Ted Anton
River Forest
'98

→ Medawar: 109
interdisc: 115
→ cf. of synthesis -Mertis
Ted Wilson, sociobiology
128

→ 140 Bohr intro to Internatl
Encyclopedia of Unified Science

The Scientific Imagination

The Scientific Imagination

With a New Introduction

GERALD HOLTON

HARVARD UNIVERSITY PRESS

CAMBRIDGE, MASSACHUSETTS
LONDON, ENGLAND
1998

Printed in the United States of America

First Harvard University Press paperback edition, 1998

Library of Congress Cataloging-in-Publication Data

Holton, Gerald James.
The scientific imagination / Gerald Holton.
p. cm.
Originally published : Cambridge, [Eng.] ; New York : Cambridge University Press, 1978.
Includes bibliographical references and index.
ISBN 0-674-79488-5
1. Science—Methodology—Case studies.
2. Physics—History—Sources. I. Title.
Q175.H775 1998
502.8—dc21 98-8463

CONTENTS

Preface

I

Considering the progress made in the sciences themselves over the past three centuries, it is remarkable how little consensus has developed on how the scientific imagination functions. Speculations concerning the processes by which the mind gathers truths about nature are among the oldest and still most prolific and controversial cognitive productions. Unless the inevitable distortion of near perspective is misleading me, it appears that only in the relatively recent period have proposals been made that have long-range promise.

The chief aim of this book is to contribute concepts and methods that will increase our understanding of the imagination of scientists engaged in the act of doing science. These chapters are therefore a continuation of the series of case studies which I published a few years ago in *Thematic Origins of Scientific Thought: Kepler to Einstein.*[1] As was the case there, my approach may be characterized by four aspects.

First, I try to make a detailed examination of the nascent phase of the scientist's work, and to juxtapose his published results, on the one hand, with firsthand documentation (correspondence, interviews, notebooks, etc.), on the other. In such a pursuit, one must be ready for the unexpected. Thus, in the studies on Einstein in my earlier book, the documents forced a reevaluation of the supposed genetic role of the Michelson experiment in Einstein's original formulation of relativity theory, and revealed that this role was small and indirect – contrary to the standard accounts and to the sequence given in practically all physics texts dealing with the matter (including a text I myself had published).

Similarly, in the new case studies presented here, the documents help us to account for the motivation of Fermi and his collaborators as we follow the historical development of the processes that led to another major scientific discovery, in this case, induced radioactivity by neutron bombardment. Both the study of how Robert A. Millikan dealt with the experimental data on which he based the published value of the electronic charge, and the comparison of his actual laboratory notebooks with his publications, require us to introduce a notion familiar from literary analysis but new to the analysis of scientific works. In my view, it is in the fine structure, in the detail of documented case studies, that one may hope to find the material with which to shape and test a theory of the scientific imagination – even if that task will not be completed easily or soon.

Second, I tend to look on any product of scientific work, whether published or not, as an "event" that stands at the intersection of certain historical trajectories, such as of the largely private or personal scientific activity; of the shared, "public" scientific knowledge of the larger community; of the sociological setting in which a science is being developed; and indeed of the cultural context of the time. In the previous book, I have described Niels Bohr's debt to works in philosophy and literature, and Einstein's interaction with epistemological currents. In this volume, I examined the effects of national resources and national styles on the organization and achievement of a laboratory team (Chapter 5).

Third, a particular concern in my studies is to find the extent to which, on certain crucial occasions, the imagination of a scientist may be guided by his, perhaps implicit, fidelity to one or more *themata*. Adherence to such preconceptions may help or impede the scientist; as Einstein once wrote to de Sitter, "Conviction is a good mainspring, but a bad regulator." The thematic structure of scientific work, one that can be thought of as largely independent of the empirical and analytical content, emerges from the study of the options that were in principle open to a scientist. It can play a dominant role in the initiation and acceptance of, or controversy over, scientific insights. Function and types of themata were discussed in detail in the case studies given in *Thematic Origins*, and are developed further here, in the cases to which the first half or so of these pages are explicitly devoted.

Last but not least, I am drawn to consider the practical consequences of such findings, for the development of scholarship in the history and philosophy of science, for the better understanding of the place of science in our culture, and for educational programs. This is the function of the last set of chapters.

II

Of these four related areas of chief interest, the topic of thematic analysis, although still an unfinished research subject, has perhaps the largest share of any claim to novelty. Judging from the commentaries on this work published in the past few years,[2] there may be several reasons for the interest that has been shown so far:

1. Thematic analysis allows discernment of some constancies or continuities in the development of science, of relatively stable structures that extend across supposed revolutions and among apparently incommensurable rival theories. Further, in this period of reactions against the philosophy that views science as a suprahistorical and culturally transcendent method of investigation, some scholars are attracted to the finding that a basic feature of the work of many seminal scientists is their acceptance of only a small number of themata, and that their debates frequently involve antithetical dyads or triplets of themata – for example, atomicity/continuum, simplicity/complexity, analysis/synthesis, constancy/evolution/catastrophic change. Such posits help to explain the formation of traditions or schools, and the course of controversies.

2. Although practically all my case studies so far have been concerned with the physical sciences, some results will be found to be applicable to the other sciences. There is evidence of this possibility, for example, with respect to recent studies in the history of biology,[3] in early biochemistry,[4] in sociology,[5] and in psychology.[6]

3. Techniques analogous to the thematic analysis that I have

applied to science have worked well before in other fields, for example, in content analysis, linguistic analysis, and cultural anthropology. It appears therefore that the work of mapping and classifying themata can lay bare basic commonalities between scientific and humanistic concerns that are not equally likely to become evident through other means. Thus in Harry Wolfson's *Philo*[7] the eloquent passage about the purpose of studying the work of the philosopher illuminates also the purpose of studying the work of the scientist in the mode here proposed. Wolfson wrote:

> No philosopher has ever given expression to the full content of his mind. Some of them tell us only part of it; some of them veil their thought underneath some artificial literary form; some of them philosophize as birds sing, without being aware that they are repeating ancient tunes. Words, in general, by the very limitation of their nature, conceal one's thought as much as they reveal it; and the uttered words of philosophers, at their best and fullest, are nothing but floating buoys which signal the presence of submerged, unuttered thoughts. The purpose of historical research and philosophy, therefore, is to uncover these unuttered thoughts, to reconstruct the latent processes of reasoning that always lie behind uttered words, and to try to determine the true meaning of what is said by tracing back the story of how it came to be said, and why it is said in the manner in which it is said.

4. The investigation of preconceptions in and concerning science connects rather directly with a number of other modern studies, including that of human cognition and perception, learning, motivation, and even career selection (as discussed in Chapter 7). Moreover, one may hope that a more sophisticated idea of the working rationality of scientists – with its full set of antithetical components, including preconceptions on the one hand and objective techniques on the other – will help to deflate foolish and dangerous ideas about science that, as noted in Chapter 3, have characterized some of the popular conceptions of science. As we are entering a period of an increasing number of externally imposed restrictions on and directions of scientific research, it is well for scientists and other scholars to ensure that

the conditions under which scientific originality can flourish are studied, more widely understood, and protected.

III

The deep attachment of some scientists to certain overarching themata may well be one of the chief sources of innovative energy (parallel to that of the instrumentalist or utilitarian thrust in science). It seems to me otherwise difficult to understand a key fact about the sciences, namely, that again and again they have been regarded as verging on a charismatic activity rather than being thought of as, say, merely one of the more successful but fundamentally pedestrian activities of mankind.

To underline, and at least briefly emphasize this sometimes neglected point, we cannot do better than reflect on a work standing at the very beginning of modern science. Copernicus, like any other good astronomer, relied of course on observation and calculation, and he greatly advanced mathematical astronomy in the technical sense. But one must look deeper to find the chief reason why he came to write the work for which he has been honored, or to explain its power. Nature, he held, is God's temple, and he implied that human beings can, through the study of nature, discern directly both the reality and the design of the creator. This was a daring and dangerous idea, and it is significant that when Copernicus's book was put on the *Index* of "Books to be Corrected," this implication was one of the relatively few deletions which were insisted upon as necessary; for it was clear both to Copernicus and to his opponents that when the purpose of science is perceived large enough, it can rival the claims of all other reality systems.

From the first sentence of *De Revolutionibus*, one senses the source of energy of a major scientific idea. It is not some pedestrian piecing together of a corner of the puzzle. Nor does the work give us merely better astrometry and applications such as calendar corrections, valuable though these are. Rather, his discovery is on a scale that produces an expansion of human consciousness, a change in cultural evolution – and it was so perceived by those who were converted to Copernicus's idea.

In his work, two themata predominate, and the mutual accommodation of theory and data that they produce seem to me to account for the quasi-aesthetic conviction in his followers that the system must be right. These themata are those of simplicity and necessity. They appear in a stern manner that became basic to all science since. In a well-known passage, Copernicus proudly writes that the heliocentric scheme he has found for the system of planets has the property that "not only their phenomena follow therefrom, but also the order and size of all the planets and spheres and heaven itself are so linked together that in no portion of it can anything be shifted without disrupting the remaining parts and the universe as a whole."

The power of this solution was precisely its restrictiveness. There is nothing arbitrary, no room for the smallest ad hoc rearrangement of any orbit, as had been quite possible in pre-Copernican work. Copernicus's system, as a whole, revealed a sparse rationale, a necessity that binds each detail to the whole design. Hence it carries the conviction that we understand why the planets are disposed as they are, and not otherwise. One is reminded here again of Einstein's remark to his assistant Ernst Straus: "What I'm really interested in is whether God could have made the world in a different way; that is, whether the necessity of logical simplicity leaves any freedom at all."

This kind of terminology, and the attitude behind it, are now rare and even somewhat embarrassing to most scientists. There are good sociological, psychological, and even political reasons why this should be so, why our usual list of motivations for scientific work tends to stress the Baconian side of the heritage of modern science – the discovery of cures, the perfection of machinery, the strengthening of the state's security, or simply the provision of a decent way to spend one's days on this earth. But while the Baconian ethos has become a necessary component of the total scientific and engineering enterprise, it would not be sufficient to sustain science, and by itself does not help us understand the nature of high discovery.

No one would argue that personal testimonies such as those referred to should be introduced into our current scientific papers. However, a quiet underground current of this cosmological tradition still exists. It comes in a somewhat disguised form, but the thematic content of simplicity and necessity as warrants of deeper truths are still among the most prized.[8] Steven Weinberg, on receiving the Robert Oppenheimer Memorial Prize, said:

> Different physicists have different motivations, and I can only speak with certainty about my own. To me, the reason for spending so much effort and money on elementary particle research is not that particles are so interesting in themselves – if I wanted a perfect image of tedium, one million bubble chamber photographs would do very well – but rather that as far as we can tell, it is in the area of elementary particles and fields (and perhaps also of cosmology) that we will find the ultimate laws of nature, the few simple general principles which determine why all of nature is the way it is . . .
>
> The reason I take such an optimistic view of where we are now is that relativity and quantum mechanics, taken together but without any additional assumptions, are extraordinarily restrictive principles. Quantum mechanics without relativity would allow us to conceive of a great many possible physical systems. Open any textbook on non-relativistic quantum mechanics and you will find a rich variety of made-up examples – particles in rigid boxes, particles on springs, and so on – which do not exist in the real world but are perfectly consistent with the principles of quantum mechanics. However, when you put quantum mechanics together with relativity, you find that it is nearly impossible to conceive of any possible physical systems at all. Nature somehow manages to be both relativistic and quantum mechanical; but those two requirements restrict it so much that it has only a limited choice of how to be – hopefully a very limited choice.[9]

All scientists since Copernicus have understood the attractiveness of a system having such qualities. And although any individual attempt of this sort is an act of intellectual and professional risk taking – for the thematic choices themselves are neither verifiable nor falsifiable, and the antithetical thema of complexity, for example, deserves a more detailed analysis than it has been given so far – no other, less cosmological, approach is likely to lead to the truth, least of all to a truth having the exalting sweep that historically has helped give the scientific enterprise its intellectual mandate.

IV

In pointing to these uses to which themata have been put, I do not mean to imply that they are the only ones of significance. On the contrary, themata have been and, I expect, ever will be used by scientists of the most opposing attitudes and interests. Moreover, I do not believe that giving attention to thematic analysis requires one to adopt a label (certainly neither "positivistic" nor "antipositivistic"), or otherwise forces one to take sides in the battles currently preoccupying some sectors in the history and philosophy of science – the more so as many of the divisions are themselves along thematically opposing conceptions about the history or philosophy of science. The campaign flags, or accusations, read "objectivity" versus "subjectivity" versus "anything goes"; "logical" versus "empirical" versus "psychologistic" studies; "rules of reason" versus "mystical conversion"; "rational" versus "irrational"; "relativism" versus "absolutism"; "analytical-reductionistic" versus "holistic"; and even "reason" versus "imagination." But to paraphrase a seminal paper that changed the state of physics in the early years of this century, the understanding of the process of scientific innovation that is implied in these antagonistic schemes is characterized by polarities which do not appear to be inherent in the phenomena, that is, in the actual work of the scientists as it reveals itself to us in archival material, oral histories, and of course in the actual participation in research.

At the very least, the present, opposing positions seem to me to lack the flexibility and the ability to accommodate themselves to the human activities – with all their natural ambiguities – that we are trying to map and study. The heat and ideological clamor emanating from some of the encounters in fields that make science their raw materials of observation are strangely incongruous with respect to the state of affairs in the sciences themselves. Ironically, even those who actually work in the "hardest" sciences now are often satisfied with claiming no more than "good reasons" and probable knowledge. Most of them are not afraid to accept humanistic interpretations of their work, and are likely to sympathize with Henry A. Murray's perceptive definition that "science is the creative product of an engagement between the scientist and the events to which he is attentive."[10]

V

The search for models of the scientific imagination, at this stage of research, must of necessity be largely inductive and empirical. It must be committed to painstaking attempts at historical accuracy and cautious scholarship based on the available evidence, but it must also possess the imaginative freedom to produce new conceptual tools with which to study well-guarded areas such as the working of the minds of scientists. Adopting a kind of ethological approach to the study of scientific activity and bringing in whatever is needed – now the state of science as understood at the time, now findings on the psychodynamics of scientists or the social forces on them – seem to me a strategy preferable to casting the accounts of achievements into formalistic structures. As we are only just beginning to gather the chief elements from which theories of the scientific imagination may be fashioned, schemes that promise certainties must be held at arm's length.

One recalls here a story told by the architect LeCorbusier.[11] Having invented his "modulor," a measure-system for fixing the dimensions of architectural space, of urban design, and of plastic arts, and believing in its necessity and power with fervor, he was arguing intensely for its wide adoption. LeCorbusier even journeyed to Princeton to convince Einstein. However, instead of the hoped-for endorsement of the system, he obtained a much milder and more appropriate judgment: Einstein told him the scheme would be quite satisfactory if it only served to make the bad more difficult, and the good easier.

Acknowledgments

In addition to sponsors, institutions, and persons to whom I have expressed my gratitude on many of the pages that follow, I wish to acknowledge especially a supporting grant for research in the history and philosophy of science, received from the National Science Foundation, and the hospitality of the Center for Advanced Study in the Behavioral Sciences at Stanford during a research leave in 1975–76. None of my work would have been possible without the generous assistance from the Albert Einstein Archives at Princeton, the Center for the History of Physics at the American Institute of Physics in New York, the archives of the American Philosophical Society in Philadelphia, the Robert A. Millikan Archive at the California Institute of Technology, the Ernst Mach Archive in Freiburg, and the archives at the Accademia dei Lincei in Rome and of the Domus Galilaeana in Florence.

Among those whose technical expertise helped me negotiate the huge distance between the mere thought and the printed word, I wish to thank Joan Laws for her patient assistance over many years, and Marcel Chotkowski La Follette for therapeutic editorial help with the printer's manuscript.

I have presented several of these cases for discussion in my seminar on the history of science at Harvard, and once more thank my students for their thoughtful responses. When a chapter is based on a previously published essay – usually in somewhat revised form – the publishing history appears in the chapter notes and the Acknowledgments; my appreciation to those publications and their editors is also hereby recorded gladly.

Introduction: How a scientific discovery is made: The case of high-temperature superconductivity

Setting priorities for research, choosing which projects are to be supported and which abandoned, triggers epic battles at the highest levels. It requires predicting which paths and which mixture of policies might best advance science and lead to fruitful technologies. Yet in such debates little attention is given to one of the most fundamental questions: What can historical cases teach us about how the scientific imagination works, and hence what it takes to make a scientific discovery? There are many popular ideas abroad, often based on oversimplified textbook accounts of famous discoveries and on charming anecdotes. They have little to do with the unruly complexity of the events themselves, and can only mislead science scholars and science policymakers.

For this reason, it will be revealing to find lessons in specific case studies of the kind to which this book is dedicated. To start off with an example of the scientific imagination at work, a good choice is the discovery of high-temperature superconductivity, not least because it is recent enough to simplify the process of reconstructing its context. Our investigation, which included interviews with Karl Alex Müller and Johannes George Bednorz, who discovered the first high-temperature superconductors, also throws light on a set of problems of intense current interest: How does "curiosity-driven" or basic science interact with "strategic" research and engineering? How important are both planning and serendipity in discovery? What laboratory culture makes success more likely? How deeply are the roots of crucial

This essay was prepared in collaboration with Hasok Chang and Edward Jurkowitz.

ideas and apparatus buried in the soil of history? How important is the practice of borrowing across traditions and disciplines? What role do the private style and presuppositions of the individual play in research?

As will be shown, in a typical case scientific innovation depends on a mixture of basic and applied research, on interdisciplinary borrowing from current as well as from old resources, on an unforced pace of work, and on personal motivations that lie beyond the reach of the administrator's rule book. While many of these findings may be generally familiar to students of scientific creativity, our operational mode of analysis, which is roughly comparable to the methods of genealogical research, makes them more precise, testable, and generally applicable. As such, they may serve as an empirical complement to some of the untested assumptions that inform policy discussions, not least in the debates in Washington and in corporate boardrooms over the relative merits of applied and basic research.

A brief history

Superconductivity—the loss of electrical resistance below a critical, or transition, temperature (T_c) characteristic of the material—was first discovered in mercury by the Dutch physicist Heike Kamerlingh Onnes in 1911. Mercury becomes superconducting at just 4.2 degrees above absolute zero (4.2 degrees Kelvin). Teams large and small worked for decades in the hope of finding electrical conductors with higher critical temperatures, which would be easier and cheaper to keep resistance-free. There beckoned the rewards both of new theories to explain the phenomenon and of practical applications to exploit it; among the latter was the possibility of enormous new efficiencies in the transmission and use of electricity.

But for a long time nature yielded little hope for real progress. By 1973, fully sixty-two years after the discovery of the phenomenon of superconductivity, all efforts had stalled at a T_c of 23.3 Kelvin, the critical temperature of a niobium-germanium compound (Nb_3Ge). After years of frustrating failures to boost T_c into a region where there were realistic prospects for commercial use, high-temperature superconductivity was no longer

Figure 1. Superconductivity, the loss of electrical resistance in a material, was discovered by the Dutch physicist Heike Kamerlingh Onnes (*right*, shown with his colleague G. J. Flim and their helium-liquefying apparatus in 1922). (Photograph courtesy of the Deutsches Museum, Munich.)

considered a promising area. Some theories held that no higher T_c could be expected. Bernd Matthias, a highly respected Bell Laboratories physicist who, together with collaborators, had discovered hundreds of new superconductors, challenged his peers to give up "theoretically motivated" searches, or "all that is left in this field will be these scientific opium addicts, dreaming and reading one another's absurdities in a blue haze" (quoted in Bromberg 1995).

All this changed virtually overnight in 1986, with the publication of a set of papers by Karl Alex Müller and his former student Johannes Georg Bednorz, two investigators at the IBM Zurich Research Laboratory in Rüschlikon, Switzerland. Unlike most of the previously discovered superconductors, the new compound was a ceramic, a mixed oxide of barium, lanthanum, and copper

Figure 2. In 1987, Karl Alex Müller *(far left)*, 60, and his former student Georg Bednorz, 37, were awarded the Nobel Prize for the discovery of high-temperature superconductors. (Photograph courtesy of IBM.)

(La_2CuO_4, or lanthanum cuprate, doped with a small amount of barium). It not only had a remarkably high T_c—in the neighborhood of 30 Kelvin—but was also relatively easy to prepare by ceramic techniques and to modify by chemical substitution. Whereas ground-breaking discoveries often involve new technology, in this instance the means to create and to measure the phenomenon had been available for decades.

The discovery became an academic and popular sensation, especially after Paul C. W. Chu's group at the University of Houston and Mao-Ken Wu's team at the University of Alabama jointly announced in February 1987 that they had achieved superconductivity at about 90 Kelvin with materials related to the Bednorz-Müller compound, a temperature well within the range of the inexpensive coolant liquid nitrogen. Now high school students could demonstrate the phenomenon. A climax of excitement was reached at the so-called Woodstock of Physics, a panel discussion on high-temperature superconductivity held at the American Physical Society's annual meeting on March 18, 1987, in New York City. Roughly 3,500 physicists crowded into the hotel where the meeting was held, some lingering so long after the session ended that they had to be ejected from the rooms by the hotel staff at 6 A.M. (Khurana 1987b; Robinson 1987; Schechter 1989).

Here was everything a physicist could wish for: a new class of materials with great potential for generating new theories and new technologies. Above all, the discoveries provided that most rare and most desired moment, a glimpse of vast unexplored scientific territory. Like Cortez's men on the peak in Darien, the physicists at the meeting "look'd at each other with a wild surmise." Or as a reporter put it: "One could have felt as if one were a part of a ceremonial gathering to affirm a new cult" (Khurana 1987b).

After the tumultuous emergence of high-temperature superconductivity—even President Ronald Reagan hailed the "new age" of superconductivity as a welcome "revolution" having great promise for new products—Müller and Bednorz were awarded, with the maximum possible speed, the Nobel Prize for physics

Figure 3. The American Physical Society's 1987 March meeting, later named the Woodstock of Physics, was an impromptu crash course following the discovery of high-temperature superconductors. According to one account, the meeting was "an insane physics demonstration . . . Suddenly, over three and a half thousand physicists (with twice that many elbows) seemed hell-bent on proving that two bodies *could* occupy the same place at the same time." (Photograph courtesy of the American Institute of Physics.)

for 1987. Their discovery unleashed the energies of dozens of teams, and laboratories all over the world rushed to synthesize other potential oxide superconductors. Indeed, it did not take long for the critical temperature to be raised to 125 Kelvin, and to even higher temperatures at high pressure. Today the record stands at about 164 Kelvin, with isolated observations, often transient, being reported of critical temperatures well above 200 Kelvin (the freezing point of water is 273.16 K). The opium addicts' blue haze has dissipated, and some physicists have even dared to hope again that room-temperature superconductors will eventually be found.

A first-level analysis

The main outlines of this discovery are well known, but our interviews and correspondence with Müller and Bednorz turned up essential details. Here we set forth an account of the discovery of the first high-temperature superconductors as Müller and Bednorz experienced it, with particular attention to the resources, either intellectual or material, on which the discovery depended. Then we put forward a systematic analysis based on this narrative, a schema designed to help answer the more general question of what it takes to make a scientific advance.

Müller, who was born in Basel, Switzerland, in 1927, graduated from the Swiss Federal Institute of Technology (ETH) in Zurich in 1958. ETH was the home base of the physicist Wolfgang Pauli, who had continued to teach there after being awarded the Nobel Prize in 1945. Müller said of Pauli, "he formed and impressed me." As will be shown, the student had learned more than physics from his teacher. By 1963, Müller had joined the research staff of the IBM Zurich Research Laboratory, and in 1972 he was put in charge of its physics group. In 1982 he was promoted to IBM Fellow, becoming one of a handful of that corporation's distinguished scientists who were free to work on anything they pleased. He was to use that opportunity well.

Previously Müller had worked for almost fifteen years on a series of problems in condensed-matter physics, many of which had links back to his doctoral research. His Ph.D. thesis, done under Georg Busch at the ETH, was on the identification of the electron paramagnetic resonance lines of iron ions, a subject quite unrelated to superconductivity. But as it happened, the material in which the iron ions were present as impurities was the then recently synthesized oxide strontium titanate ($SrTiO_3$), and that fact, in a way nobody could have foreseen, would turn out to be Müller's first step toward research on high-temperature superconductivity.

Indeed, Müller's use of strontium titanate was entirely accidental. Initially, he had set out to map the paramagnetic resonance spectrum of impurities in tin. When that worked out poorly, Müller explained in an interview, he went "by chance into Professor Heine Gränicher's office," looking for crystals of other materials. Gränicher offered Müller some samples, and among them was strontium titanate. It was a fateful moment. Strontium titanate not only helped Müller get his doctorate, but also led him to study the crystallographic literature about the class of materials to which this compound belongs, and set him on a road whose destination would become apparent only years later.

When he was made an IBM Fellow in 1982, Müller felt that, having passed the age of fifty, he was ready for an entirely new challenge. Perhaps he remembered the advice of an old supervisor, H. Thiemann, whose byword had been "One should look for the extraordinary." In any case, Müller chose extraordinary conductivity as his next challenge.

At the time, superconductivity was not a promising field of research. Not only had the incremental progress toward higher T_c apparently stalled, but IBM had had to abandon the effort to produce a computer using Josephson junctions—electronic devices made of superconducting materials that can switch states faster than devices made of semiconductors—despite the enormous investments the company had made in this project.

Müller was aware of all this. In 1978 he had spent an eighteen-month leave at IBM's Thomas J. Watson Research Center in Yorktown Heights, New York. John Armstrong, the vice president in charge, wisely gave Müller discretion to pursue any subject he wished while he was there. That encouraged him to look into the troubled field of superconductivity, about which he then knew very little. As he put it, he started "from page one of Michael Tinkham's book," *Introduction to Superconductivity* (1975).

But thorough study throughout the available literature turned up no theory that would lead beyond the usual materials to new substances and higher critical temperatures. Müller then "decided I just don't talk to the theoreticians. They just held me back."

After he returned to Zurich, Müller continued to work on superconductivity, first alone, and then from 1983 with Bednorz. Born in Neuenkirchen, Germany, in 1950, Bednorz was highly trained in crystallography, solid-state chemistry, and physics. Müller and Gränicher were supervising his Ph.D. thesis at the ETH; not surprisingly, Bednorz's first experimental work was on the growth and characterization of strontium titanate.

Bednorz was an ideal partner for Müller. As Bednorz explained in their joint Nobel lecture, he had become interested in superconductivity in 1978, when he was invited by the IBM Zurich laboratory to improve the superconductive properties of strontium titanate single crystals. In this he quickly succeeded by adding trace amounts of niobium to the crystal. Yet the highest achievable T_c was still only 1.2 Kelvin, so his supervisor had lost interest, and Bednorz had returned to the institute to work on his thesis.

But the seeds of fascination with the field had been sown, and when in 1983 Müller asked Bednorz to join him in the search for a new superconductor, Bednorz accepted with such alacrity that it took Müller by surprise. Still, at the outset the two men spent a fruitless couple of years with nickel-based oxides. Right up to its culmination, their research was typically in the "little-science" style, meaning on a relatively small budget. Bednorz later described it as one step on a "long and thorny path." Moreover, they worked in self-imposed isolation. Müller admitted that they kept their early work completely to themselves, not informing even the IBM managers, in part because superconductivity research was not then a popular subject with management. This decision, made possible by Müller's status as an IBM Fellow, was also taken so that if they failed, they could quietly give the project "a burial in very restricted family circumstances, in order not to jeopardize Bednorz's career."

The breakthrough came when they decided to look for superconductors among copper-containing oxides, a class of materials fundamentally different from those that had been ransacked by the pioneers in the field, such as Matthias. During a literature search on these types of compounds, Bednorz happened on a 1985 paper by Claude Michel, L. Er-Rakho, and Bernard Raveau of the Université de Caen that described barium-doped lanthanum cuprate. But the authors were chemists and had concentrated on catalytic rather than superconducting properties.

The decision of the Swiss team to investigate the oxides, which is a key turn in this story, "took every condensed-matter physicist by surprise" (Chakravarty 1994). Indeed, one physicist recently confessed that when he and his colleagues had heard that Müller "was searching for high T_c in oxide, we thought he was crazy." Up to that time searches had concentrated on intermetallic compounds; ceramic oxides were generally thought to be insulators, not conductors, much less superconductors. But according to their joint Nobel Prize lecture, the team's "aim was primarily to show that oxides could do better in superconductivity than metals and alloys."

They had several reasons for striking out in this new direction. Some oxides, including strontium titanate, had previously been found to be superconductors of the traditional sort, although with T_c's no higher than 14 Kelvin, they ranked well below the niobium compounds. As Müller and Bednorz later noted, they were also reasoning from the then-standard theory for superconductivity—the BCS theory, named after the physicists John Bardeen, Leon Cooper, and Robert Schrieffer—and the Jahn-Teller theorem devised by the physicists H. A. Jahn and Edward Teller. Müller had read a paper by Karl-Heinz Höck and H. Nickisch of the Technische Hochschule in Darmstadt, Germany, and H. Thomas of the Universität Basel in Switzerland (Höck, Nickisch, and Thomas 1983) that led him to think that a material that met the Jahn-Teller criterion and was metallic at high temperatures might have an unusually high T_c. The lanthanum cuprate met those criteria as well. Yet among the interesting ironies of the story is that the BCS theory is now thought to have only limited applicability to high-temperature superconductivity, and that the Jahn-Teller effect has little to do with establishing superconductivity in the high-temperature superconductors. Furthermore, as Müller later put it, "You always need a kick of luck": Bednorz had prepared the lanthanum cuprate differently (by coprecipitation) than had the French team, thereby favorably altering the structure, as it turned out. (Dordick 1987).

As Müller explained, however, there was another, very crucial factor: the attraction that he in particular felt toward the substance described by Michel, Er-Rakho, and Raveau. It had a perovskite-type structure. This structure had special meaning for Müller; indeed, he had an "atavistic type of feeling that it might work for superconductivity." When he and Bednorz came to deliver their joint Nobel lecture, they gave it the significant title "Perovskite-Type Oxides—The New Approach to High-T_c Superconductivity."

Perovskites, named in 1830 in honor of the Russian amateur geologist Lev Aleksevitch von Perovski, are a class of ceramics that have a particular atomic arrangement. In their ideal form perovskites, which can be described by the general formula ABX_3, consist of cubes that are decorated with three elements. The *A* cation (positively charged ion) lies at the center of each cube, the *B* cations occupy all eight corners, and the *X* anions (negatively charged ions) lie at the midpoints of the cube's twelve edges. The cubic structure has particular appeal because, of the seven possible crystal systems, it is the one with the highest degree of symmetry.

As noted, Müller and Bednorz had each devoted their graduate research largely to the perovskite strontium titanate. Müller later wrote that "the perovskite structure determined, even dominated, my scientific efforts for many years" (Müller 1988). Indeed, in Müller's extensive bibliography, perovskites recur in widely varying studies, ranging from paramagnetic resonance to sound attenuation and heat capacity, from structural phase transitions to photochromism. For example, Müller gave special attention to perovskites in a decade-long investigation of the manner in which the Jahn-Teller effect can lead to structural phase transitions (Thomas and Müller 1972).

As Müller emphasized, perovskites "always worked" for him. This highly symmetric structure became for him a thematic guide, quite different from and supplementary to the elements traditionally considered central to the logic of scientific research. It will become apparent in the following chapters that, in this tendency, Müller joins many other scientists who found themselves being led by thematic commitments, at least during the early, rather private stages of their projects. Einstein, for example, had a predilection for symmetry, continuity, and classical causality, whereas Heisenberg embraced discontinuity and abandoned classical causality. In Müller's case the thematic influence on the scientific imagination was just as compelling.

After reading the French paper about barium-doped lanthanum cuprate, Bednorz and Müller prepared the compound and showed that it became superconducting at a temperature of about 30 Kelvin, a T_c substantially above that of any material previously studied. They also confirmed that the sample exhibited another important indicator of superconductivity, the Meissner effect: when a superconductor in a magnetic field is cooled to the temperature at which it loses resistance, all or part of the magnetic flux within the material is expelled. Still, Bednorz and Müller initially found that their reports were "met by a skeptical audience" (Bednorz and Müller 1987a). But soon the confirmations came pouring in, and research groups grew explosively the world over.

A second-level analysis

For a serious study of what it took to make this discovery and what lessons it implies, we must press on beyond this sketch of events to a deeper level, where the resources that history had prepared for the success of the team lie hidden. A complete analysis would take into account in more detail the personal research trajectories of Müller and Bednorz, evaluate the influences of encounters with other researchers, and thoroughly explore the educational systems through which they passed, as well as the universities and corporate institutions that employed them. Here we will concentrate on the vast treasury of *intellectual and material resources* that these two scientists were able to exploit. On the basis of this case history, we then generalize, proposing a structured description of how new scientific work is rooted in and nourished by previous achievements, some from the distant past.

It is often said that scientific work is "based on" earlier work,

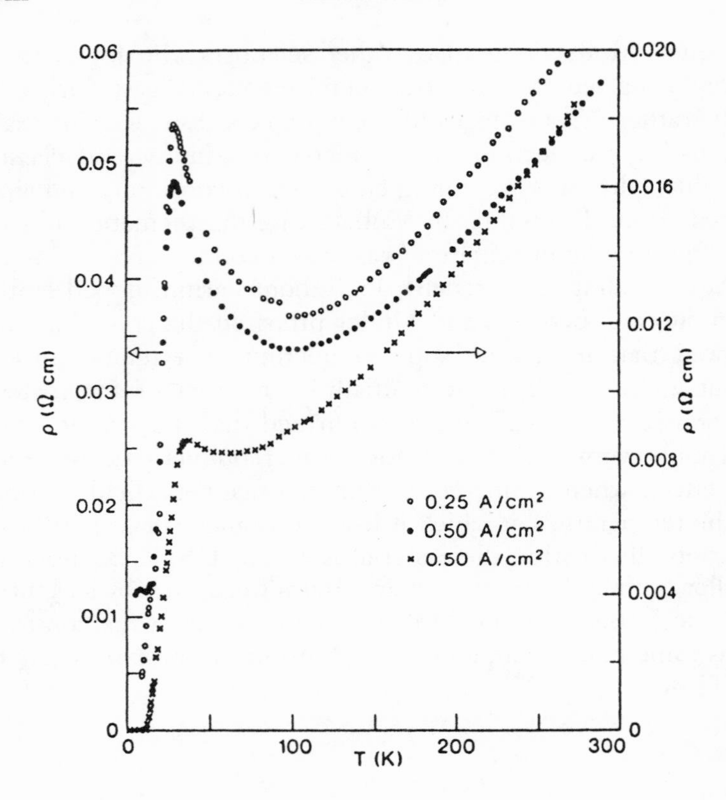

Figure 4. Data indicating the discovery of the first high-temperature superconductor, barium-doped lanthanum cuprate, published in *Zeitschrift für Physik B.* The crucial figure shows the temperature dependence of the resistivity (ρ) of lanthanum cuprates doped with varying amounts of barium. The figure's shock value lay in its abscissa: the materials underwent the transition of zero resistivity at temperatures well above absolute zero. (Courtesy of Springer-Verlag.)

or that earlier work "gave rise to" later work; there is also much talk of "traditions," "influences," and "connections." These notions must be made more precise to be useful. In particular, one must seek to operationalize them, that is, to define them in terms of identifiable and repeatable operations.

By resources we mean mathematical techniques, physical laws, analytical instruments, factual information, and the like. Although an investigator may well create some such resources on the spot, more frequently they are derived from previous work done by others. Indeed, the most important relation among scientific research efforts is that of adapting, assimilating, transforming—or in general "borrowing"—whether consciously or not.

Such borrowing leaves identifiable traces, just as one can discern one's ancestors' traits in one's own makeup. To pursue this suggestive metaphor, it should be possible in principle to reveal many "generations" of "ancestors" that lie behind a scientific work. In short, "influence" can be operationalized by attempting to tease out the genealogy of a work by looking for documentable facts that are equivalent to a line of inheritance.

One way to locate the resources used in a given piece of published scientific work is to trace the citations, in footnotes or in the text, to other publications or to private communications it contains. To be sure, citations cannot be used blindly, because they may be merely pro forma, intended to acknowledge the existence of related projects in the same field, or to serve other, largely social purposes. Another problem is that not all important resources are explicitly cited. Many resources will be considered generally known and silently assumed. The scientific genealogist, therefore, must rely on his or her own scientific and historical background knowledge to find implicit citations in the target paper or in the ancestral papers. Finally, any genealogical exercise is open-ended; how far back to trace the connections is a pragmatic decision. We have found that it is not necessary to go back further than three or four "generations" to test interesting hypotheses about scientific innovation.

A genealogical analysis of Bednorz and Müller's main scientific publications announcing the discovery of high-temperature superconductivity shows the need to distinguish among the various types of resources on which the team drew. They made use of at least four types of resources: initial, motivating theoretical framework and ideas (schema); experimental techniques and material resources (production); means of gathering and analyzing data (observation); and theoretical concepts for interpreting the results (interpretation). Figure 5 indicates the kinds of items that make up these four main components.

As Figure 5 shows, analysis of the original five papers that

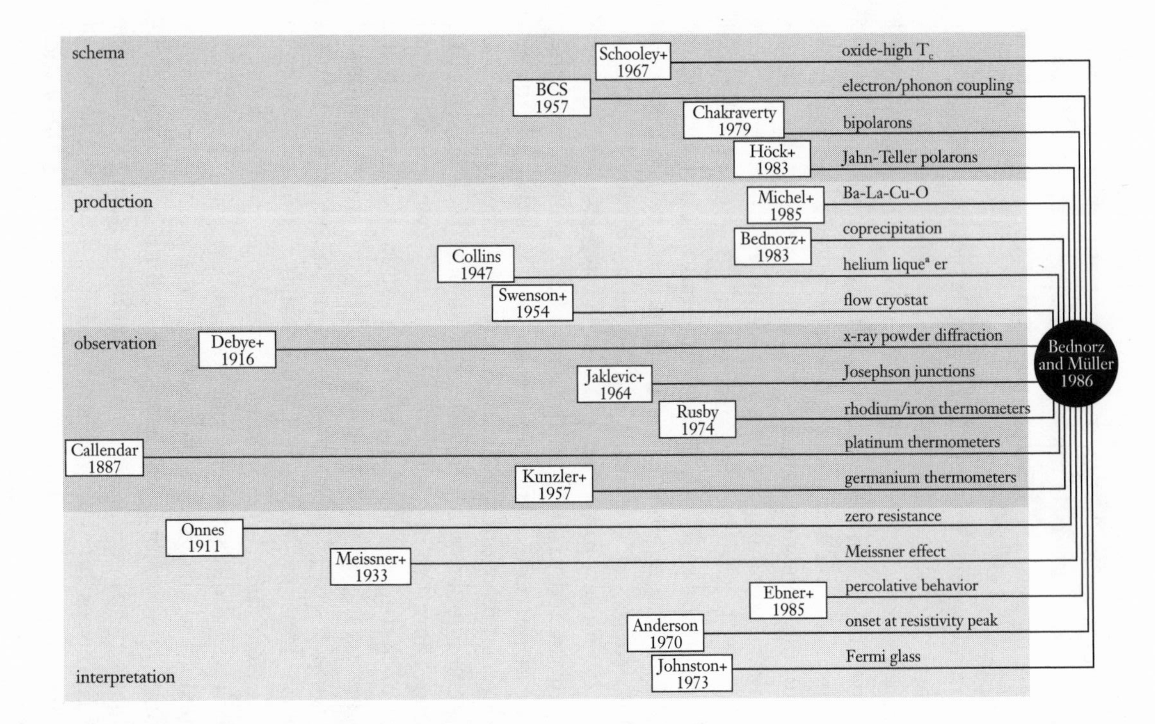

Figure 5. To determine the immediate intellectual and material "ancestry" of the discovery of high-temperature superconductors, we traced implicit and explicit citations ("+" indicates coauthors) appearing in five papers that form the essential core of the Bednorz-Müller work. This genealogical analysis quickly showed that at least four types of resources fed into the discovery, as marked along the left edge of each "layer."

constituted the announcement of their breakthrough quickly revealed a number of silent resources that Müller and Bednorz put to use. For example, among the tools for observation were several standard techniques whose origins are no longer referred to explicitly in research reports: x-ray powder diffraction for analyzing the structure of the sample, and various electrical resistance thermometers, among others. Similarly, the theoretical resources needed to interpret the experimental results included some long considered commonplace and whose original sources were not cited, such as the criteria for identifying superconductivity: zero electrical resistance and the Meissner effect.

Figure 5 indicates how each resource can be connected either to one of the two-dozen publications explicitly cited in the basic Bednorz-Müller papers or to a publication implicitly referred to in their papers. For example, the passing mention of the Meissner effect implicitly refers to the 1933 publication describing the effect by the German physicists Walther Meissner and Robert Ochsenfeld. Similarly, the platinum thermometers the teams used imply reference to an 1887 publication by Hugh L. Callendar of the Cavendish Laboratory in Cambridge, England, that ushered in the resistance thermometer as a practical means of measuring temperature.

It doesn't take long to see that the Bednorz-Müller work harbors a broad and intricate system of ancestors. Our search revealed many cross-references *among* early resources and also quickly took us back to work published a century or more ago. Unwittingly but documentably, the stage for the 1986 discovery was set by scientists, many long in their graves. For instance, the apparatus used to liquefy helium stems from a liquefier developed by the MIT engineer Samuel C. Collins in 1947; its predecessor was the Russian physicist Pyotr Kapitza's 1934 liquefier, which in turn made use of two principles of cooling first laid out by the British physicists William Thomson (Lord Kelvin) and James P. Joule in the 1850s and by the French chemists Nicolas Clément and Charles-Bernard Desormes in 1819.

Obviously, the network of ancestors extends to yet earlier generations (Figure 6). Moreover, if one focused on any one node—say, the BCS theory of superconductivity—it would reveal a broad and intricate network of its own (Figure 7). That, of

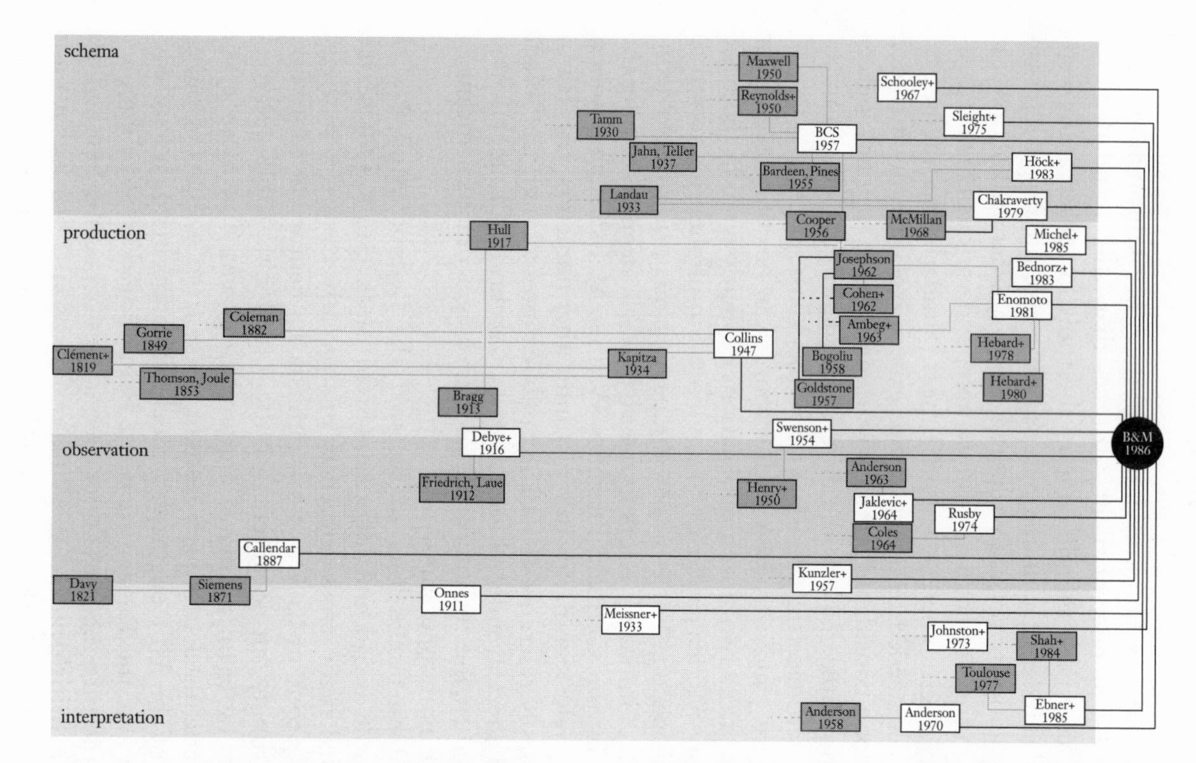

Figure 6. More complete genealogical analysis of the Bednorz-Müller achievement uncovered a broad network of intellectual and material ancestors, grouped here generally as in Figure 5. For illustrative purposes, the first-generation ancestors *(heavy lines)* are shown in more detail than earlier generations *(light lines)*. The genealogical tree has been arbitrarily truncated.

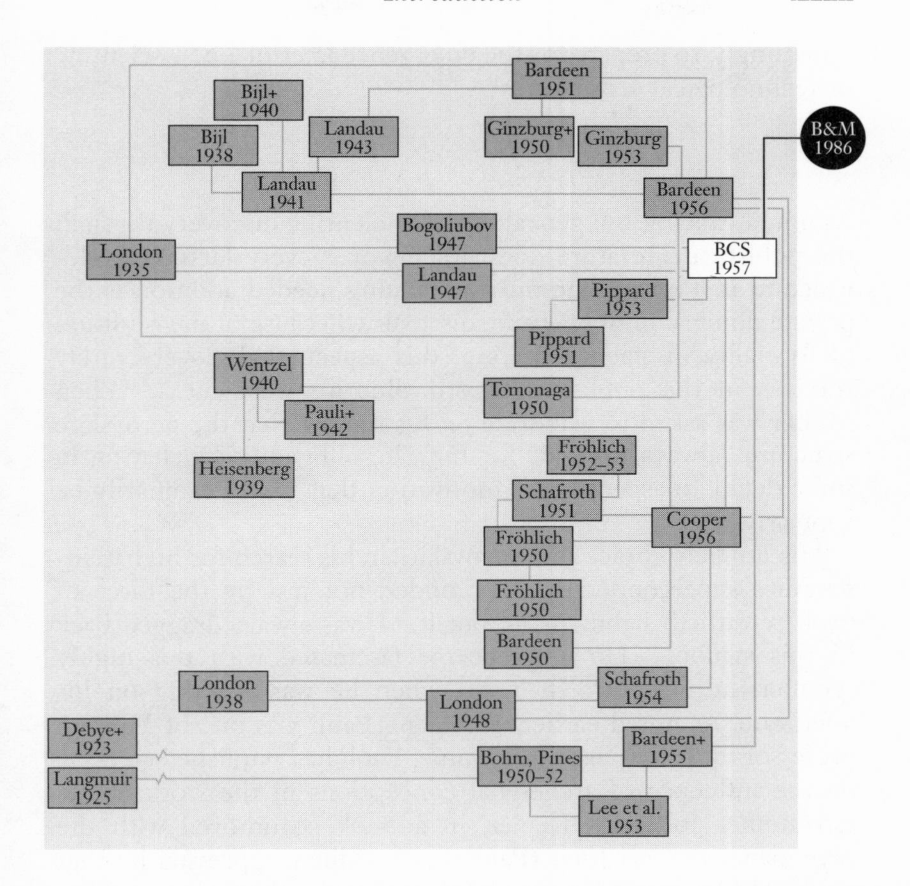

Figure 7. Density of connections that characterizes genealogical analysis of scientific work is demonstrated by focusing on one node in Figure 6—the BCS theory for low-temperature superconductivity—which explodes to reveal its own dense network of ancestors. This figure also illustrates a quixotic aspect of scientific discovery: the BCS theory in its original form is now widely thought not to apply to high-temperature superconductors, but ideas do not have to be correct to be fruitful.

course, is the point: in an operationally meaningful sense we begin to perceive "what it took" to discover high-temperature superconductivity. We can generalize that any significant advance relies on a large but identifiable set of earlier contributions. Some may be famous and profound; many more are much less significant in themselves. But all have served, almost always unwittingly, to prepare for the emergence later of a new scientific or technological achievement.

The private dimension

As noted, tracing the genealogy of a scientific discovery through the published literature does not uncover every factor of relevance to it. Perhaps the most intriguing needed addition is the private dimension of scientific discovery. Because of the tradition of formality in science writing, this aspect of discovery rarely survives in the published record. But we were lucky. When Müller was asked to elaborate on his remark that the perovskite structure "always worked" for him, he obliged us by sharing in some detail an aspect of his motivation that would ordinarily be kept private.

His unlikely choice of a perovskite in his search for high-temperature superconductors was guided not just by the force of (well-rewarded) habit. As he put it: "I was always dragged back to this symbol." He first became fascinated with this highly symmetrical structure in 1952, when he was working on his doctorate. As noted earlier, Wolfgang Pauli was one of Müller's professors at ETH; just at that time, Pauli had published an essay on the influence of archetypal conceptions in the work of the astronomer Johannes Kepler, in a book coauthored with the psychoanalyst Carl Jung (Pauli 1952). Much impressed by that essay, Müller started to read Kepler avidly, thus encountering Kepler's deep commitment to the guidance of three-dimensional structures of high symmetry—the five Platonic solids—in his work on planetary motion (Figure 8).

Müller continued, "If you are familiar with Jung's terminology, the perovskite structure was for me, and still is, a symbol of—it's a bit high-fetched—but of holiness. It's a *Mandala*, a self-centric symbol which determined me . . . I dreamt about this perovskite symbol while getting my Ph.D.; and more interesting about this is also that this perovskite was not just sitting on a table, but was held in the hand of Wolfgang Pauli, who was my teacher." At the time, Müller had divulged this aspect of his inspiration only to friends and to Pauli's last assistant, Charles P. Enz. He has since discussed it in an introspective essay (Müller 1988) illustrated with the Dharmaraja Mandala (Figure 9).

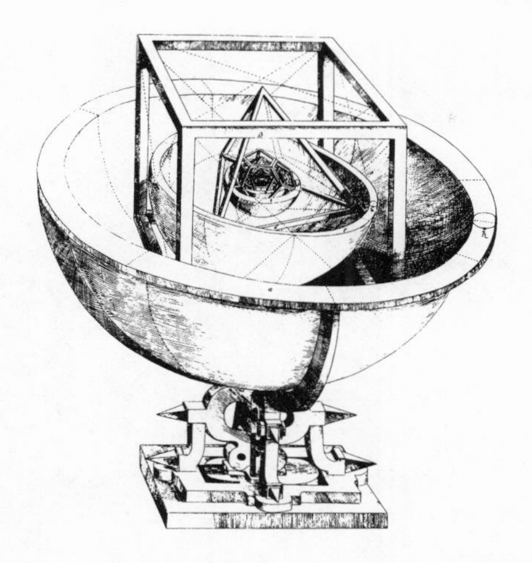

Figure 8. Like many scientists, Müller was also influenced by older scientific ideas. Wolfgang Pauli, his teacher, had written an essay on archetypal conceptions in the work of Johannes Kepler, including a model of the planetary orbits as a concentric structure defined by the enclosed highly symmetrical solids. (Reproduced from Kepler, *Mysterium Cosmographicum*, 1596.)

To the historian this is familiar ground. Scientists from Kepler to Kekulé, from Newton to Crick and Watson, were guided in the early stages of their research by a visually powerful, highly symmetric geometrical design. In faithfulness to Müller's self-report, our genealogy should therefore include a new type of resource, *personal thematic presuppositions*, and with it a new line of inheritance, reaching in this case back first to Pauli and Jung and then to the works of Johannes Kepler, four centuries earlier. This added intellectual resource played as big a role in motivating the 1986 discovery as any of the other resources we have mentioned. Other case studies will reinforce the fact that personal thematic presuppositions of various sorts were essential motivators in major advances throughout the history of science.

Some testable hypotheses

What can we learn about scientific discovery in general from this genealogical analysis of a particular advance? Four hypotheses offer themselves that may be found to hold generally for modern

Figure 9. Müller chose the Dharmaraja Mandala *(above)* to illustrate his source of inspiration. (Courtesy of Joachim Baader, Galerie für tibetische Kunst; Munich, Germany.)

science. Although students of scientific discoveries will not find them surprising, we would contend that our genealogical method of analysis has allowed these hypotheses to be put in a more testable, and therefore more useful, form.

1. Borrowing of resources routinely takes place between different traditions within a conventionally defined discipline. For instance, among the theoretical ancestors of the Bednorz-Müller work are ideas from thermodynamics, statistical mechanics, the old quantum theory, quantum mechanics, and quantum field theory. Even when a given work superficially appears to be the result of a narrow line of research, it is likely to have a deep and broad ancestry. When scientists borrow from different subfields within a discipline, these can blend together or be transformed in an alchemical process that turns them into gold. One may also note that the unpredictable way scientists reach back to earlier research done in a different part of the discipline suggests it would be futile to attempt to "rationalize" or "direct" this process, but argues for making a scientific education as wide-ranging as possible.

2. *Borrowing of resources also routinely occurs across traditional boundaries between disciplines.* The Bednorz-Müller work borrowed directly or indirectly from a wide variety of different disciplines, each with its own professional societies and journals. They included physical chemistry, material science, crystallography, metallurgy, electronics, and low-temperature techniques. This feature, most obvious in experimental projects but also found in theoretical ones, has, we suggest, become more and more characteristic of modern scientific work.

3. *Basic research borrows resources from applied research, and applied research borrows resources from basic research.* A good example of this symmetrical exchange is Bednorz and Müller's use of a SQUID (superconducting quantum interference device) to measure changes in magnetic fields. The initial pursuit of superconductivity itself can be regarded as basic, or "curiosity-driven." So can Brian Josephson's prediction that superconducting currents can tunnel across an insulating film. But then the Josephson effect led to the production of SQUIDs, making possible exquisitely sensitive magnetic susceptometers, whose development is considered a piece of applied work. The susceptometers, however, proved useful in further basic research into superconductivity, including Bednorz and Müller's. In short, to use a metaphor from physics, the exchange of energy between pure and applied research resembles the exchange of energy between a pair of coupled pendulums. Such feedback effects can be observed even within discipline-oriented research lines.

4. *Our findings emphasize the great importance for scientific research of unintended interactions or applications.* Most borrowed resources had been developed by others in research with a goal quite different from that of the eventual borrower. Moreover, the

research of the borrower also often ends up somewhere other than the intended destination; for example, Onnes's initial discovery of superconductivity was based on ideas of Kelvin's, which predicted "exactly the opposite of what was found eventually" (Meijer 1994). As noted, Bednorz and Müller discovered high-temperature superconductivity by studying a compound that had been synthesized and researched by others for unrelated purposes. Nor could the team have predicted that now there exist more than a hundred high-temperature superconductors, as well as a growing set of industrially promising applications—motors, transformers, thin films, and power cables—some of them already on a production basis. What is more, all of this has transpired in the continued absence of any consensus about the mechanism of high-temperature superconductivity.

The Müller-Bednorz story, replete with unpredictable turns of events and rife with unintentionality, has yet another twist. The perovskite structure inspired Müller and Bednorz to gamble on investigating the oxides in the first place. The barium cuprate compound they prepared contained well-separated planes of copper and oxygen atoms, and these layers turned out to be a universal property of high-temperature superconductors. Moreover, these layers exist because the compound is not, after all, a true perovskite; because of the way its unit cells stack, it has orthorhombic rather than cubic symmetry. As Müller said to us in this connection, although Kepler was initially dedicated to decomposing planetary orbits into perfect circles, he was eventually led to ellipses instead—but thereby helped prepare for Newton's *Principia*.

Unity in science

If these four hypotheses are more generally confirmed, they will have the effect of providing support for the old assumption that there is some underlying unity in science, perhaps not of the Theory-of-Everything variety but of a different, operational kind.

A distinguished and vocal minority of scientists (including Philip W. Anderson) has asserted that we should not look for unifying theories emerging from the study of elementary particles, and that each area of science, such as biology or fluid dynamics, has its own laws, which cannot be derived from something more fundamental. Those arguments, whether right or wrong, do not touch our idea of unity, which is exemplified in the ceaseless borrowing that connects diverse traditions and disciplines. In principle, any two research efforts, however removed in time, subject, or purpose, may well turn out to be genealogically connected. And in the limit, the whole of natural science may be represented as one thickly linked continuum, which can be divided into distinct disciplines and traditions only in a more or less arbitrary way. However they may differ, the multitudinous projects of science at any given moment share in and emerge from a common history.

Implications for science policy

This study has significant implications for science policy. It suggests, first of all, that far more attention should be paid to the history of actual advances, which demonstrate, in operational terms, that major accomplishments in science depend on healthy systems of education and research administration that nurture a mixture of basic, applied, and instrument-oriented developments. The traditions and management styles of laboratories and their parent institutions can greatly advance or hinder research. At the Zurich laboratory, Müller and Bednorz benefited from access to highly trained machine and glassblowing technicians, schooled in the traditions of excellence and craftsmanship that can be traced back to the guilds of previous centuries. Then too, it was probably not an accident that their discovery, which is basically an advance in the science of materials, occurred at a laboratory with a long-standing commitment to this science, most notably to the study of ferroelectricity.

But the most striking feature of the culture at the Zurich laboratory was the willingness to give good people the freedom to pursue projects with long gestation periods. This was rewarded twice in quick succession. The year before Bednorz and Müller won the Nobel Prize in physics, the prize had been awarded to Gerd Binnig and Heinrich Rohrer, also of the Zurich laboratory, for their patient development of the scanning tunneling microscope. The stories of the transistor and the laser also suggest that the chance of serendipitous encounters with key ideas is increased by permitting research to proceed at an unforced pace.

We recognize here the well-known phenomenon of the self-amplification of confident, successful, high-quality cultures. They exhibit what Robert K. Merton of Columbia University, who pioneered the modern sociology of science, has memorably termed the Matthew effect (a reference to the text: "Unto every one that hath shall be given, and he shall have abundance . . ." Matthew 25:29).

It is equally important that the system of research administration encourage the flexibility that promotes borrowing within and across disciplines and between basic and applied research. The culture of the laboratory, including its financing, should allow both a natural, unforced pace of work and a degree of self-direction of the scientific imagination that allows researchers to draw on the personal sources of inspiration on which administrative rule books and traditional science texts are so silent.

But above all, our research suggests that the current debate about the relative merits of and support warranted for basic and mission-oriented research is oversimplified. Historical study of cases of successful modern research has repeatedly shown that the interplay between initially unrelated basic knowledge, technology, and products is so intense that, far from being separate and distinct, they are all portions of a single, tightly woven fabric (Mort 1994; Ehrenreich 1995). Even research that narrowly targets a specific application sooner or later must rely on results from a wide spectrum of research areas. For example, it is sometimes said that Irving Langmuir looked into blackened light bulbs and so created modern surface chemistry. But of course his achievement did not spring full-fledged from his brow. Its genealogy, if traced back as carefully as we have traced the genealogy of Bednorz and Müller's discovery, would quickly reveal the crucial role of many types of research in earlier generations and in different fields.

If we wish to achieve noteworthy science, even if noteworthy is defined to mean only science with an economic payoff, we have no alternative but to support the seamless web of research.

I gladly acknowledge advice on early drafts received from J. Georg Bednorz, George Benedek, Henry Ehrenreich, Theodore H. Geballe, Marc A. Kastner, K. Alex Müller, and Michael Tinkham. I am also grateful for support of this study from the Andrew W. Mellon Foundation.

G. H.

BIBLIOGRAPHY

Bednorz, J. G., and K. A. Müller. 1986. Possible high T_c superconductivity in the Ba-La-Cu-O system. *Zeitschrift für Physik B* 64:189–193.

———. 1987a. Perovskite-type oxides—the new approach to high-T_c superconductivity (Nobel lecture). In *Les Prix Nobel 1987*. Stockholm: Almqvist & Wiksell International, pp. 65–98. (Also printed in *Reviews of Modern Physics*, 1988, 60:585–600.)

———. 1987b. A road towards high T_c superconductivity. *Japanese Journal of Applied Physics* 26 (Supplement 26–3):1781–1782.

Bednorz, J. G., M. Takashige, and K. A. Müller. 1987a. Preparation and characterization of alkaline-earth substituted superconducting La_2CuO_4. *Materials Research Bulletin* 22:819–827.

———. 1987b. Susceptibility measurements support high-T_c superconductivity in the Ba-La-Cu-O system. *Europhysics Letters* 3:379–386.

Bromberg, J. L. 1995. Experiment vis-à-vis theory in superconductivity research: The case of Bernd Matthias. In *Physics, Philosophy, and the Scientific Community*, ed. K. Gavroglu et al. Boston: Kluwer Academic Publishers.

Chakravarty, S. 1994. Cuprate superconductors: A broken symmetry in search of a mechanism. *Science* 266 (21 October): 386–387.

Dahl, P. F. 1992. *Superconductivity: Its Historical Roots and Development*. New York: American Institute of Physics.

Dordick, Rowan. 1987. The field of superconductivity heats up. *IBM Research Magazine* 25(1):15–17.

Ehrenreich, H. 1995. Strategic curiosity: Semiconductor physics in the 1950s. *Physics Today* 48(1):28–34.

Felt, U., and H. Nowotny. 1992. Striking gold in the 1990s: The discovery of high-temperature superconductivity and its impact on the science system. *Science, Technology & Human Values* 17:506–531.

Höck, K-H., H. Nickisch, and H. Thomas. 1983. Jahn-Teller effect in itinerant electron systems: The Jahn-Teller polaron. *Helvetica Physica Acta* 56:237–243.

Khurana, A. 1987a. Bednorz and Müller win Nobel prize for new superconducting materials. *Physics Today* 40 (12):17–19.

———. 1987b. Superconductivity seen above the boiling point of nitrogen. *Physics Today* 40 (4):17–23.

Matricon, J., and G. Waysand. 1994. *La Guerre du Froid: Une Historie de la Supraconductivité.* Paris: Editions du Seuil.

Meijer, P. H. E. 1994. Kammerlingh Onnes and the discovery of superconductivity. *American Journal of Physics* 62:1105.

Mort, J. 1994. Xerography: A study in innovation and economic competitiveness. *Physics Today* 47(4):32–38.

Müller, K. A. 1988. Äussere und innere Forschungserfahrung und Erwartung. *Technische Rundschau* (44):2–4.

Müller, K. A., and J. G. Bednorz. 1987. The discovery of a class of high-temperature superconductors. *Science* 237:113.

Müller, K. A., M. Takashige, and J. G. Bednorz. 1987. Flux trapping and superconductive glass state in La_2CuO_4–yBa. *Physical Review Letters* 58:1143–1146.

Pauli, W. 1952. Der Einfluss archetypischer Vorstellungen auf die Bildung naturwissenschaftlicher Theorien bei Kepler. In *Naturerklärung und Psyche*, C. G. Jung and W. Pauli. Zurich: Rascher Verlag. (Translated into English by R. F. C. Hull and P. Silz as *The Interpretation of Nature and the Psyche*, published in 1955 by Pantheon Books.)

Robinson, A. L. 1987. Research news. *Science* 235:1571.

Schechter, Bruce. 1989. *The Path of No Resistance: The Story of the Revolution in Superconductivity*. New York: Simon and Schuster.

Thomas, H., and K. A. Müller. 1972. Theory of a structural phase transition induced by the Jahn-Teller effect. *Physical Review Letters* 28:820–823.

Tinkham, Michael. 1975. *Introduction to Superconductivity.* New York: McGraw Hill. Second edition 1996.

Part I: On the thematic analysis of science

1

Themata in scientific thought

When the historian, philosopher, sociologist, or psychologist of science studies a product of scientific work – a published paper, a laboratory record, a transcript of an interview, an exchange of letters – he is usually dealing primarily with an *event*.[1] We can distinguish at least eight different facets of such events, each facet corresponding to a different type of interesting research question.

First is, of course, the understanding of the scientific content of the event at a given time, both in contemporaneous terms and, separately, in terms of what we now believe to be the case. What did the scientist claim was at issue? What was he in fact confronted with? For this we try to establish his awareness (within the area of public scientific knowledge at the time of the event) of the so-called scientific facts, data, laws, theories, techniques, and lore. I would include under this heading the larger part of historical research on what are called scientific world views, exemplars, and research programs. Chiefly, however, historians and scientists are still concerned with digging out the concepts and propositions embodied in the events studied and with rendering them in empirical and analytical language.

The second is the time trajectory of the state of shared (that is, "public" rather than "private") scientific knowledge that led up to and perhaps goes beyond the time chosen for the event. Establishing this means, so to speak, the tracing of the World Line of an idea or a subject of research, a line on which the event (E) is a point. Whether we are studying the problem of falling bodies from Kepler to Newton, or the flowering of quantum electrodynamics from Feynman to the last issue of *Physical Review Letters*, under this heading we are dealing with antecedents, parallel developments, continuities and discontinuities, and the like. This tracing of conceptual development and of the "context of justification" is the most frequent and the strongest activity of historians of science and historically inclined science educators.

Third is the more ephemeral personal aspect of the activity in which *E* is embedded. Here we are in the context of discovery, trying to understand the "nascent moment," which may be poorly documented and not necessarily appreciated or understood even by the agent himself. Except for work on a few figures such as Kepler or Einstein, scientists (and philosophers) until recently have been rather impatient with such studies. The very institutions of science – the methods of publication, the meetings, the selection and training of young scientists – are designed to minimize attention to this element. The success of science itself as a sharable activity seems to be connected with this systematic neglect of what Einstein called the "personal struggle." Moreover, the apparent contradiction between the often "illogical" nature of actual discovery and the logical nature of well-developed physical concepts is perceived by some as a threat to the very foundations of science and rationality itself.

The alternative path is not easy. In one of his interviews, Einstein urged historians of science to concentrate on comprehending what scientists were aiming at, "how they thought and wrestled with their problems." But he pointed out that the scholar would have to have sufficient insight, a kind of educated sensitivity both for the content of science and for the process of scientific research, as solid facts about the creative phase are likely to be few; and that, as in physics itself, the solution to historical problems may have to come by very indirect means, the best outcome to be hoped for being not certainty but only a good "probability" of being "correct anyway."[2]

A fourth component of historical research is indeed the establishment of the time trajectory of this largely "private" scientific activity – the personal continuities and discontinuities in development, or science in the making as experienced in the scientist's own personal struggle. Now the event E at time t begins to be seen as the intersection of two trajectories, of two World Lines, one for "public science" (let us call it S_2), and one for "private science" (S_1), to use a shorthand terminology which is useful enough if not pushed too hard.[3]

Fifth, parallel to the trajectory of S_1 and shading into it as one of its boundaries is a band tracing the psychobiographical development of the person whose work is being studied. We are dealing here with the new and tantalizing field which explores the relation between a person's scientific work and his intimate style of life.

Sixth is unavoidably the study of the sociological setting, conditions, or influences which arise from colleagueship, the dynamics of teamwork, the state of professionalization at the time, the institutional means for funding, for evaluation and acceptance, and quantitative trends. Here we deal with the fields of science policy studies and sociology of science in the narrower sense.

Seventh, a similar band, parallel to and shading into the trajectories of S_1 and S_2 deals with cultural developments outside science that influence science or are influenced by it – with questions concerning the feedback loops in the entities science-technology-society, science-ethics, and science-literature.

Finally, there is the logical analysis of the work under study. In my own development, first as a student of P. W. Bridgman and Philipp Frank and later as their colleague, interest in and respect for a valid analysis of the logic of science in fact preceded work in the analysis of the more strictly historical aspects of a case.

These eight areas of study are not separated by hard barriers. To be sure, each has invited its own specialization and thereby its own operational self-definition. For each we could quickly put forward the names of heroes and the shape of future hopes of development – though we might now all agree, with various intensities of regret, that the resolution of a real case in the history of science (in all its ambiguities and interdisciplinary connections) in separable components is, after all, a reductionistic strategy which our human limitations force or doom us to employ.

Toward thematic analysis

The method of dealing with complex entities by resolution or reduction found its use in science itself very early – for example, in the passage in the Second Day of Galileo's *Dialogo,* where Salviati and Simplicio are discussing the motion of an object released from the mast of a moving ship. Simplicio refuses Salviati's proposal to resolve the motion into a horizontal and a vertical component, one for free fall straight to the center of the earth, the other with constant velocity in the direction of initial motion. Perhaps we should credit Simplicio's resistance to a premonition that the whole method of resolution and reduction is precarious and has no more necessity than any other methodological thema – that is, it is neither verifiable nor falsifiable, and its usefulness depends entirely on how soon you are satisfied with your results.

As we now know, Salviati was grossly exaggerating. Resolving the motion of the falling object into two components in order to understand motion and its causes is only the first step in an essentially infinite chain of resolutions. If one wants more detail about the motion, other laws enter. The appearance of the Coriolis force is responsible for an eastward deflection of the object. The laws for falling bodies in real media at various Reynolds numbers have to enter to calculate the effect of friction and turbulence. The more detail one wants to know, the more resolutions become necessary. The process would have become infinitely regressive if an Occam's Razor had not been invented in our century for cutting off all side effects below a certain limit. Quantum physics did give us a way to stop, owing to the uncertainty principle and the finite size of Planck's constant; they extinguish the meaningfulness of all further questions.

And there is another lesson. The two components Salviati chose, while they were plausible enough and even turned out to be useful, were not endowed with any provable necessity over any other set of two or more components of motion that might have been imagined. I mention this to acknowledge that my list of components is not to be taken as the recital of an unchangeable, sacred Eightfold Way. On the contrary, one reason for making the list was to be able to conclude that it is incomplete in an important respect. In other words, there remains a set of questions that is irresistible (to me, at any rate); that cannot be handled naturally in this eightfold scheme at all; and that lay bare a link between scientific activity and humanistic studies, a link that few have studied so far.

Any listing of such questions must include these: What is constant in the ever-shifting theory and practice of science – what makes it one continuing enterprise, despite the apparently radical changes of detail and focus of attention? What elements remain valuable in science long after the theories in which they are embodied have been discarded? What are the sources of energy that keep certain scientific debates alive for decades? Why do scientists – and for that matter also historians, philosophers, and sociologists of science – with good access to the same information often come to hold so fundamentally different models of explanation? Why do some scientists, at enormous risks, hold on to a model of explanation, or to some "sacred" principle, when it is in fact being contradicted by current experimental evidence?

Why do scientists often privately not acknowledge a dichotomy between the context of verification and that of discovery, and yet publicly accept it? If it is true, as Einstein believed,[4] that the process of formulating laws purely by deduction is "far beyond the capacity of human thinking," what may be guiding the leap across the chasm between experience and basic principle? What is behind the obviously quasi-aesthetic choices that some scientists make – for example, in rejecting as merely ad hoc a hypothesis that to other scientists may appear to be a necessary doctrine? Are the grounds from which such choices spring confined to the scientific imagination, or do they extend beyond it?

To handle such questions I have proposed a *ninth* component for the analysis of a scientific work – that is, thematic analysis (a term familiar from somewhat related uses in anthropology, art criticism, musicology, and other fields). In many (perhaps most) past and present concepts, methods, and propositions or hypotheses of science, there are elements that function as themata, constraining or motivating the individual and sometimes guiding (normalizing) or polarizing the scientific community. In the scientists' own public presentations of their work, and during any ensuing scientific controversy, these elements are usually not explicitly at issue. Thematic concepts do not usually appear in either the index of textbooks or in so many words in the professional journals or debates. Such traditional discussions concern chiefly the empirical content and analytical content, that is, the repeatable phenomena and the propositions concerning logic and mathematics. By way of a very rough analogy, I have suggested that those two elements be considered the x and y coordinates of a plane within which the discussion seems chiefly to proceed, since the "meaningfulness" of concepts is tested by the resolution of concepts or propositions into those elements – "meaningful" in the sense that agreed-upon rules generally exist for the verification or falsification of statements made in that language.

Thus (as we shall see in Chapter 2), in R. A. Millikan's famous oil drop experiment, the question of whether or not the electric charges on small objects always come in multiples of some fundamental constant (called the charge of the electron) could in principle have been resolved quickly by coming to terms on how and what was being observed through the telescope or ultramicroscope when a particle was seen to move in the view field, and whether and how to amend the equation for Stokes's law for the fall of small objects by extrapolation of a correction term. If that were all, the lengthy debate about the existence of a postulated "subelectron" would never have happened. But in 1910, and continuing for some years afterward, the controversy between Millikan and his opponent was joined – at the intersection, as it were, of two sets of World Lines. Analysis of the expressed motivations, and of the ever-hardening attitudes of the protagonists on opposite sides of the question shows here, as in other cases, the strong role of an early, unshakable commitment by the opponents to different themata.

The themata that appear in science can, in our very rough

analogy, be presented as lying along a dimension orthogonal to the *x-y* plane in which verification and falsification can take place, hence somewhat like a *z* axis rising from it. Although the *x-y* plane does suffice for most discourse within science as a public, consensual activity, the three-dimensional (*xyz*) space is required for a more complete analysis – whether historical, philosophical, or psychological – of scientific statements, processes, and controversies. (My argument is not to introduce thematic discussions or even a self-conscious awareness of themata into the practice of science itself. It is indeed one of the great advantages of scientific activity that in the *x-y* plane many questions – for example, concerning the "reality" of scientific knowledge – cannot be asked. Only when such questions were ruled out of place in a laboratory did science begin to grow rapidly.) It is fruitful to make distinctions between three different uses of themata: the *thematic concept*, or the thematic component of a concept (examples I have analyzed are the use of the concept of symmetry and of the continuum); the *methodological thema* (such as the preference for expressing the laws of science where possible in terms of constancies, or extrema, or impotency); and the *thematic proposition* or *thematic hypothesis* (exemplified by overarching statements such as Newton's hypothesis concerning the immobility of the center of the world, or the two principles of special relativity theory).

The attitude I have taken in the task of identifying and ordering thematic elements in scientific discussions is to some degree analogous to that of a folklorist or anthropologist who listens to the epic stories for their underlying thematic structure and recurrence. Although the analogy leaves much to be desired, there are more than superficial relations. For example, the awareness of themata which are sometimes held with obstinate loyalty helps one to explain the character of the discussion between antagonists far better than do scientific content and social surroundings alone. The attachment of physicists such as H. A. Lorentz, Henri Poincaré, and Max Abraham to the old electromagnetic world view and their discomfort with Einstein's relativity theory become a great deal more understandable when the ether is thought of as operating as the embodiment of thematic concepts (for example, of the absolute and the plenum). Thus in their obituary for Abraham, Max von Laue and Max Born wrote perceptively:

> [Abraham] found the abstractions of Einstein disgusting in his very heart. He loved his absolute ether, his field equations, his rigid electron, as a youth loves his first passion whose memory cannot be erased by any later experience . . . His opposition was grounded in physical, fundamental persuasions to which he, purely in accord with his feelings, held on as long as possible . . . [As Abraham himself once said] against the logical coherences he had no counterarguments; he recognized and admired them as the only possible conclusion of the plan of general relativity. But this plan was to him thoroughly unsympathetic, and he hoped that the astronomical observation would disconfirm it and bring the old, absolute ether again into honor.[5]

A finding of thematic analysis that appears to be related to the dialectic nature of science as a public, consensus-seeking activity is the frequent coupling of two themata in antithetical mode, as when a proponent of the thema of atomism finds himself faced with the proponent of the thema of the continuum. Antithetical ($\Theta\bar{\Theta}$) couples – such as evolution and devolution, constancy and simplicity, reductionism and holism, hierarchy and unity, the efficacy of mathematics (for example, geometry) versus the efficacy of mechanistic models as explanatory tools – are not too difficult to discern, particularly in cases that involve a controversy or a marked advance beyond the level of common work.

I have been impressed by how few themata there are – at least in the physical sciences. I suspect the total of singlets, doublets, and occasional triplets will turn out to be less than 100. The appearance of a new thema is rare. Complementarity in 1927 and chirality in the 1950s are two of the most recent such additions in physics. Related to that is the antiquity and persistence of themata, right through scientific evolution and "revolution." Thus the old antithesis of plenum and void surfaced in the debate early this century on "molecular reality" – indeed, it can be found in the work of contemporary theoretical physicists. One may even predict that, no matter how radical the advances will seem in the near future, they will with high probability still be fashioned chiefly in terms of currently used themata.

The persistence in time, and the spread in the community at a given time, of these relatively few themata may be what endows science, despite all its growth and change, with what constant identity it has. The interdisciplinary sharing of themes among various fields in science tells us something about both the meaning of the enterprise as a whole and the commonality of the ground of imagination that must be at work.

An illustration

To illustrate some of these points and to show that current as well as historical cases are amenable to this analysis, I want to focus on an example in one of the liveliest fields of physics today, as embodied in publications of Steven Weinberg.[6] The line tracing the development of Weinberg's thoughts intersects the trajectory of a stream of developments in quantum electrodynamics initiated by Enrico Fermi in 1934 and now basing itself on techniques started independently in the late 1940s by R. P. Feynman, Julian Schwinger, Freeman J. Dyson, and Sinitiro Tomonago. Other points on the trajectory include discoveries by groups at CERN, Argonne Laboratory, and the National Accelerator Laboratory. In thematic terms, the "event" we shall study is only the latest in a very old sequence of attempts, reaching past many revolutions and heady victories back to the first scientist of recorded history; for the main preoccupation is the fundamental constituent of which all matter is presumed to be made.

To put it briefly, Weinberg, his collaborators, and other groups have been working on the problem of finding common ground between the four types of interaction ("forces") that are now believed to account for all physical phenomena: the gravitational interaction that all particles experience; the electromagnetic force that accounts for phenomena involving charged particles and the interaction of light with matter; the "strong" nuclear force that acts between members of the large family of elementary particles called hadrons;[7] and the "weak interaction" postulated to describe extremely short-range interactions of some elementary particles (such as the scattering of a neutrino by a neutron, and the radioactive decay of a neutron into a proton, an electron, and an antineutrino).

In 1967, Weinberg (and, independently, Abdus Salam of Trieste) proposed that the electromagnetic force and the weak interaction are essentially connected. Each of the four types of interaction has been considered to be the result of processes analogous to radiation or absorption between two interacting objects, the particle radiated or absorbed being characteristic for each of the interactions. Thus electromagnetic phenomena are due to the exchange of the massless photon, whereas the gravitational interaction is thought to be due to the exchange of particles called gravitons. The weak interaction is mediated by the so-called intermediate vector boson (IVB) which, if it is found to exist, will have to be exceedingly massive.[8] Weinberg's proposal was that the massless photon and the very massive IVB are close relatives – that the IVBs are by and large members of the photon family but get their mass (the appearance of their difference) by virtue of being associated with broken gauge symmetry groups.

At the time Weinberg proposed the theory, "there was," he notes,[9] "no experimental evidence for or against it, and no immediate prospects for getting any." To this day the IVBs cannot be produced directly (for instance, in accelerators), but indirect evidence for their existence has been reported. In a paper published under the names of fifty-five investigators from seven institutions in a Pan-European collaboration at the CERN laboratory,[10] two events were found in which a mu-neutrino was scattered by an electron, and several hundred events in which a mu-neutrino was scattered by a neutron or a proton. (The latter reaction showed up nicely also on more recent experiments at Argonne National Laboratory and the National Accelerator Laboratory.) This is evidence that the "neutral current" reaction, a new kind of weak interaction predicted by Weinberg involving the postulated neutral IVB, may be taking place,[11] and so indirectly supports the theory which makes these particles a member of the same family as the photons.

Moreover, strong interactions also become amenable to calculations with the same methods as are used for weak and electromagnetic interactions. It is possible, therefore, that the strong interactions are caused by exchange of particles that belong to the same family as the photon and the IVB. "If these speculations are borne out by further theoretical and experimental work," Weinberg says in the last sentence of his recent survey, "we shall have moved a long way toward a unified view of nature" (p. 59; see n. 9).

Now let us go to the beginning of this same report, which is entitled "Unified Theories of Elementary-Particle Interaction," and look at it through eyes alert to themata. What, then, are the thematic conceptions, methodological themata, and thematic suppositions that inhere in this search for the IVBs and their photonlike family membership?

We can list a few of the more evident themata when we scan just the first page (see Figure 1.1) of the article.[12] It begins:

> One of man's enduring hopes has been to find a few simple general laws that would explain why nature, with all its seeming complexity and variety, is the way it is. At the present moment the closest we can come to a unified view of nature is a description in terms of elementary particles and their mutual interactions. All ordinary matter is composed of just those elementary particles that happen to possess both mass and (relative) stability: the electron, the proton and the neutron. To these must be added the particles of zero mass: the photon, or quantum of electromagnetic radiation, the neutrino, which plays an essential role in certain kinds of radioactivity, and the graviton, or quantum of gravitational radiation . . .

What strikes us at once is the acknowledgment that "one of man's enduring hopes has been to find a few simple general laws" and thereby obtain a theory that is "unified" (the first word of the title). Unification or synthesis, with its promise of increased understanding through increased economy of thought, is a member of a connected set of themata, one of its opposing aspects being multiplicity (or complexity, variety), but the chief antithetical thema being one we discussed before, that of resolution or analysis rather than synthesis. Each of these members of the constellation has it uses (as will be further studied in Chapter 4). Here, clearly, unification is taken to be preeminent.

". . . Why nature, with all its seeming complexity and variety, is the way it is." Kepler, who asked in the preface of the

Unified Theories of Elementary-Particle Interaction

Physicists now invoke four distinct kinds of interaction, or force, to describe physical phenomena. According to a new theory, two, and perhaps three, of the forces are seen to have an underlying identity

by Steven Weinberg

One of man's enduring hopes has been to find a few simple general laws that would explain why nature, with all its seeming complexity and variety, is the way it is. At the present moment the closest we can come to a unified view of nature is a description in terms of elementary particles and their mutual interactions. All ordinary matter is composed of just those elementary particles that happen to possess both mass and (relative) stability: the electron, the proton and the neutron. To these must be added the particles of zero mass: the photon, or quantum of electromagnetic radiation, the neutrino, which plays an essential role in certain kinds of radioactivity, and the graviton, or quantum of gravitational radiation. (The graviton interacts too weakly with matter for it to have been observed yet, but there is no serious reason to doubt its existence.) A few additional short-lived particles can be found in cosmic rays, and with particle accelerators we can create a vast number of even shorter-lived species [*see top illustration on page 52*].

Although the various particles differ widely in mass, charge, lifetime and in other ways, they all share two attributes that qualify them as being "elementary." First, as far as we know, any two particles of the same species are, except for their position and state of motion, absolutely identical, whether they occupy the same atom or lie at opposite ends of the universe. Second, there is not now any successful theory that explains the elementary particles in terms of more elementary constituents, in the sense that the atomic nucleus is understood to be composed of protons and neutrons and the atom is understood to be composed of a nucleus and electrons. It is true that the elementary particles behave in some respects as if they were composed of still more elementary constituents, named quarks, but in spite of strenuous efforts it has been impossible to break particles into quarks.

For all the bewildering variety of the elementary particles their interactions with one another appear to be confined to four broad categories [*see bottom illustration on page 52*]. The most familiar are gravitation and electromagnetism, which, because of their long range, are experienced in the everyday world. Gravity holds our feet on the ground and the planets in their orbits. Electromagnetic interactions of electrons and atomic nuclei are responsible for all the familiar chemical and physical properties of ordinary solids, liquids and gases. Next, both in range and familiarity, are the "strong" interactions, which hold protons and neutrons together in the atomic nucleus. The strong forces are limited in range to about 10^{-13} centimeter and so are quite insignificant in ordinary life, or even on the scale (10^{-8} centimeter) of the atom. Least familiar are the "weak" interactions. They are of such short range (less than 10^{-15} centimeter) and are so weak that they do not seem to play a role in holding anything together. Rather, they are manifested only in certain kinds of collisions or decay processes that, for whatever reason, cannot be mediated by the strong, electromagnetic or gravitational interactions. The weak interactions are not, however, irrelevant to human affairs. They provide the first step in the chain of thermonuclear reactions in the sun, a step in which two protons fuse to form a deuterium nucleus, a positron and a neutrino.

From this brief outline one can see that a certain measure of unification has been achieved in making sense of the world. We are still faced, however, with the enormous problem of accounting for the baffling variety of elementary-particle types and interactions. Our prospects for further progress would be truly discouraging were it not for the guidance we receive from two great products of 20th-century physics: the development of quantum field theory and the recognition of the fundamental role of symmetry principles.

The Necessity of Fields

Quantum field theory was born in the late 1920's through the union of special relativity and quantum mechanics. It is easy to see how relativity leads naturally to the field concept. If I suddenly give one particle a push, this cannot produce any instantaneous change in the forces (gravitational, electromagnetic, strong or weak) acting on a neighboring particle because according to relativity no signal can travel faster than the finite speed of light. In order to maintain the conservation of energy and momentum at every instant, we say that the pushed particle produces a field, which carries energy and momentum through surrounding space and eventually hands some of it over to the neighboring particle. When quantum mechanics is applied to the field, we find that the energy and momentum must come in discrete chunks, or quanta, which we identify with the elementary particles. Thus relativity and quantum mechanics lead us naturally to a mathematical formalism, quantum field theory, in which elementary-particle interactions are explained by the exchange of elementary particles themselves.

Figure 1.1. First page of Steven Weinberg's article, *Scientific American* 231, no. 1 (July 1974): 56.

Mysterium Cosmographicum why the planets are at the distances they are, of the number and with the motions which we find them to have, "and not otherwise," would have agreed with this description of one of man's enduring hopes. So would most scientists since. The second sentence, however, bares a preconception which not all scientists will share. We find here a new thematic commitment, that of constructing the desired unified view of nature out of "elementary particles and their mutual interactions." We hear the echo of Democritus's "all is atoms and void." But as we shall note shortly, not all physicists in our time have subscribed to this belief. Nor of course would biologists, psychologists, or social scientists be satisfied with this particular unified view of *nature* in terms of particles and their interactions. A choice has been made here, though a choice that promises indeed a breathtaking unification of this part of nature.

What is being conveyed in Weinberg's opening passage by "elementary"? A few sentences later (col. 1, bottom) it is defined to mean that there is not now "any successful theory that explains the elementary particles in terms of more elementary constituents." Some day, to be sure, one may find "still more elementary constituents, named quarks" (col. 2, top); but until that time, so long as "strenuous efforts" make it "impossible to break particles," they are elementary.

This quality of being elementary anchors the whole arrow of explanation, upward from these presumed elementary particles to the antithetical entities, constructs (such as nuclei [col. 1, bottom], atoms, or ordinary matter, all of which are "composed" of elementary matter). The antiquity of that quest, from Thales to Prout to J. J. Thomson to our day, is evident. These elementary particles, then, are today's true "atoms" in the sense of the Greek *atomos*. They form one leg of another triplet of themata, the second being the *construct* made of and explained by these atoms or elementary quanta, and the third being the notion of the *continuum*, the indefinitely cuttable.[13]

The list of elementary particles then consists of the electron, the proton, and the neutron. "To these must be added the particles of zero mass: the photon . . . the neutrino . . . and the graviton" (col. 1, one-third down). We are here clearly in a world of particulate discreteness; although the wave property that inheres in such particles is of course not doubted, it simply is not part of the image that has captured attention and primacy.[14]

The number and variety of elementary particles, Weinberg says, are "bewildering." But there are ways of retaining sanity and gaining insight by mastering the bewildering variety. The ordering of chaos by means of the concept of hierarchy or levels of categories – a manageable few, just four – comes to the rescue as a methodological theme. The division into four categories – gravitation, electromagnetic interaction, strong interactions, and weak interactions – is not merely a separation into separate pigeonholes for very different birds. There is a real hierarchy here which orders the subsections, showing a sequence of ranges of interactions, from infinity to much less than 10^{-14} centimeter.

Already one can see from this brief outline that, as Weinberg puts it, "a certain measure of unification has been achieved in making sense of the world" (col. 3, top). Helping to make sense of the world in a way not possible through the demands of logicality alone is indeed one of the chief functions of a thema. "We are still faced, however, with the enormous problem of accounting for the baffling amount of elementary-particle types and interactions" (col. 3, top). Methodologically, the theory evokes more than an echo of an older scheme of fourfold categories, one so magnificently successful that it helped to rationalize the observable phenomena for some 2,000 years: the four Elements, with their own internal hierarchy, from lightest to heaviest, and their own rules of interaction. However, the new unification through hierarchical ordering promises among its many advantages that two and perhaps three of the forces in the four categories "have an underlying identity."

The way to discover this identity is through analogies in behavior which would collapse the superficially different entities to a state in which they share something more than membership in a hierarchical order. This quest for something more is answered by turning to the conception of family (for example: "Our hopes of perceiving an underlying identity in the weak and electromagnetic interactions lead us naturally to suppose that there may be some larger gauge symmetry that forces the photon and the intermediate vector boson into a single family" [p. 55; see n. 9]). The chief explanatory tool on the road to greater simplicity is this "family" connection, existing despite an "appearance" of greater differences – for example, the difference between the photon's zero mass and the necessarily very large mass of the intermediate vector boson. Throughout Weinberg's article, and many others in this field, one of the recurring conceptions is precisely this splendid one of groups, families, and superfamilies ("superfamilies of eight, ten, or even more members").[15] The familial relationships between the elementary particles are far more profound than in the ad hoc families that were discovered in the chemical periodic table in the past century or in, say, the work of Linnaeus. But the methodological use as a tool of explanation is not qualitatively different.

Let me take the occasion of the surfacing of this fine anthropomorphic word to go back to Weinberg's opening page, where reference is made to "a few additional short-lived particles" and we are told that "we can create a vast number of even shorter-lived species." Elementary-particle physics is sometimes wryly referred to as zoology. Indeed, it is shot through and through with themes that may well have their origins in a part of the imagination that was formed prior to the conscious decision of the researcher to become a scientist. The technical report of, say, the analysis of a bubble chamber photograph is cast largely in terms of a life-cycle story. It is a story of evolution and devolution, of birth, adventures, and death. Particles enter on the scene, encounter others, and produce a first generation of particles that subsequently decay, giving rise to a second and perhaps a third generation. They are characterized by relatively short or relatively long lives, by membership in families or species.[16]

Listening to these village tales told by physicists, one is aware that the terminology may initially not have been meant "seriously." Yet the life-cycle thema works, and so do a number of other themata imported into the sciences from the world of human encounters. It has always seemed curious to me how strenuously the psychologists of the period around the turn of this century tried to gain added respectability by borrowing concepts from physics for the description of human relationships. Evidently they were unaware that they were reimporting conceptual tools when they themselves were closer to the real thing. One is reminded of the story of the bank building in Athens, under the Acropolis, which looked like a particularly bad copy of a Greek temple. It turned out that the architect had not taken as his model one of the great temples right at hand, but had gone to a much more fashionable source. His design was based on that of a bank in Berlin which in turn had been derived from a distant, third-rate copy of an idealized Greek temple.

We have not yet finished with Weinberg's first page. Several other magnificent themata begin to show themselves: *isotropy* and *homogeneity* (for example, particles of the same species are, as far as we now know, "absolutely identical, whether they occupy the same atom or lie at opposite ends of the universe" [col. 1, two-thirds down]); *symmetry* (col. 3); and *conservation* ("of energy and momentum at every instant" [col. 3]).

In later pages we encounter the following additional themata, among others: the efficacy of geometrical representation (such as Feynman diagrams), the efficacy of integers as explanatory tools (the debt of modern quantum mechanics to the holiest precept of Pythagoras), again conservation (of charge), infinity and finiteness (of mass), more on symmetry principles,[17] and, above all, models (p. 57; see n. 9). The word "model" is probably one of the most frequently used words in the writings by theoretical physicists.

In this manner we are brought to the last sentence in the paper. It has been quoted previously, but we can now look at it in a somewhat new light: "If these speculations are borne out by further theoretical and experimental work [meaning, by analytical or formalistic as well as empirical content, or by y and x axis representations] we shall have moved a long way toward a unified view of nature" – that is, toward the fulfillment of one of man's enduring hopes, hopes that find expression in his themata, some new and many ancient. In this case, the hope rests on the Democritean thematic commitment to a corpuscular or atomistic point of view to explain physical phenomena – not on its opposite, the thema of the primacy of the continuum, as in the work of the theorist who explained matter as singularities or vortexes in a fluid or field, and who could not believe quantum discreteness to be truly basic. Most modern physicists are thematically Democriteans, but Einstein, Erwin Schrödinger, and others to whom the fundamental tool of explanation was the continuum, passionately disagreed; if discreteness had to be adopted as basic in atomic processes, one of them asserted, he would prefer giving up being a physicist.

Between such thematic opposites, there is no simple way to arrive at consensus. Werner Heisenberg was one of those who tried to convince Einstein.[18] He reported: "It was a very nice afternoon that I spent with Einstein, but still when it came to the interpretation of quantum mechanics I could not convince him and he could not convince me. He always said, 'Well, I agree that any experiment the results of which can be calculated by means of quantum mechanics will come out as you say, but still such a scheme cannot be a final description of Nature.' " Heisenberg understood the impossibility of resolving such basic preconceptions by appealing to the kind of arguments that work so well to bring about scientific consensus on other matters. He added, "I doubt whether the unwillingness of Einstein, Planck, von Laue, and Schrödinger to accept [quantum-mechanical descriptions as basic] should be reduced simply to prejudices. The word 'prejudice' is too negative in this context and does not cover the situation."

As if to demonstrate the truth of his own remark, Heisenberg went on to reveal that, contrary to most contemporary scientists, he himself could no longer agree with the thrust of current theory which makes the notion of "elementary particle" a basic reference point for explanation. Since elementary particles can be generated by collision of other particles, *they*, he felt, are really the complications that require explanation themselves; "or to formulate it paradoxically: Every particle consists of all other particles." The search for "*really* elementary particles" on which to base a theory of matter goes "back to this philosophy of Democritus," but it is an "error" (or, in our terms, at least a commitment to a thema abhorrent to him).

His commitment was in another direction: "What then has to replace the concept of a fundamental particle? I think we have to replace this concept by the concept of a fundamental symmetry . . . And when we have actually made this decisive change . . . then I do not think that we need any further breakthrough to understand the elementary – or rather non-elementary – particle." Elsewhere Heisenberg explains "the 'thing-in-itself' is for the atomic physicist, if he uses the concept at all, finally a mathematical structure." This is a thematic choice that aligns Heisenberg with the great Platonic tradition: One cannot build matter out of matter, but must seek the base in formal, mathematical principles; for "our elementary particles are comparable to the regular bodies of Plato's *Timaeus*. They are the original models, the idea of matter."[19]

To be sure, anyone who has studied the rise and fall in the acceptance of a thema will wonder whether it is not premature to believe the ancient opposition between Democritean and Platonic approaches has been settled once and for all, in our day, in favor of one rather than the other. We are dealing here not with resolvable puzzles, but with the raw material of the scientific (and not only the scientific) imagination.

A second example

Having examined the rich texture of themata in a particular imaginative publication by a major contemporary scientist, we can look at the same matter in another way – namely, by following a particular thema–antithema couple through the history of modern science.

To give an example: The whole tradition in physics which was founded in Newton's time held that any evidence of chaos or of uncertainty must rest on, and be explained by, an underlying layer of order and certainty, even as the seemingly erratic observable motion of planets in Greek science had been understood as the complex results of many simple and orderly motions superposed on one another. This prototype for explanation (classical causal sequences account for observed accident or disorder) is a thematic commitment. It is not an experimental or logical necessity. Indeed, it seemed endangered by the introduction in the mid-nineteenth century of imagery of the opposite kind, originating in kinetic theory. Now it turned out that a good way to understand cases of simple order was to imagine them to be the result of underlying chaos. Thus a balloon filled with gas under pressure and observed to be at rest on the table is understood by saying that an immense number of gas particles, all with different speeds and directions, is unceasingly colliding with the inside surface of the balloon. The hail of collisions just cancels in all directions, and results in the object remaining at rest. Simple order on the visible level was thus explainable by chaos on the invisible level.

It was, however, not merely an accident that it fell on Einstein, in his 1905 paper on Brownian motion, once more to reverse the direction of the arrow of explanation – Einstein (who did not believe that God played at dice) reestablishing the ontological priorities that went back to Newton (who had written that God was a "God of Order"). Einstein's success was to explain the erratic, eternal dancing motion of tiny but visible particles of dust seen with a microscope. The seemingly accidental motions of that visible microscopic world were, Einstein found, entirely explainable by positing that the simple Newtonian laws which guide the motion of two colliding billiard balls also explain the action of the invisible, submicroscopic molecules bombarding the dust particle. A Newtonian order could be taken to be at the bottom of things, after all.

With the development of quantum physics, however, it became more and more clear that the appearance of Newtonian order among colliding particles was itself explained best by considering that order (on the scale of measurement which had been satisfactory for the purpose so far) was merely the apparent result of a large sum of atomic events, each of which individually is subject to the laws of chance – even as earlier the quiescence of a blown-up balloon on the table could be regarded as the result of canceling accidents and agitations within. With his Uncertainty Principle, Heisenberg was saying that the fundamental explanatory thema is after all not the simple, causal, point-by-point sequence typical of, say, the progress of a satellite orbiting around a central planet, but the probabilistic sequence of a random-number generator or a game of chance. The ontological ladder was once more turned around.

And once more, attempts were and are still being made to reverse it yet again. Einstein himself, and a small but hardy group that followed him, never accepted as verified the primacy of the thema of fundamental probabilism in physical nature.

Their hope has been to show that beneath the layer at which the Uncertainty Principle operates, there is yet another level of nature where hitherto inaccessible, hidden mechanisms act, on classical principles, to yield the appearance of randomness in atomic processes – chaos out of order, not vice versa.

Caveats

I have not come as John the Baptist; and would indeed like to avoid his fate. Let me therefore end with a list of limitations I see in the thematic analysis of scientific work.

1. While themata can have a strong grip on the scientist or the community, and can be the most interesting aspect of a given case, there exist important parts of the history of science and of current work where themata do not seem to enter prominently. In studying the case of the work of Enrico Fermi and his group (cf. Chapter 5), I found it no great help to think of it thematically.

2. Even if this were not true, I would not like it to be thought that the themata in a scientific work are its chief reality. Otherwise, work in the history of science would degenerate into descriptivism, and scientific findings would seem to be on a par with the tales of the old men in the hills of Albania, to whom today's story is just about as good or as bad as yesterday's. There is in science evidently a sequence of refinements, a rise and fall, and occasionally the abandonment or introduction of themata. But also there undoubtedly has been on the whole a progressive change to a more inclusive, more powerful grasp on natural phenomena.

3. The study of the role of themata in the work of scientists can be equally interesting whether the work led to "success" or to "failure" – the commitment to a set of themata does not make a scientist necessarily right or wrong. In any case, attempts to "purge" one's self of themata to improve one's science are probably futile. However, a conscious examination of the possible merits of themata opposite to one's own might well have some healthy effect.

4. We need to know more about the origins of themata. It is rather clear to me that an approach stressing the connections

between cognitive psychology and individual scientific work is a proper starting point.

I have already expressed my belief that much, perhaps most, of a scientist's thematic imagination is fashioned in the period before he becomes a professional. Some of the most fiercely held themata are evident even in childhood.[20] This is undoubtedly an area worthy of further research.

5. Once formed, the thematic commitment of a scientist typically is remarkably long lived. But it can change. Examples are Wilhelm Ostwald, who first turned against atomism and then reversed himself once more; Planck; Einstein; and a few others. Moreover, embracing a thema such as atomism in one field of physics occasionally has not prevented the embrace of the opposite thema by the same person for another field of physics. A case in point is Millikan's championship of the "atom" in electricity, even while he was struggling fiercely against the quantum of light. Poincaré was conservative and ether-bound when it came to relativity theory, but quite oppositely directed in quantum theory.

6. While the individual scientist is the primary repository of themata, they are also shared with minor variations by members of a community. Some themata have a career that can be conveniently understood in life-cycle terms; that is, their wide acceptance can rise or atrophy and fade away. Explanatory devices such as macrocosmic-microcosmic correspondence, inherent principles, teleological drives, action at a distance, space-filling media, organismic interpretation, hidden mechanisms, and absolutes of space, time, and simultaneity – all once ruled strongly in physics. Detailed study of the mechanisms of such rise and decay is much needed.

7. There is always the danger of confusing analysis with something else: with Jungian archetypes, with metaphysics, with paradigms and world views. (It might well be that the latter two contain elements of themata; but the differences are overwhelming. For example, thematic oppositions persist during "normal science," and themata persist through revolutionary periods. To a much larger degree than either paradigms or world views, thematic decisions seem to come not only from the scientist's social surrounding or "community," but even more from the

individual.) Although thematic analysis may be limited by the requirement of some firsthand experience with the scientific material, the rewards of doing more specific work on real cases seem to me far more evident than those to be had from current fashions such as comparisons between historiographic schools or the invention of speculative "rational reconstructions."

8. Finally, there is a need for self-awareness. The search for answers in the history of science is itself imbued with themata, just as is the search for a unified theory of elementary particles. Therefore, we must be prepared for the criticisms of those who are afflicted, not with our themata, but with their antithemata; and we must be ready to run up against the limitations within which we necessarily work – as Einstein did in his frank way when he said, "Adhering to the continuum originates with me not in a prejudice, but arises out of the fact that I have been unable to think up anything organic to take its place."[21] His own work is, of course, testimony to the fact that one can turn such inherent limits of the scientific imagination into strength, rather than merely deploring or neglecting them.

2

Subelectrons, presuppositions, and the Millikan-Ehrenhaft dispute

Introduction

Peter Medawar is one of the few first-rank research scientists still concerned with the problem of knowledge – the sources, warrants, and degrees of certainty of scientific findings, the interplay between fact and belief and between perception and understanding. In *The Art of the Soluble* he asks: "What sort of person is a scientist, and what kind of act of reasoning leads to scientific discovery and the enlargement of the understanding?"[1] He finds the usual approaches too limited: "What scientists *do* has never been the subject of a scientific, that is, an ethological inquiry . . . It is no use looking to scientific 'papers,' for they not merely conceal but actively misrepresent the reasoning that goes into the work they describe . . . Only unstudied evidence will do – and that means listening at a keyhole."[2]

Medawar proposes that to study scientific activity one should live in the laboratory or in the theoretician's workroom and observe the work as it is carried out. To approach Medawar's aim when dealing with historical problems, historians and sociologists regularly make use of unself-conscious evidence such as letters, autobiographical reports cross-checked by other documents, oral-history interviews conducted by trained historians, transcripts of conversations that took place in the heat of battle at scientific meetings, and, above all, laboratory notebooks – firsthand documents directly rooted in the act of doing science, with all the smudges, thumbprints, and bloodstains of the personal struggle of ideas.

These sources can help us in understanding the belief structure and activity of some scientists and how they dealt with new ideas when systematic tests of them, if available at all, were difficult to believe or apply. In this study I treat the period after the earliest phase of discovery, when the stirrings of a new conception are difficult to document, but before the new work has been absorbed into the mainstream of science through the mechanisms of justification. In this period one may hope to find evidence of the fragile and obscure process of science in the making which has been explicitly avoided by Reichenbach[3] and Popper,[4] among others.

"*The most fundamental question of modern physics*"

This study centers on events, in the years around 1910, that led two physicists into exactly opposite directions – one to "success" (and the Nobel Prize), the other to "failure" (and eventually a broken spirit). Failures are not remembered in science, and they are rarely analyzed in histories of science. Hence today this controversy is virtually forgotten.

Initially, the protagonists seemed not well matched. Robert A. Millikan was a practically unknown professor at the new University of Chicago, a man over forty years old, with few scientific publications. Felix Ehrenhaft, at the venerable University of Vienna, was regarded as an accomplished physicist, eleven years younger than Millikan, and having a dozen publications.[5] Their disagreement was about the value of the smallest electric charge found in nature. Both men recognized that the subject of their experimental research, as well as the import of their controversy, went to the foundations of science.

Millikan's first major paper begins:

> Among all physical constants there are two which will be universally admitted to be of predominant importance; the one is the velocity of light, which now appears in many of the fundamental equations of theoretical physics, and the other is the ultimate, or elementary, electrical charge, a knowledge of which makes possible a determination of the absolute values of all atomic and molecular weights, the absolute number of molecules in a given weight of any substance, the kinetic energy of agitation of any molecule at a given temperature, and a considerable number of other important physical quantities.

> While the velocity of light is now known with a precision of one part in twenty thousands [thanks largely to R. A. Millikan's patron and colleague at Chicago, Albert A. Michelson], the value of the elementary electrical charge has until very recently been exceedingly uncertain.[6]

Since Michael Faraday's time it was known that during electrolysis 1 gram-atomic weight of univalent material would be released at the electrode if about 10^5 coulombs of charge pass through the electrolyte. If one assumed that this quantity of charge was carried by N ions of charge e each (where N is Avogadro's number), then $Ne \doteq 10^5$ coulombs. If e is now measured independently with accuracy, N, the number of atoms per gram-atomic weight of any substance, is also known with accuracy, and as a result many other fundamental constants may be calculated. At the beginning of the twentieth century, e was identified by many physicists with the magnitude of the charge of the electron. Poor values for e put into doubt the value of N and all that followed from it.

The controversy between Ehrenhaft and Millikan, often called "The Battle over the Electron," erupted in the spring of 1910. Only a year earlier, Ehrenhaft had published measurement results for the value of the "elementary quantum of electricity" along the same general lines as Millikan. But now he suddenly announced finding electric charges much smaller than the charges on the electron. Millikan wrote later that Ehrenhaft's new claim "raises what may properly be called the most fundamental question of modern physics."[7] In a series of increasingly lengthy and detailed articles, Ehrenhaft and his students claimed to find "subelectrons." That is, they found droplets of liquid, metal particles, and other small objects to have charges with a value much smaller than that of the electron. In the course of time Ehrenhaft found charges of a half, a fifth, a tenth, a hundredth, a thousandth that of the electron. As his work progressed there seemed to be no reason to assume that Ehrenhaft would find any lower limit to exist for the electric charge associated with matter. On the other hand, in almost no laboratories other than those of the Vienna group were these results obtained. At the same time Millikan and his students, among others, were assiduously refining and publishing evidence for the unitary electron.

This controversy reverberated for years in the scientific community. The number of articles devoted to it multiplied. Discussants at scientific meetings included Max Planck, Jean Perrin, Albert Einstein, Arnold Sommerfeld, Max Born, and Erwin Schrödinger. Periodically, the evidence was reviewed in depth.[8] In 1927, three years after Millikan had received the Nobel Prize in part for his work on the charge of the electron, the respected physicist O. D. Chwolson still called the fight a "delicate case"; and he added, "It has already lasted 17 years, and up to now it cannot be claimed that it has been finally decided in favor of one side or the other, i.e., that all researchers have adopted one or the other of the two possible solutions of this problem. The state of affairs is rather strange."[9]

To appreciate today the seriousness of Ehrenhaft's claims, we must guard against some ahistorical impressions. First, anyone familiar with the beautiful "Millikan oil drop experiment," now routinely assigned in elementary physics classes, may be inclined to dismiss contrary findings. Such class activities, however, are really only pedagogic exercises to bolster belief in the electron, rather than serious tests of belief. Even so, it is quite difficult to obtain "good data." According to one instructor's recent analysis of class experience: "In spite of the improvements in the Millikan oil-drop apparatus . . . the experiment remains perhaps the most frustrating of all the exercises in the undergraduate laboratory."[10]

Second, the existence of a kind of "subelectronic" charge has been postulated in recent years as part of the quark model in elementary-particle physics. In that model, objects having one-third or two-thirds the magnitude of the charge of an electron are assumed; but theory and experiments so far agree that it is highly improbable that such fractional charges due to quarks would show up outside a nuclear particle (on the droplets and similar objects Ehrenhaft and Millikan were watching) even once, let alone time after time. In his February 1910 paper, Millikan did make a passing remark that Ehrenhaft later seized: "I have discarded one uncertain and unduplicated observation, apparently upon a singly charged drop, which gave a value of the charge on the drop some 30 percent lower than the final value of *e*."[11] The evidence is overwhelming that this one anomaly in Millikan's published data (and in all publications by others on the same matter, except by Ehrenhaft and his school) is the result of an error of measurement. This explanation is the one Millikan proposed, and it has been generally adopted.[12]

Third, we must neither superimpose on the early phases of the development our present perception of the opposing points of view, nor apply criteria more appropriate for the final justification of a theory. It could not be known around 1910 that the results of Millikan's research on the electron would ultimately be so stimulating, not only in physics but also in chemistry, astronomy, and engineering. It could also not be foreseen that the amazing results of Ehrenhaft would stimulate nothing usable at all, unlike the experiments of, for example, Henri Becquerel, who initiated the study of radioactivity with an experiment he quite misinterpreted. It is only in retrospect that the "risks" of Millikan and Ehrenhaft in the early period seem greatly different.

Fourth, the controversy was of added interest at the time because it concerned not only the nature of electric charges but also the behavior of the small particles that carried them. Recent improvements in microscopy, as well as the basic work of Einstein, Marian von Smoluchowski, and Perrin, had made more accessible what Wolfgang Ostwald called "the World of Neglected Dimensions." It was widely thought that research on the colloidal state (the dispersed state of matter where particle dimensions are between 10^{-4} and 10^{-7} cm) was a great frontier for both pure and applied science, one that might bridge organic and inorganic matter. This field seemed filled with promise for medical-biological research as well as for industry.[13] Against this background, the discussions on the charges carried by small particles gained further importance.

The protagonists

In their work, Millikan and Ehrenhaft interacted with each other and with the trajectory of public science (canonical knowledge, institutions for development of controversy or consensus, etc.) around 1910. Biographical details and some awareness of cultural and social contexts may therefore be expected to be of some help in understanding the encounter.

Robert Andrews Millikan was born in 1868 in Illinois and died in 1953 in Pasadena, California.[14] At the height of his career he was perhaps the most renowned and influential scientist in the United States: physicist, administrator, educator, and policy maker. As shown in his autobiography, Millikan's origins were humble. Like many other American scientists of his generation, he was the son of a small-town minister. His parents, Silas Franklin Millikan and Mary Jane Andrews, brought up six children in a tradition that had little use for pretensions. Robert's grandfather had been among the earliest farming settlers of the Mississippi River country in western Illinois; it is said that in 1825 he had walked alongside the covered wagon as the family was moving from the Berkshire Hills in the East to the frontier, the "Western Reserve." As a boy, Robert Millikan led a life recognizable from the stories of Mark Twain – steamboats on the Mississippi, family farm work in their one-acre yard, the swimming hole, the barefoot existence, the rural, simple, pragmatic, direct, and fundamentally pious background.

In 1886, Millikan went to Oberlin College, where he registered for only one physics course ("a total loss"). He discovered his interest and aptitude in the subject only when a professor asked him to help teach physics. For graduate work he went to Columbia University and studied under Michael Pupin for two years as the only graduate student in physics. A. A. Michelson, whom he met in Chicago in 1894, made suggestions that helped Millikan form his experimental thesis work. When he obtained a Ph.D. from Columbia in 1895, he found that there was no satisfactory job available. With a loan from Pupin, he went to Germany in May 1895 for additional study. It was the best moment to arrive. Within a few months, the work of Wilhelm Conrad Röntgen came up like a storm, followed by that of Becquerel, and the field of physics burst into excitement. In 1896, Millikan accepted an invitation from Michelson to join the physics department at the University of Chicago, and, from 1908, Milllikan's articles on the charge of the electron came from Chicago's Ryerson Laboratory. Eventually, Millikan achieved a large and varied research output, first at Chicago and, after 1921, while shaping and heading the California Institute of Technology. He listed nine fields of research in the second edition (1910) of *American Men of Science*, and twenty fields in the fifth edition (1933). In Isaiah Berlin's terminology he was a fox rather than a hedgehog.

His scientific breadth from the beginning is evident in the archival material he left.[15] There is a revealing notebook, probably starting in 1897 or 1898, entitled "References to Important Articles." Orderly entries list recently published articles in physics, under headings such as "Zeeman Effect" (1897–1907), "Brownian Movement" (from 1905), and "Blondlot's N-Rays." These reading lists were probably at least in part connected with Millikan's duties at Chicago. He writes in his autobiography:

> I soon found myself responsible for the weekly seminar in physics, which Professor Michelson asked me to take off his hands . . . Furthermore, I soon began to give advanced courses on the electron theory, on the kinetic theory, and on thermodynamics . . . [After 1900, Michelson was so absorbed in his research] he asked me to assign research problems to three of the prospective candidates for the doctor's degree . . . and to take the whole responsibility for supervising their work, so that by 1902 and 1903 I had quite a group of problems going in addition to my own . . .[16]

One set of pages in this notebook is entitled "Electron Theory of Matter," apparently compiled by adding entries from time to times over the years. It begins with "m/ε Zeeman effect, *Phil. Mag. 43*, p. 226, 1897," and "Cathode Rays, J. J. Thomson, *Phil. Mag. 44*, p. 293, '97," and continues with significant articles over the following few years, including the early determination of *e* by Thomson, J. S. Townsend, and H. A. Wilson. Evidently, Millikan was keeping a careful eye on this work as it was developing.

The last page and inside cover of the notebook are entitled "Research Subjects," with entries dated from 1898 to 1914. The first of the twenty-seven entries, May 21, 1898, is, "Resistance of air in its relation to the velocity of the (falling) moving body" – a major component of the problem that was to be treated a decade later in Millikan's work on the charge of moving droplets. The ninth entry, probably made in 1903, reads, "Stokes law for size of water particles in clouds, see J. J. T. articles on size of ε and Barus, *Phil. Mag. 4*, p. 24/ 1902." Evidently, during the years prior to 1908, when Millikan was preoccupied with teaching and with his first investigations of the emission of electrons from metals by incident light and by high-intensity electric fields, he was laying the conceptual grounds for his later work on the electronic charge.[17]

We leave Millikan for a time, at the verge of starting his big work, and turn to the other protagonist, Felix Ehrenhaft. He was born in Vienna in 1879 into a professional family; his father was *Obermedizinalrat;* his mother was the niece of a student of J. B. L. Foucault. He studied at both the University and the Institute of Technology at Vienna.[18] In his earliest work (1900) he was one of the first to produce and study inorganic colloids. In 1903 he became assistant to Victor von Lang at the University of Vienna. Accepted as privatdocent in 1905, he was teaching statistical mechanics by 1909. Among his colleagues were Felix Exner, Friedrich Hasenöhrl, Stefan Meyer, Egon von Schweidler, Karl Przibram, and Ernst Lechner. Ehrenhaft was called to an associate professorship at the University of Vienna in 1912 and became director of the Third Physical Institute in 1920. He found photophoresis and named the effect in 1918. After the Nazis took over in Austria in 1938, he came to the United States as a refugee. He returned in 1946 to Vienna to resume his position, and he died there in 1952.

By 1909 he was already known for his experimental study of Brownian motion in gases, building as he did on the theoretical ideas of Einstein and von Smoluchowski. For this work he received the Lieben Prize of the Vienna Academy of Sciences in 1910. He was a genial person whose house was always open to scientists from all corners of the world. According to Philipp Frank,[19] Einstein found Ehrenhaft congenial and would stay with him when passing through Vienna. From about 1920, and particularly after his claimed discovery in the mid-1930s of magnetic monopoles, Ehrenhaft's life centered on unresolvable controversies concerning the interpretation of complex physical phenomena. He made some thirty presentations on monopoles before skeptical audiences of the American Physical Society in the period 1940–46.[20] When he is still remembered, it is usually in this context.

Antiatomism and a faculty vacancy

The European Continental tradition of physics first entered Millikan's training through his teacher Michael Pupin, who had received his doctorate in Berlin. Millikan reports in his autobiography that Pupin's course on optics and electromagnetism was an eyeopener, and he came to admire and respect Pupin greatly. But Millikan was amazed by Pupin's attitude to atomism at that time. Pupin had been impressed with the teachings of the schools of energetics and antiatomism, and he had once told Millikan that he did not believe in the kinetic theory at all. The importance or truth of the atomic theory was still being argued in 1904, when it was a chief subject of debates at the scientific congress at St. Louis.[21] The first Solvay Congress, late in 1911, was to a large degree concerned with fundamental, persisting impasses in a physics based on the atomic hypothesis, and a few critics such as Pierre Duhem scoffed at the hypothesis as late as 1913.

Many students absorb the epistemology of their honored teacher. In the case of Millikan, nothing of the sort happened. Let us recall what Millikan resisted, despite Pupin's example. Ostwald, Mach, Stallo, Helm, and others around the turn of the century hoped to erect science on a purely phenomenological base, without "unnecessary hypotheses" such as atomism, to provide a frequently given example. Despite triumphs of atomic theory such as Maxwell's proof of the independence of the viscosity of gases from density, there really was little direct evidence from phenomena for the reality of atoms and molecules, that is, for the necessity of discreteness itself. Scientists would not see particle tracks in cloud-chamber photographs until around 1912.[22] Not until these supported the more indirect evidence of scintillation screens and Geiger counters did the individual flashes or clicks of those instruments become persuasively associated with individual atomic events. Until that point was reached, the scientists were working with average values, not individual atomic entities.

One of the best short descriptions of the school of thought of which Ernst Mach was the most powerful proponent has been given in a biography of Mach written by the physicist Anton Lampa. Because Lampa, too, will soon enter this story, I shall use his account. Lampa points out that Mach's research interests were in very widely scattered specialities and that Mach sought one unifying position which he could adopt in doing any research. That basis he found in the world of elementary sensations that precedes the world of scientific construction:

> In trying to find a point of view which required no change when Mach went from [problems in] physics to physiology and psychology, he started from a natural world picture which everyone, without conscious effort, finds within oneself upon one's intellectual awakening. Mach analyzed that natural world picture. The result of the analysis is his Theory of Elements. The physical findings can be resolved into elements that hitherto are not further resolvable: colors, sounds, pressures, warmth, spaces, times, etc. These elements turn out to be dependent upon circumstances both outside the spatial limits of one's own body and within those limits. Insofar, and only insofar, as the latter is the case, we call these elements also experiences [impressions, *Empfindungen*]. The physical and the psychological [world] thus contains shared elements . . . The natural world picture designates as corporeal objects relatively durable element-complexes of colors, sounds, heat, pressure, etc. . . .
>
> The complex of all elements forms the world . . . The pseudoproblems arise with the formation of the conception of substance (matter, soul); such problems can be solved only if one analyses the complexes and goes back to the elements.[23]

One consequence of this position relates to what was called "atomistics." In looking for the ideal of a phenomenologic physics, Mach refused to give the atom a fundamental basis in physics, but instead asked that it be considered at most as a heuristic device for research. (Under proper conditions and safeguards, he would, however, tolerate a far more daring, speculative use of atomistic ideas than is customary, for he proposed using more than three dimensions to represent the structure of molecules.)

Making atomic entities the subject of research, whether in physics, chemistry, or electricity, was considered by the Machists a false and even dangerous metaphysical hypothesis. The liberation of science from all metaphysical bonds was Mach's lifelong ambition. Hence he acted not only as a productive physicist and an influential philosopher but also as a powerful figure in the politics of academic life.[24] He kept in touch with his students and followers and saw to it that his point of view would be represented in journals and on faculties.

As it happened, the year 1910, when the Millikan-Ehrenhaft dispute first arose, was characterized by two other events relevant to this study. One was the culmination of the widely observed epistemological battle involving Ernst Mach. In fact, the same volume of the *Physikalische Zeitschrift* that carried Ehrenhaft's first detailed account of his discovery of subelectrons also contained the heated, painful, often *ad hominem* and polemical articles that were being exchanged between Mach and Max Planck.[25]

A second event in the same year added to the sense of urgency felt in Mach's circle: A vacancy became available in the physics faculty of the German University in Prague, where Mach himself had been active for nearly three decades. Two members of the faculty there, Anton Lampa and Georg Pick, began at once the search for proper candidates. Pick formerly had been an assistant of Mach; Lampa had been a disciple of Mach, then an assistant to Victor von Lang at the University of Vienna, and from 1904 taught there until his move to Prague in 1909. Lampa was a physicist as well as an idealistic fighter for the reform of education. As Philipp Frank, later his colleague in Prague, put it: "Lampa saw it as his life's chief goal, to propagate Mach's views and to find adherents for them."[26]

Lampa and Pick looked for someone who could be relied upon to carry on physics in accord with Mach's views. A chief candidate was Gustav Jaumann of Brno. To obtain Mach's approval, Lampa wrote to Mach in a letter of February 9, 1910:

> I need not reassure you that Jaumann's high talent seems to me beyond doubt and that his whole cast of thought is sympathetic. I consider the ideal of theoretical physics to be the purely phenomenological presentation [*Darstellung*], as lies at hand for example in thermodynamics. Jaumann proceeds from the wish to build up such a phenomenological presentation for electricity and all that can be connected with it. He therefore rejects the theory of atoms and of electrons . . .

Lampa ended by sharing some worries about Jaumann and by announcing his visit to Mach in Vienna "in a few weeks." Evidently, Mach sent his approval speedily, for in a letter of February 18, 1910, Lampa thanked Mach for the reply, stating that all qualms were laid to rest and that he would intervene warmly on behalf of Jaumann.[27] Yet the selection process went on for many months more. Another candidate was Albert Einstein of the University of Zürich, who was still regarded by the Machists to be of their persuasion.[28] He was just then corresponding with Mach and, indeed, signed one of his letters, "Ihr Sie verehrender Schüler."[29] Einstein was finally called to the chair in Prague in March 1911.[30]

Seeing electrons

Let us leave these Europeans for a time to their philosophies and academic negotiations, and turn to Millikan, who was unaware of these events or of their future implications. Around him was a very different atmosphere. Millikan confesses to an unsophisticated, pragmatic, straightforward point of view of his own, one element of which is seeking direct explanation in terms of concrete visualization. The words "concrete visualization" recur in his writings, possibly to counter the charge that he engaged in making up hypotheses. When Millikan wrote about the electron in his early years, he did not of course think of a particle that has magnetic moment, angular momentum, wavelength, intrinsic self-energy, or any of the properties that we now think of as being associated with and defining the electron. He thought of the electron as a discrete corpuscle[31] of unitary electric charge, whose action one can see with one's own eyes. Indeed, he was not far from asserting that one can see the electron itself: "*He who has seen that experiment,*" he writes in his autobiography about the oil drop experiment, "*and hundreds of investigators have observed it,* [*has*] *in effect SEEN the electron.*" And again, with even less qualification: "*But the electron itself, which man has measured . . . is neither an uncertainty nor a hypothesis. IT IS A NEW EXPERIMENTAL FACT that this generation in which we live has for the first time seen, but which anyone who wills may henceforth see.*"[32]

Because the autobiography was published when Millikan was over eighty, it may invite a suspicion about the reliability of some statements. There are passages where this is a valid concern. However, the autobiography is really a patchwork of new and old writings. One can gather by inspecting the materials in the Millikan archives that the publication probably was assembled with the aid of an editor under Millikan's supervision.[33] Large portions of the published book are repetitions of earlier publications. This is the case with the preceding passages, which come directly from Millikan's Nobel Prize acceptance speech in 1924.[34]

Other passages in his documents and publications further elaborate the anthropomorphic metaphor that Millikan adopted to deal directly with the experimental situation. He writes, for example, that when the small oil droplet was "moving upward [in the electric field, against the gravitational pull] with the smallest speed that it could take on, I could be certain that just one isolated electron was sitting on its back. The whole apparatus then represented a device for catching and essentially seeing an individual electron riding on a drop of oil."[35]

Sometimes, while watching a charged oil droplet held in the electric field, he observed it change its motion suddenly, when the droplet encountered a charged molecule (ion) in the air. This observation was even more important; for the *discontinuity* in the observable phenomenon – new at the time – fitted splendidly with the hypothesized discontinuity in the concept of quantized charge. Here was a great new fact, and the image that helped interpret it was directly at hand: ". . . one single electron jumped upon the drop. Indeed, we could actually see the exact instant at which it jumped on or off."[36] Earlier documents contain the same metaphor. In an early draft of his autobiography,[37] Millikan wrote that he "could actually see the exact instant at which [the electron] jumped on or off." He also provides other visual images; for example, "I had seen a balanced drop suddenly catch an ion."[38] Millikan had the same power of visualization as other distinguished scientists.[39] Thus in his brief essay on Ernest Rutherford, Millikan quoted with approval what he called "a very characteristic Rutherfordian remark": "Ions are jolly little beggars, you can almost see them."[40]

At about the time Millikan began to "see" his electrons, Jean Perrin in France was battling for the atomicity of matter with the same strength of preconception and consequent focusing of vision that characterized Millikan's determination to demonstrate the atomicity of electric charge. Mary Jo Nye writes of Perrin: "Perrin's primary goal from the very beginning of his scientific career was to prove the reality of the invisible atom, to eliminate as "puerile anthropomorphism" those strictures which seemed logically necessary to many others . . . One student wrote of Perrin . . . 'He "sees" atoms – there is no doubt at all – as Saint Thomas saw seraphim.' "[41]

The way that Millikan launched his research on the charge of the electron illustrates three related factors: (1) his capacity of looking with fresh, clear eyes at what was going on; (2) his powers of visualization as an aid in drawing conclusions; and (3), behind all these, almost unconfessed and certainly unanalyzed, a preconceived theory about electricity which gave him eyes with which to look and interpret.

On the road to the electronic charge – Method I

Millikan described frankly the series of accidents that set him on his way. At one of the weekly seminars in physics at Chicago, he presented a review of J. J. Thomson's great paper of 1897 on cathode rays. Millikan later wrote:

> [it] put together in matchless manner, the evidence for the view that the "cathode rays" consist not of ether waves, as Lenard and the Germans were maintaining, but rather of material particles carrying electric charges, each

> particle possessing a mass of about a thousandth of that of the lightest known atom and therefore constituting the most minute known masses in existence. He called these particles "corpuscles". . . . [This paper] impressed me greatly and started me on the researches which have been life work.[42]

However, for the next ten years Millikan's researches did not go well. Up to 1907, he had published only an article on his (1895) thesis, two short notes in 1897 and one in 1906, a translation of Paul Drude's *Optics*, and five introductory textbooks. In 1907 he published with George Winchester two articles on photoelectricity which received some notice.[43] In his autobiography Millikan hints that he was rather dissatisfied with himself at that point. He uses phrases such as "this apparently fruitless work"[44] and "my own research failures"[45] in describing his research. He may well have been concerned about his chances as a research scientist. In 1908, for some reason that one wishes to know more about, he "kissed textbook writing good-bye . . . and [while aware of the risk of further failures] started intensively into the new problem"[46] – the magnitude of the elementary charge *e*.

There were four obvious merits in Millikan's choice of this particular subject. One was that "everyone was interested in the magnitude of the charge of the electron,"[47] then known only to low accuracy and with widely varying results, depending on the method used. Another merit was that the best experimental method to use seemed to Millikan quite obvious and rather simple, although this turned out not to be the case. Third, measuring a basic constant with greater accuracy (rather than looking for daring new things in physics as Ehrenhaft was to do) was quite in keeping with Millikan's talents and temperament and with the tradition Michelson had set. Fourth, the theoretical basis or epistemological assumptions needed for the work seemed quite clear to Millikan: "Being quite certain that the problem of the value of the electric charge (Franklin's fundamental atom of electricity – apparently invariant and indivisible – the assumed unit building block of the electrical universe) was of fundamental importance, I started into it."[48]

Not for him all the turmoil and bitter debate raging in Europe concerning the "reality" of molecules, atoms, and electrons, or the admissibility of discrete rather than continuous entities! The electronic charge existed, and it was of "fundamental importance" to find the value of the charge. If Millikan needed philosophical underpinnings for his work, he found them, appropriately enough, in the work of the great American folk hero, statesman, and scientist, the sensible Benjamin Franklin.

Millikan consistently refers to Franklin as the first to formulate a granular structure and material reality for the "electrical particle or atom,"[49] and he quotes frequently a sentence he ascribed to Franklin: "The electrical matter consists of particles extremely subtle, since it can permeate common matter, even the densest, with such freedom and ease as not to receive any appreciable resistance"[50] Franklin is the father of the subject, "for there are no electrical theories of any kind which go back of our own Benjamin Franklin,"[51] The result of all modern research has been merely "to bring us back very close to where Franklin was in 1750, with the single difference that our modern electron theory rests upon a mass of very direct and convincing evidence."[52] In 1948, looking back on the recent fiftieth-anniversary celebration of J. J. Thomson's "unambiguous establishment of the electron theory of matter," Millikan remarked that since Franklin had begun his experiments in 1747, one should have also been celebrating the bicentenary of "Franklin's discovery of the electron."[53] Even before Millikan turned seriously to his work on the charge on the electron, an account of Franklin's accomplishments (and his full-page portrait) could be found in some of the early school texts Millikan coauthored. A book published in 1908 describes Franklin's "so-called one-fluid theory," adding, "A modern modification . . . has recently come into prominence through . . . Lord Kelvin and J. J. Thomson," featuring "very minute negatively charged corpuscles, or electrons."[54]

Millikan was not the only one to see a connection between Franklin's ideas and the modern theory of electricity. To give only two examples: Rutherford had pointed it out in an address in Philadelphia in 1906 at the bicentennial celebration of Franklin's birth,[55] and some years earlier Lord Kelvin had developed it in a paper that concentrated on Aepinus's elaboration of

Franklin's theory.[56] Yet when Millikan began his work in the first decade of the new century, one did not have to accept the atomistic view of electricity, let alone subscribe to a theory associated with Franklin. If Millikan had followed Pupin's example, he could have supported a rival theory of electricity, based on the thematic concept of the continuum rather than on the thematic concept of atomism. Maxwell's theory of electricity, while outwardly agnostic on the nature of electricity, permitted electricity to be more easily thought of in terms of continuous displacement, a motion within the electromagnetic ether, than in terms of an atomistic structure. Maxwell noted in 1873 in his *Electricity and Magnetism* that electrolysis seems to invite conceptualizing a definite value for an electric charge: "For convenience in description we may call this constant molecular charge (revealed by Faraday's experiments) one molecule of electricity." He added, however, this convenient terminology, "gross as it is and out of harmony with the rest of this treatise," this should not mislead us to ascribe reality to granules of electricity:

> . . . the theory of molecular charges may serve as a method by which we may remember a good many facts about electrolysis. It is extremely improbable, however, that when we come to understand the true nature of electrolysis we shall retain in any form the theory of molecular charges, for then we shall have obtained a secure basis on which to form a true theory of electric current and so become independent on these provisional hypotheses.[57]

The view that the atomicity of electricity was only a heuristic device was widespread in England and on the Continent before the successes of the corpuscular view through the work of Pieter Zeeman, H. A. Lorentz, and J. J. Thomson. Arthur Schuster wrote of the early 1880s: "The separate existence of a detached atom of electricity never occurred to me as possible, and if it had, and I had openly expressed such heterodox opinions, I should hardly have been considered a serious physicist, for the limits to allowable heterodoxy in science are soon reached."[58] In 1897, Lord Kelvin still thought that careful consideration should be given to the idea that "electricity is a continuous homogeneous liquid."[59] Max Planck confessed that as late as 1900 he did not fully believe in the electron hypothesis.[60] Even where an atomistic hypothesis of electricity seemed persuasive, it did not have to imply a unitary charge for the electron. Millikan later noted that the possibility of the electronic charge being merely a "statistical mean" was one that some physicists were supporting at the time.[61] According to all available documents, however, neither at the start of his work on the electron nor later did Millikan subject this possibility to any detailed test.

With hindsight, it is easy to see evidence that should have clinched for everyone the argument in favor of the particle theory of unitary electric charge, even prior to Millikan's work: J. J. Thomson's measurement of the constant charge-to-mass ratio of cathode rays; Rutherford's measurement of the charge on α particles; the charge on cloud droplets of various liquids determined by J. J. Thomson, his student Townsend, and H. A. Wilson.[62] Yet even where the error margins were not enormous, the methods all shared one fatal flaw with the calculation of the unit charge exchanged in electrolysis: They represented the determination of an *average* charge from observations made on very many hypothetical individual charges as the same time. At best, these were indirect measurements of *the* charge e; at least e would be the statistical mean value of a distribution of unknown shape. Nobody before Millikan had measured the charge of an individual object, and found it to be equal to one or two or any small multiples of a unit of electricity, much less watched a charged object changing its charge discontinuously by 1, 2, 3, . . . units of charge.

Millikan also did not have the slightest hope of doing this when he set out to measure the value of the electronic charge. When Millikan began this work with his student L. Begeman, he used H. A. Wilson's method essentially unchanged. Clouds of droplets were produced in an expansion cloud chamber between the parallel, horizontal plates of a charged condenser. They observed the slowly falling top layers of the clouds containing the smallest droplets. One set fell under gravity (at speed v_1), and another set fell faster, with the additional aid of an electric field set up across the condenser (at speed v_2). Assuming, first,

Stokes's law to hold for the droplets, second, each of the droplets to have formed on a singly charged ion, and not to shrink noticeably owing to evaporation and, third, that the different clouds in succession were all similarly formed, one could quickly obtain the charge of the hypothetical unit of electricity in terms of the observables (speeds of fall v_1 and v_2; electric field strength E; density of drop δ; and viscosity of the gas μ).[63] The average charge per droplet would thus be given by

$$q = \frac{4}{3}\pi \left(\frac{9\mu}{2g}\right)^{3/2} \frac{g}{E\delta^{1/2}} \left(v_2 - v_1\right) {v_1}^{1/2}$$

This method (to be called *Method I*) was full of unsatisfactory features, both theoretically and practically. Wilson's published measurements had shown a spread of values in determinations for e from 2.0×10^{-10} esu to 4.4×10^{-10} esu, with a mean of 3.1×10^{-10} esu. (Earlier in 1903, J. J. Thomson had obtained $e = 3.4 \times 10^{-10}$ esu by a similar method.) However, Millikan's plan in 1907–8 was to make only minor changes in the procedure, to improve accuracy. Thus Millikan and Begeman used radium instead of x-rays to ionize the moist gas prior to the expansion that formed the cloud. Their results of ten sets of observations for e spread from 3.66×10^{-10} esu to 4.37×10^{-10} esu, with the mean given as 4.03×10^{-10} esu. It was evidently an improvement over Wilson's results – although it shared with those the implicit assumption to rule out a statistical distribution of divergent values of electric charges occurring in nature.

The accidental discovery of an experiment – Method II

The joint paper by Millikan and Begeman was read at the American Physical Society meeting in Chicago in early January 1908. A one-page abstract was published that February.[64] Almost immediately, prominent attention was drawn to this maiden effort – by none other than Ernest Rutherford.[65] Millikan's result had come just in time to help Rutherford and Geiger in their major new work: They had measured the magnitude of the α-particle's charge to be 9.3×10^{-10} esu, and assumed it should be equal to $|2e|$. Hence e should be 4.65×10^{-10} esu. The values for e found earlier by Thomson and Wilson had been 30 percent lower, but the new ones by Millikan and Begeman were only 15 percent lower. Although the work of Millikan and Begeman appeared to be the best of the three, Rutherford implied that it, too, could be improved and the gap closed. Rutherford suggested that a failure to allow adequately for the evaporation of the droplets in those methods caused the estimate of the number of ions (droplets) present to be too large, and hence the value of e too small. Pending such improvements of the method of others, the Manchester group associated with Rutherford continued to use his value $e = 4.65 \times 10^{-10}$ esu confidently until Bohr's early work of 1912; it "had been gospel at Manchester since the measurement of Rutherford and Geiger in 1908."[66]

With the incentive of Rutherford's suggestion, if indeed it was needed, Millikan's strategy was now clear: The error owing to evaporation had to be eliminated.[67] Millikan planned to work in his typically gradualist way, by arranging the electric field to hold steady the top surface of the charged cloud, keeping it suspended to permit studying its rate of evaporation. The work was apparently done in the spring and summer of 1909[68] and seemed at first to require essentially only small modifications of existing techniques, chiefly using an exceptionally large (10,000-volt) battery to set up a stronger electric field, now in *opposition* to the effect of gravity.

When Millikan turned on the electric field, something happened that at last allowed him to orient and gather his immense energy, his talent as an observer and a researcher, his ability to use students, his instinct for recognizing important and basic problems, and his great eye for the accident that opens an unsuspected door. He chanced upon a sequence of accidents that he described consistently and frankly in the resulting publication of 1910, and the 1939 draft autobiography, and in the published autobiography. As he put it, the accident "*made it possible for the first time to make all the measurements on one and the same individual droplet, and . . . made it possible to examine the attracting or repelling of properties of an individual isolated electron . . .*"[69]

When he turned on the switch, the cloud, far from being held stationary, dissipated instantaneously and completely. The strong field, acting on the variously (not, as had always been assumed, equally) charged droplets, cleared them out, and thus there was no top surface of the cloud left on which to make measurements. Indeed, the decade-long technique of measuring *e* by cloud watching came to an abrupt end; I have found no evidence that anyone used it again. Millikan wrote that the dispersal "seemed at first to spoil my experiment. But when I repeated the test, I saw at once that I had something before me of much more importance than the top surface . . . For repeated tests showed that whenever a cloud was thus dispersed by my powerful field, *a few individual droplets remain in view*,"[70] those with just the right charge relative to their mass to allow balancing their weight in the electric field.[71]

It was the first time that one of the cloud experimenters concentrated on the individual, charged droplet instead of the whole cloud. Indeed, Millikan had stumbled on a new instrument, a very sensitive balance for holding an object of the order of 10^{-13} to 10^{-15} grams in view. It marked the change from Method I (falling cloud of water droplets) to Method II (balanced droplets of water, later also of alcohol). While that was only an intermediate stage before he arrived at Method III (nonsuspended oil drops), his perception had guided him to a tool for opening up a new experimental field, the more so as serendipity struck a second time:

> I chanced to observe . . . on several occasions on which I had failed to screen off the rays from the radium [for ionizing the air before producing the cloud] that now and then one of [the balanced drops] would suddenly change its charge and begin to move up or down in the field . . . This opened the possibility of measuring [later] with certainty, not merely the charges on individual droplets as I had been doing, but the charge carried by a single atmospheric ion by comparing the speeds in the electric field of one drop before and after it chanced to catch an ion.[72]

Watching a water droplet suspended in a field made it easier to allow for evaporation and so increased the precision of measurement.[73] This situation was a direct response to Rutherford's challenge. The rest of Millikan's work would soon follow fairly naturally – from the replacement of water by a liquid with much lower vapor pressure, to the long labors of removing or narrowing the source of uncertainty, for example, by the modification of Stokes's law for small droplets. Even in the first months, in the summer of 1909, Millikan claimed the "*charges actually always came out, easily within the limits of error of my stopwatch measurements,* 1, 2, 3, 4, or some other exact multiple of the smallest charge on a [water] droplet that I ever obtained. *Here, then, was the first definite, sharp, unambiguous proof that electricity was definitely unitary in structure . . .*"[74]

The importance of the discovery should not divert attention from the fact that Millikan did not design or devise the experiment from which his early fame sprang; rather, he discovered the experiment.[75] The character of this discovery differs somewhat from that of, say, the existence of Uranus or America. No one had doubted the existence of individual droplets. Anyone could have put together existing equipment a good decade or more earlier if one had only thought of watching a drop instead of a cloud. The equipment used by Millikan in 1909 was a simple structure made of available materials; even making a big enough battery had been no major challenge for a long time. It is not altogether clear why Thomson, Townsend, and Wilson, among many others, did not think of determining the electric charge in the first place with the far simpler method of individual droplets than with the rather complex cloud experiments. The stranglehold on the imagination exerted by the tradition of work on clouds appears to have yielded only to Millikan's accident.

Ehrenhaft in 1909 – the paths converge

Millikan's first reports on individual droplets were not published until December 1909 in *The Physical Review*, and in February 1910 in the *Philosophical Magazine*. Before analyzing the papers or describing the setting in which the scientific community first heard of Millikan's verbal report of his discovery in August 1909, I turned to the work that Ehrenhaft, quite independently, was doing at about the same time – for the trajectories of the scientific work of the two protagonists are about to intersect. Millikan had been led to the single-object determination of charge, starting from his cloud work, in line with research techniques developed in England and the United States. Meanwhile, Ehrenhaft had been progressing toward the same research problems by techniques more characteristic of work on the Continent, namely preparation of colloids and the ultramicroscopic Brownian-movement observations of individual fragments of metal (e.g., from the vapor of a silver arc) and of cigarette smoke. In this period, Ehrenhaft's atomistic preference was as clear and explicit as Millikan's. He ended one of his papers of 1907 with the hope that the work would be "a new support for the molecular-kinetic hypothesis."[76]

Ehrenhaft's first report on a new method to measure the charge on small particles in order to determine what he called the "*elektrische Elementarquantum*," was dated March 4, 1909, and appeared as a one-page summary in the *Anzeiger* of the Academy of Sciences of Vienna.[77] He explained he had noted that colloidal metal particles occasionally showed an electric charge, as indicated by their motion in a *horizontal* electric field. (It is plausible that he happened upon this effect during earlier studies on Brownian movement.) Measuring the motions of particles with and without an electric field, and applying Stokes's law to obtain their mass, he could thus measure the charges on the particles. In short, he was very nearly doing what Millikan did, but he did not use a vertical electric field. Two weeks later, in a longer report,[78] Ehrenhaft announced that he intended to arrange it in just that way; however, the results did not appear until a year later. Consequently, in 1909 his method suffered from obvious shortcomings. Different particles were needed, one set observed when moving with a horizontal component in the presence of the electric field, the other when moving vertically without the electric field; e, therefore, cannot be the charge determined on a single object but is an average. Nevertheless, Ehrenhaft's three papers, submitted between March 4, 1909 and April 10, 1909, are the first in the literature in which the paths and motions of *individual, charged particles* were followed and used to compute a value of e.[79] Moreover, the value for e that Ehrenhaft here obtains (4.6×10^{-10} esu)[80] is far closer to Rutherford's (4.65×10^{-10}) and Planck's (4.69×10^{-10}, from black-body radiation) than that of Millikan and Begeman

(1908) had been, and Ehrenhaft did not hide this fact. Indeed, in view of the subsequent debate, it is ironic that Ehrenhaft here notes that Millikan's measurement results of 1908 were lower than Rutherford's, in other words, that Millikan's effort "yielded values of the elementary quantum that are too small."

At the Winnipeg meeting, August 1909

In attempting to gain acceptance for one's work, early communication in an excellent forum is a great advantage. Here again Millikan was extraordinarily lucky. He finished the measurement of *e* by the balanced-waterdrop method in the late summer of 1909, just before the British Association for the Advancement of Science was to hold its seventy-ninth meeting in Winnipeg, Canada. It was too late for him to get on the printed program,[81] and to the last moment Millikan did not know if he would be called on to present his results during the sessions.[82] It must have been a heady meeting for this relative newcomer to scientific research. The presidential address was given by J. J. Thomson, who chiefly devoted his time to discussing three research frontiers; the structure of electricity ("We know that negative electricity is made up of units all of which are of the same kind . . ."); the ether ("The ether is not a fantastic creation of the speculative philosopher; it is as essential to us as the air we breathe . . . The study of this all-pervading substance is perhaps the most fascinating and important duty of the physicist"); and radioactivity. Among the physicists and astronomers listed as reading papers at the large meeting were C. V. Boys, A. S. Eddington, A. S. Eve, E. Goldstein, Otto Hahn, W. J. S. Lockyear, Oliver Lodge, Percival Lowell, A. E. H. Love, Theodore Lyman, D. C. Miller, J. H. Poynting, Lord Rayleigh, E. Rutherford, A. Schuster, and G. J. Stoney.[83]

The ground for Millikan's presentation was prepared not only by the attention directed to the field of research by J. J. Thomson, but also by Rutherford's address on August 26 as president of the Section on Mathematical and Physical Science.[84] Rutherford's aim was to summarize how recent progress in physics strengthened the credibility of the atomic theories of matter and of electricity. As he had also done in 1906,[85] Rutherford wasted little time opposing directly the antiatomists who were still active, particularly on the Continent. Rather, he made short counterthrusts in sentences such as "The negation of the atomic theory has not and does not help us make discoveries," casting doubt on the atomic theory "is quite erroneous," and, in the very last sentence of the paper, "In the light of these and similar direct deductions, based on a minimum amount of assumption, the physicists have, I think, some justification for their faith that they are building on the solid rock of fact, and not, *as we are often so solemnly warned by some of our scientific brethren, on the shifting sands of imaginative hypothesis.*"[86]

Rutherford's main attention was devoted to reviewing the developments favorable to the atomistic point of view, which he acknowledged to appeal particularly "to the Anglo-Saxon temperament." Nevertheless, on the list of findings he was reporting were efforts by three Austrians, Exner and Richard Zsigmondy (determination of mean velocity of particles in various solutions, from Brownian-movement calculations), and Ehrenhaft (1907, experimental determination of Brownian movement by small particles suspended in gases). Rutherford's own recent work with Geiger on the charge of α particles was further support, "showing that this radiation is, as the other evidence indicated, discontinuous, and that it is possible to select by a special electric method the passage of a single α particle . . ."[87]

As in the paper he had written with Geiger during the previous year, Rutherford now cited the work of J. J. Thomson, Townsend, Millikan, and Begeman on clouds (Millikan's report on using individual droplets had not yet been delivered), as another indication that "electricity, like matter, is supposed to be discrete in structure." He added: "This method is of great interest and importance," although the exact determination of e in this manner was "beset with great experimental difficulties."[88] Rutherford lauded recent work by Ehrenhaft (1909) on the charge carried by ultramicroscopic dust particles of metal, and grouped Ehrenhaft's value for e with Rutherford's and Geiger's as one of "the most recent measurements by very different methods which are far more reliable than the older estimates."[89]

The implication was that these values were more reliable than those of Thomson, Townsend, and Wilson. It now is no longer reasonable, he concluded, "to believe that such concordance [in the experimental values of e and N, based on different theories] would show itself if the atoms and their charges had no real existence"; hence doubts concerning the atomic theory of matter are "quite erroneous."[90]

Rutherford did voice one regret: ". . . it has not yet been possible to detect a single electron by its electrical or optical effect, and thus count the number directly, as in the case of α-particles." This was precisely the missing link. Although he could not have known that Millikan had already found it, Rutherford was optimistic: ". . . there seems to be no reason why this should not be accomplished by the electrical method." Rutherford evidently had in mind the possibility of using scintillations produced by β rays for this purpose. But Millikan was on hand at that very meeting, waiting to give his paper a few days later in which he would show just how one might go about detecting the single electron by another method, that is, by its effects on the observed motion of a small droplet of liquid.

At last, Millikan's turn to speak came, and apparently he was well received. Millikan later singled out Joseph Larmor as having been "intensely interested in my paper,"[91] and as having suggested that Millikan should look into the limits of Stokes's law, promising to do the same himself from the theoretical side.[92] After this presentation on August 31, 1909, Millikan must have thought the end of the quest to establish the unitary nature of the electric charge was in clear sight. The subject had been acknowledged at the highest level to be at the very frontier of urgent research; his own, earlier results, even before his recent improvements, had been believed and cited with respect; they fitted well with the other pieces in the jigsaw puzzle of physical theory; and he had been able to present his new method and new results just as soon as the need for them was announced by Rutherford. He later recalled that even his final, major improvement of technique, that of using oil drops to avoid all the problems caused by evaporation, occurred to him suddenly while he was riding the train back to Chicago from the Winnipeg meeting.[93]

Millikan's first major paper, February 1910

We do not have a copy of the talk Millikan delivered at Winnipeg, but a few weeks later he published a very brief account of his work,[94] and on October 9 sent the paper – his first major one – to the *Philosophical Magazine* for publication in February 1910.[95] These two contributions are the first reports in the literature of the use of single, isolated *drops*, and of the balancing-field method (Method II).

If Millikan and his readers thought that the search for the value of *e* was essentially over, they soon found that the battle was just beginning. Indeed, two likely causes for the approaching "fight over the electron" can be found directly in Millikan's *Philosophical Magazine* paper. One has to do with the well-known role which feelings about priority claims play in the actual unfolding of scientific controversies. In the section giving "The Most Probable Value of the Elementary Electrical Charge," Millikan presents his own, new mean value, $e = 4.65 \times 10^{-10}$ esu, and assigns also equal weight to "all the recent determinations of *e* by methods which seem least open to question":[96] the value obtained by Planck from radiation theory (4.69×10^{-10}), which Rutherford had mentioned with favor at Winnipeg; the value of Rutherford and Geiger (4.65×10^{-10}); E. Regener's value, obtained by a method very similar to Rutherford's (4.79×10^{-10}); and Begeman's "recent and as yet unpublished" value of 4.67×10^{-10}, obtained in Millikan's laboratory.[97] The final mean for *e*, Millikan declares, is thus 4.69×10^{-10} esu. Rutherford's objections of 1908 and 1909 have been well met. In addition, because Millikan's and Rutherford's results are so close, Millikan feels that the "results seem to constitute experimental verification of Stokes' law for these drops."

But while accepting the work of these authors, Millikan specifically rejects the values for *e* published by four others, and gives explicit reasons: Perrin's value "involves so many assumptions of questionable rigor"; Maurice de Broglie's relies on Perrin's *N*, among other difficulties; Moreau's depends on Perrin's *e*. Millikan also specifies why "uncertainties" in Ehrenhaft's results, "obtained by a method similar to the one here presented save that it involves the measurement of the velocities produced first by the action of gravity, and second by the action of an electrical field upon the charged particles thrown off by a metallic arc," make that work unacceptable to him,[98] although Millikan agrees that "this [Ehrenhaft's mean value for e, 4.6×10^{-10}] is in very good agreement" with the other accepted values.

Millikan's reservations regarding Ehrenhaft's work were reasonable, although they appear more reasonable in retrospect. Had Millikan accepted Ehrenhaft's value, it would have worked to Millikan's advantage in the sense that it would have brought the average value of all accepted determinations of e down a bit toward Millikan's own. In fact, he was rejecting a confirmatory value, one obtained by an established researcher who had used a method closer to his own than the methods of others whom Millikan was not rejecting. Millikan's decision was grounded in his suspicions, plausible but far from proved, that the value obtained by Ehrenhaft was invalidated by the method used to obtain it. As we shall soon see, Millikan was also sensitive to the obverse, to the possibility that his own results of measurement were at times unwelcome and disconfirmatory but that even without a solid analysis of the causes he could continue to accept his hypothesis and reject the apparent falsification of it.

It appears that from Ehrenhaft's point of view the glove had been thrown down before him. Beginning in his next publication, he and some of his students dedicated themselves to the "question of the elementary quantum of electricity." Ehrenhaft's own output of a dozen papers over the next four years was entirely on this subject. To be sure, Millikan's publication of February 1910 was vulnerable to criticism. With idiosyncratic frankness and detail, Millikan shows in the section "The Results" that the measurements with the new technique were still difficult to make, that he relied heavily on personal judgment, and that it was really still his first major paper. The most important raw data for five series of data on balanced water drops and ("to vary the conditions") one series of balanced alcohol drops are presented. The observer is also identified (Millikan or Begeman); but, in a move rarely found in the scientific literature, each of the thirty-eight sets of observations is then given a more or less personal rating:

> The observations marked with a triple star are those which were marked "best" in my notebook and represent those which were taken under what appeared to be perfect conditions. This means that we could watch the drop long enough to be very certain that it was altogether stationary: that we could time its passages across the cross-hairs with perfect precision, and that it showed no apparent retardation in falling through the two equal spaces. The double-starred observations were marked in my notebook "very good." Those marked with single stars were marked "good" and the others "fair."[99]

There were two "three-star" observations, seven "two-star" ones, ten with a single star, and thirteen without any. The average value of *e* obtained in each of the seven series (sets of observations) is then assigned a "weight" from one to seven, to obtain the final, weighted grand average ($e = 4.85 \times 10^{-10}$ esu, as compared with the unweighted, simple mean of 4.70×10^{-10}). Although we can discern a general relation between the total number of "stars" in a series and the weight given to the particular individual series-average, the relation is neither explained nor linear: Millikan was evidently saying he knew a good run when he saw one, and he was not going to overlook that knowledge even if it was not obvious how to quantify and share it on the record.

Equally significant was Millikan's frank admission that another seven observations had been discarded altogether, and so had not entered at all into the computation of the final average value of *e:* "First, I discarded three very good observations of my own, taken under conditions of potential and position of cross-hairs which made them uncertain in spite of the accurate timing. These observations . . . would not affect appreciably the final result if they were included."[100] Only the internal ethos of science, which prizes the fullest disclosure of data, seems to have motivated even mentioning this set of discarded observations. Millikan continues:

> Second, I have discarded three observations which I took on unbalanced drops, timing them as they rose against gravity under the influence of the field, and then again as they fell under gravity between the same cross-hairs, when

> the field was thrown off. Although all of these observations gave values of e within 2 per cent of the final mean, the uncertainties of the observations were such that *I would have discarded them had they not agreed with the results of the other observations*, and consequently I felt obliged to discard them as it was.[101]

This is an unusual statement. Nothing in the rest of his paper has prepared us to expect, as Millikan mentions casually in passing, he would discard some observations if they "disagreed with the results of the other observations." Was it enough that these three runs were on *un*balanced drops (which became his method of choice immediately afterward)? His comment that, in this instance, the omission had no practical effect on the final results either way is reassuring, but one must not overlook the more general methodological point. Such judgments are not infrequent. They often can be (and in this case were) supported by plausibility arguments which allow the experimenter to assert that he believes the discordant observations do not go to the heart of the matter, that is to say, are not grounded in a serious way in the phenomenon being studied. For just this reason, such judgments expose the researcher to a risk, one that he is willing to take in the framework of beliefs and assumptions within which the judgments are plausible acts, and one which allows him to avoid the interruptions, delays, and detailed research that might be necessary to pin down the exact causes of the discrepant observations.

At work here is the belief that, in principle, a reinterpretation can be found that would fit the neglected observations into the pattern of nature signaled by the accepted observations. Conversely, the belief is also present that an alternative picture of nature, although in principle not foreclosed – in this case, that charges exhibit a stochastic distribution about the mean value e, or that charges are made up of subelectrons or of some congealing of a continuum of charge – is so unlikely or abhorrent that it is not worth even the effort of falsifying it in detail. Millikan's confidence was explicit. He announces, "There is no theoretical uncertainty whatever left in the method,"[102] although he adds: "Unless it be an uncertainty as to whether or not Stokes' law applies to the rate of fall of these drops on the gravity." On

TABLE.

SERIES No. 1 (Bal. pos. water drops). Distance between plates ·545 cm. Measured distance of fall ·155 cm.				SERIES N Dista Measu
Volts.	*Time.* 1 space.	*Time.* 2 spaces.	*Observer.*	*Volts.*
**2285	2·4 sec.	4·8 sec.	Millikan.	2365
2285	2·4 sec.	4·8 sec.	"	**2365
**2275	2·4 sec.	4·8 sec.	Begeman.	*2365
***2325	2·4 sec.	4·8 sec.	Millikan.	*2365
2325	2·6 sec.	4·8 sec.	"	*2395
*2325	2·2 sec.	4·8 sec.	"	*2395
**2365	2·4 sec.	4·8 sec.	"	***2395
				2365
				2365
				2365
2312	2·4	4·8		2374
Mean time for ·155 cm. = 4·8 sec. $e_3=3.422\times10^{-9}\times\frac{980.3}{14.14}\times\left(\frac{.155}{4.8}\right)^{\frac{3}{2}}$ $=13.77\times10^{-10}$. $\therefore e=13.85\times10^{-10}\div3=\mathbf{4.59}\times10^{-10}$.				Mean t $e_4=3.422$ $=18.25$ $\therefore e=18.25\div$

Figure 2.1

the next page even this doubt is laid to rest: "It is scarcely conceivable that Stokes' law fails to hold for them."[103]

As to the last of the neglected observations, Millikan says simply, without apology or plausibility argument: "In the third place, I have discarded one uncertain and unduplicated observation apparently upon a single charged drop, which gave a value of the charge on the drop some 30 per cent. lower than the final value of *e*. With those exceptions all of the data recorded in our notebooks are given below."[104]

The data, presented in six tables such as the one shown in Figure 2.1, also raise questions. Just how many individual droplets were used in these observations? Each of the six series of observations may have used as many droplets as there were observations. Why, then, were the raw data (voltage, times) on all drops in each series pooled and averaged in order to obtain one preliminary average value of *e*, which in turn received a single weight for the final averaging to obtain one final value of *e*? How confident can the reader be that one integer is the correct divisor that works for all data in a given series (for example, $n = 3$ in Figure 2.1), and how were the data assembled to form a given series in the first place? Finally, since the chief point was to find *e*, what prevented Millikan from using a bigger section of the battery, or smaller spacing between the condenser plates, to obtain usable data on singly charged droplets – not to speak of testing, head on, whether fractional charges do exist?

Unfortunately, the notebooks containing the data of Millikan's work of 1909 have not been found, and we cannot watch through the keyhole what occurred in the laboratory. But soon we shall have better luck, for Millikan continued his work with great energy. It was beginning to be seen as good scientific work by the criterion of fruitfulness. For example, his 1909 value *e* was adopted in 1913 in Niels Bohr's epochal paper on the hydrogen atom,[105] and Millikan's observational method became directly applicable in the closely related efforts by his student H. Fletcher to obtain *N* and hence *e* from measurements of Brownian movement.[106]

Between the fall of 1909 and the spring of 1910, Millikan's methods of experimentation and calculation underwent a process of significant maturation (to Method III, to be examined later) by working not with balanced water drops but with falling and rising oil drops and, as well, by calculating values of elementary charge from each set of observations on a given drop. Unknown to Ehrenhaft, Millikan reported on his new method first on April 23, 1910, to the American Physical Society.[107] In the meantime, however, Ehrenhaft had committed himself.

Ehrenhaft's attack on e

It began with a note to the Vienna Academy session on April 21, 1910.[108] Ehrenhaft had been silent for a year, but now he had startling news. He used a horizontal condenser, with a vertical electric field strong enough to make particles rise against gravitation – a deployment equivalent to the one Millikan reported shortly afterward in the paper read at an American Physical Society meeting on April 23, 1910 – and he studied platinum and silver particles from arcs. Ehrenhaft reported more than 300 measurements, yielding a new and remarkable result: Particles are not only singly or doubly charged but can also have charges "between and below" these values. The twenty-two measurements of charge reproduced in Ehrenhaft's paper range from 7.53×10^{-10} esu down to 1.38×10^{-10} esu – about one-third of the value of the elementary charge he had previously measured. Ehrenhaft concluded that these findings cannot be explained away as inadequacies of method. Rather, these quantities are "in nature." If "theory presupposes the existence" of an indivisible quantum of electricity, the value of the latter will thus have to "fall considerably below" the hitherto accepted value. A counter-challenge was thus issued to all believers in *e* as the quantum of charge, for which nothing in theory or experiment seemed to have prepared the ground. Out of the blue, the subelectron had appeared on the stage.

Ehrenhaft followed his announcement of April 21, 1910,[109] with a report delivered to the Vienna Academy on May 12, 1910,[110] in which he coined the word "subelectron" and announced that his results indicated that indivisible quantities of electric charge do not exist in nature at the level of 1×10^{-10} esu or above. His subelectrons do show a propensity for aggregating; for example, he reports the total charge on gold particles to have ranged continuously from 5×10^{-11} esu to a heaping up (*Häufung*) of 1.75×10^{-10} esu, that is, up to a third of the usual charge of the electron. Although all his observations are made on very small particles observed in the ultramicroscope, he sees no reason to abandon Stokes's law in the classical form (which would, in any case, make the true charge even smaller) or to worry about Brownian movement although it made time measurements uncertain. Moreover, he assumes again that the density of the metal fragments, developed in an electric arc, are of the same density as the mother material in the electrode.[111]

More remarkable, however, is the conclusion that emerges ever more forcefully as the number and length of the articles by Ehrenhaft and his collaborators increase: These experiments do not permit them to "hold on to the fundamental hypothesis of the electron theory," namely, the indivisible electron.[112] The large spread of values for *e* which have been measured by various researchers and by different methods should be taken, they say, as a signal that one is dealing here with an aspect of natural law itself. These variations of net charge are "in nature."

If Millikan and others felt that Ehrenhaft's data could in principle be interpreted without giving up the undivided electron, they must have felt somewhat embarrassed when Ehrenhaft turned to Millikan's data in the *Philosophical Magazine* 1910 paper on balanced water and alcohol drops. He subjected the data to a devastating attack,[113] turning them against Millikan. Ehrenhaft recalculated the charge on each drop from each of Millikan's observations separately, instead of following Millikan's method of lumping several runs to obtain values of *e* from the average of measured values of voltage, and so on, measured on different droplets. The result was a large spread of values of droplet charge, from 8.60×10^{-10} to 29.82×10^{-10} esu. The case for each of these being an integral multiple of one elementary charge now did not look at all self-evident (see Figure 2.2).[114] It appeared rather that the same observational record could be used to demonstrate the plausibility of two diametrically opposite theories, held with great conviction by two well-equipped proponents and their respective collaborators. Initially, there was not even the convincing testimony of independent researchers.[115]

The oil drop experiment, 1910 (Method III)

Happily for Millikan, Ehrenhaft's attack in mid-1910 was quickly made moot by the timely publication of Millikan's new results with his new method. Millikan's documentation of his adoption of and early success with the oil drop is extensive.[116] Of interest to us here are the real advances made by Millikan over his and all other previous work.

His account in the second major paper in his career, in *Science* in September 1910,[117] verged on the euphoric. He was able to measure separately the frictional charge on an oil drop as well as the additional charges it may pick up from ions in the atmosphere

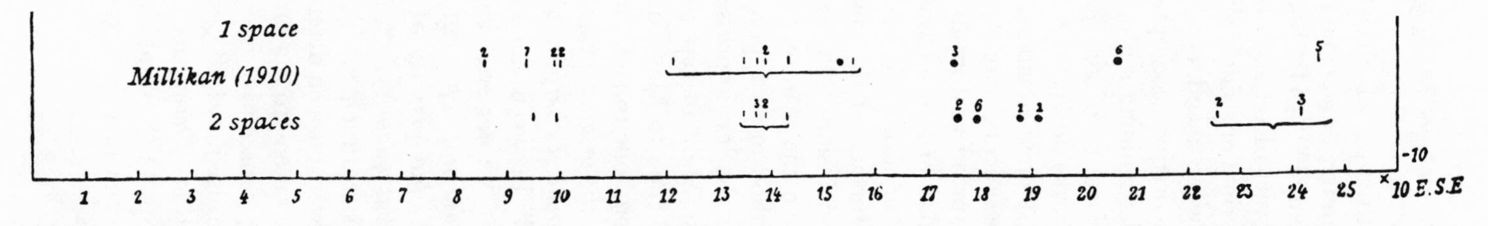

Figure 2.2 Portion of Felix Ehrenhaft's analysis of Millikan's data on the charges q carried by droplets. (*Sitzungsberichte Akad.* Wiss. [Vienna], n. 119, 1910.)

during its travel, and both types of charges were found to be quantized in the same manner, "*exact* multiples of one definite, elementary, electrical charge."[118] He boasted that he could "catch upon a minute droplet of oil and hold under observation for an indefinite length of time one single atmospheric ion or any number of such ions between 1 and 150." The method is free from "all questionable theoretical assumptions," and the limitation on the accuracy of determining e is only the accuracy with which the value of the viscosity of air (μ or η) is known. He found that Stokes's law breaks down for very small spheres, and determined a correction. A view advanced many years ago was confirmed by these experiments, namely, that "an electrical charge, instead of being spread uniformily over a charged surface, has a definite granular structure, consisting, in fact, of an exact number of specks, or atoms, of electricity, all precisely alike, peppered over the surface of the charged body." Indeed, Millikan now held "the conclusions follow so inevitably from the experimental data that even the man on the street can scarcely fail to understand the method or to appreciate the results."

Nor would the scientist be less impressed by the confidence expressed in the findings: Millikan reports that working with Fletcher from December 1909 to May 1910 on droplets of oil, mercury, and glycerin – on "one to two hundred drops" in all – they "found in every case *the original charge on the drop* [to be] *an exact multiple of the smallest charge which we found that the drop caught from the air*." Between 1,000 and 2,000 changes of charge were observed, yet "*in not one single instance has there been any change which did not represent the advent upon the drop of one definite invariable quantity of electricity, or a very small exact multiple of that quantity*."[119]

Of interest is Millikan's treatment of his data. His final value of e is the mean value of twenty-seven determinations of e on that many individual droplets, taken from a larger number "studied throughout a period of 47 consecutive days." Three other drops "have been excluded [because they] all yielded values of e from two to four per cent. too low" compared with the plotting of the values from other drops. A "natural" hypothesis concerning these three drops is that each may have been "two drops stuck together." At any rate, he adds, "After eliminating dust we found not more than one drop in ten which was irregular." The context shows that the word "irregular" means that the drop's unitary charge e deviated by as much as 4 percent from the curve plotting the other value.[120] Nevertheless, ten more drops – the four slowest and the six fastest ones – studied during that period were also eliminated from the final averaging in the 1910 *Science* article before Millikan obtained his "final mean value of e." Although these ten drops would not appreciably alter the final mean value, the probable error in each of the individual determinations is necessarily much higher than in the middle range of speeds.[121]

The publication of 1913 – Drop No. 41

With this knowledge of Millikan's use and treatment of data in his published work, we can turn to the last and most mature of his major papers in the 1909–13 period: his August 1913 *Physical Review* publication, "On the Elementary Electrical Charge and the Avogadro Constant."[122] This is the most authoritative version of the oil drop experiment to that point, and although Millikan continued to make improvements for years, all the chief elements were now assembled: a new optical system, a chronoscope (to 0.001 sec), temperature control to 0.02 °C, a more accurately calibrated voltmeter, a better value for μ, and the ability to change the gas pressure in the viewing chamber over a wide range. As a result, he could announce that "the largest departure from the mean value found anywhere in the table [of values of e, determined for fifty-eight droplets] amounts to 0.5 per cent., and the probable error in the final mean value computed in the usual way is 16 out of 61,000."[123]

We are now getting close to Medawar's keyhole, for Millikan is generous in this publication in presenting his findings. Millikan provides panels containing the critical law observations, together with sample calculations, for sixteen of the many drops he had followed. A typical example is "Drop No. 41" in his Table XV. It is reproduced here in Figure 2.3. Millikan also gives, in his Table XX, "a complete summary of the results obtained on all of

Figure 2.3

Table XV.

Drop No. 41.

t_g	t_F	$\frac{1}{t_F}$	n'	$\frac{1}{n'}\left(\frac{1}{t'_F}-\frac{1}{t_F}\right)$	n	$\frac{1}{n}\left(\frac{1}{t_g}+\frac{1}{t_F}\right)$	
24.016	42.188						
24.142	42.078	.02369			8	.009336	
24.130	42.098		1	.009380			
24.070	69.900	.01431			6	.009328	$V_i = 5065$
24.000	203.200	.004921	1	.009389	5	.009316	$V_f = 5059$
24.030	23.844	.04194	4	.009255	9	.009286	$t = 23.05°$ C.
24.046	30.606				8	.009289	$p = 19.01$ cm.
24.028	42.800						$v_1 = .04253$
23.968	42.944	.02326			7	.009276	$a = .0001816$
24.018	71.400	.01400	1	.009260	6	.009277	$l/a = .1394$
23.770			2	.009295			$e_1 = 6.097 \times 10^{-10}$
23.882	30.652	.03259				.009282	
24.008				**.009314**		**.009301**	
(1)	(2)	(3)	(4)	(5)	(6)	(7)	

Initial and final potential differences of battery (V_i, V_f)

temperature (t)

pressure in chamber (p)

(cm/sec), average speed of fall without electric field (v_1)

calculated radius of oil drop (a)

mean fall path ÷ radius* (l/a)

calculated mean value of elementary charge, before Stokes's law correction** (e_1)

*Misprint for 0.2073, as given in Millikan's Table XX.

**Misprint for 6.110×10^{-10}, as given in Millikan's Table XX.

(1) t_g column = time (sec) of drop's fall under gravity through 10.21 mm distance; twelve successive observations, and the mean value. $[t_g \propto (1/v_1)]$

(2) t_F column = time of rising when the (charged) drop is retrieved after its fall by applying the electric field. $[t_F \propto (1/v_2)]$

(3) $(1/t_F)$ column = reciprocal of some of t_F observations, hence proportional to v_2

(4) n' column = change of charge (in units of e) between successive ascents (t_F being followed by t_F'), owing to the encounters of the drop with gas ions during the fall. Since

$$e_{\text{ionic}} \propto \left(\frac{v_2' - v_2}{n'}\right), \; n' \text{ is given by } \left(\frac{1}{t_F'}\right) - \left(\frac{1}{t_F}\right).$$

In a given case n' is found by adopting that trial value, assumed to be a small integer, that assures that the product $(1/n') \cdot [(1/t_F') - (1/t_F)]$ is constant throughout the experiment with that drop.

(5) Column of values of e_{ionic} indicates (uncorrected) values of elementary charge on the gas ions encountered in five excursions.

(6) and (7) Columns showing number of elementary charges n on drop, initially owing to friction in preparation of drop (by "atomizer"). Because $e_{\text{frict}} \propto (v_1 + v_2/n)$, n is obtained similarly to (4) and (5), but now v_1 is given by $(1/t_g)$ and v_2 by $(1/t_F)$ for the ascent immediately after the descent measurement, t_g.

Note: The chief point is the determination of e_{ionic} (5) and e_{frict} (7) *and the coincidence of the two values.* They are then used together to obtain e_1, and, after Stokes' law correction, e. See ref. 123.

the 58 different drops upon which complete series of observations like the above [i.e., a table of data on one of the drops] were made during a period of 60 consecutive days." And again, after showing that no more than a single one of these fifty-eight drops gives results for *e* that deviate "as much as 0.5 per cent." from the others,[124] Millikan writes, in italics, and without repeating the previous qualification, "*It is to be remarked, too, that this is not a selected group of drops but represents all of the drops experimented on during 60 consecutive days*, during which time the apparatus was taken down several times and set up anew." In his book *The Electron*, Millikan uses the same passage and all the data of the 1913 paper in Chapter 5, "The Exact Evaluation of *e*." He adds for extra emphasis: "These [58] drops represent all of those studied for 60 consecutive days, no single one being omitted."[125]

Drop No. 41 revisited – the laboratory notebooks, 1911–12

All of these publications, and the controversy itself, take on additional significance in view of the happy circumstance that two laboratory notebooks have been found for the years 1911 and 1912, containing the data of observation and some of the data reduction which led to Millikan's 1913 *Physical Review* paper.[126] The first notebook begins with an entry dated October 28, 1911, "Density of Clockoil. By R. A. Millikan,"[127] and ends some 110 pages later, with a run dated March 11, 1912. On each page there is typically an experiment of one oil drop followed during changes of charge as it picks up ions from the air. Some experiments are lengthy and elaborate, fewer are brief, and a small fraction are aborted early. The second notebook begins with a run on March 13, 1912, and the last run, about 65 pages later, is dated April 16, 1912. Again there is usually one experiment per page. In all, there are about 140 identifiable runs during about six months.

Millikan's energy is evident in long series of runs following one another. The controversy over the existence of the electron is in full swing, and the stakes are high. Even though the work is still beset by difficulties, Millikan and his students are no longer novices. Millikan had been carrying out some form of droplet

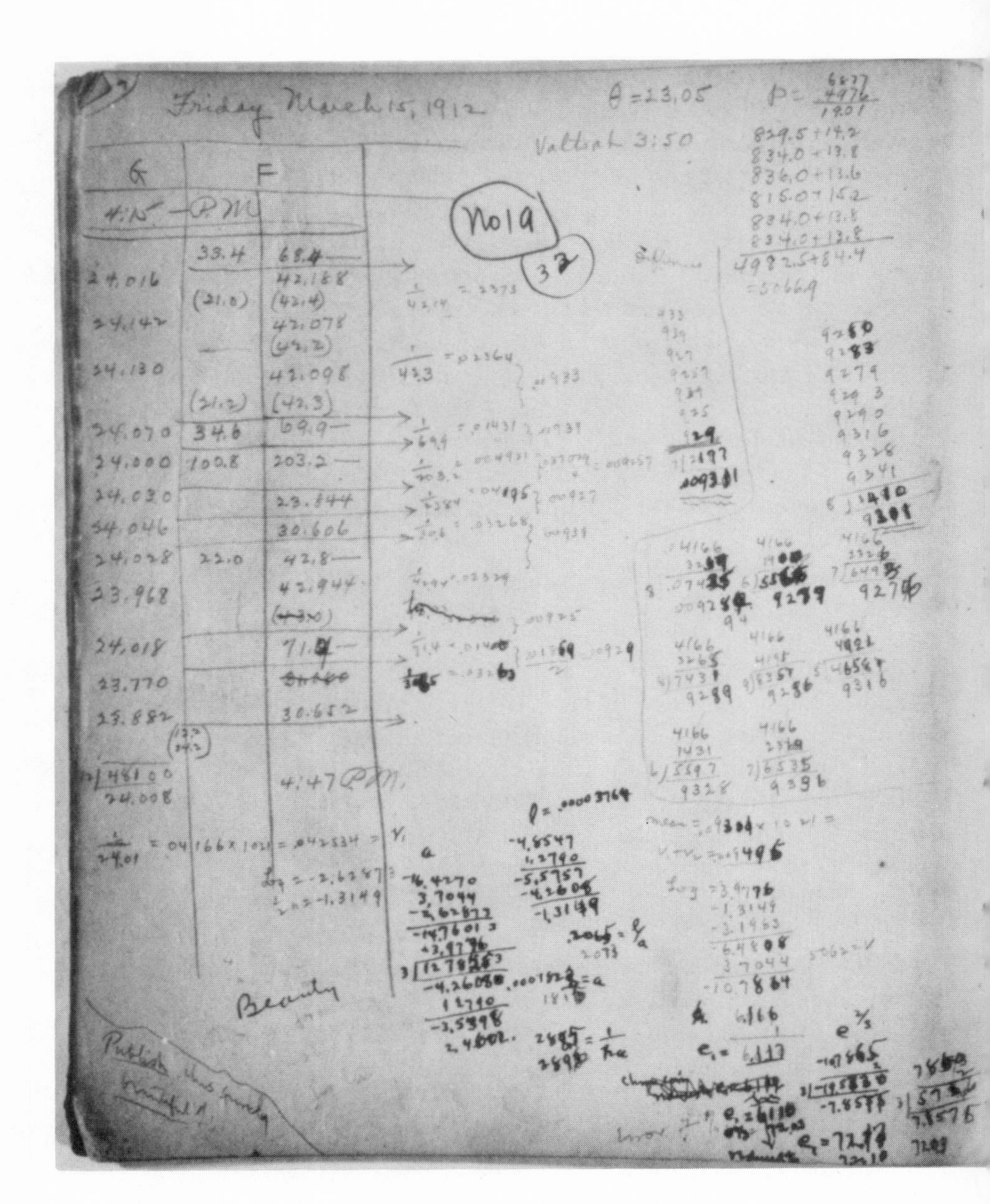

Figure 2.4. First and second observations in Millikan's Laboratory Notebook for

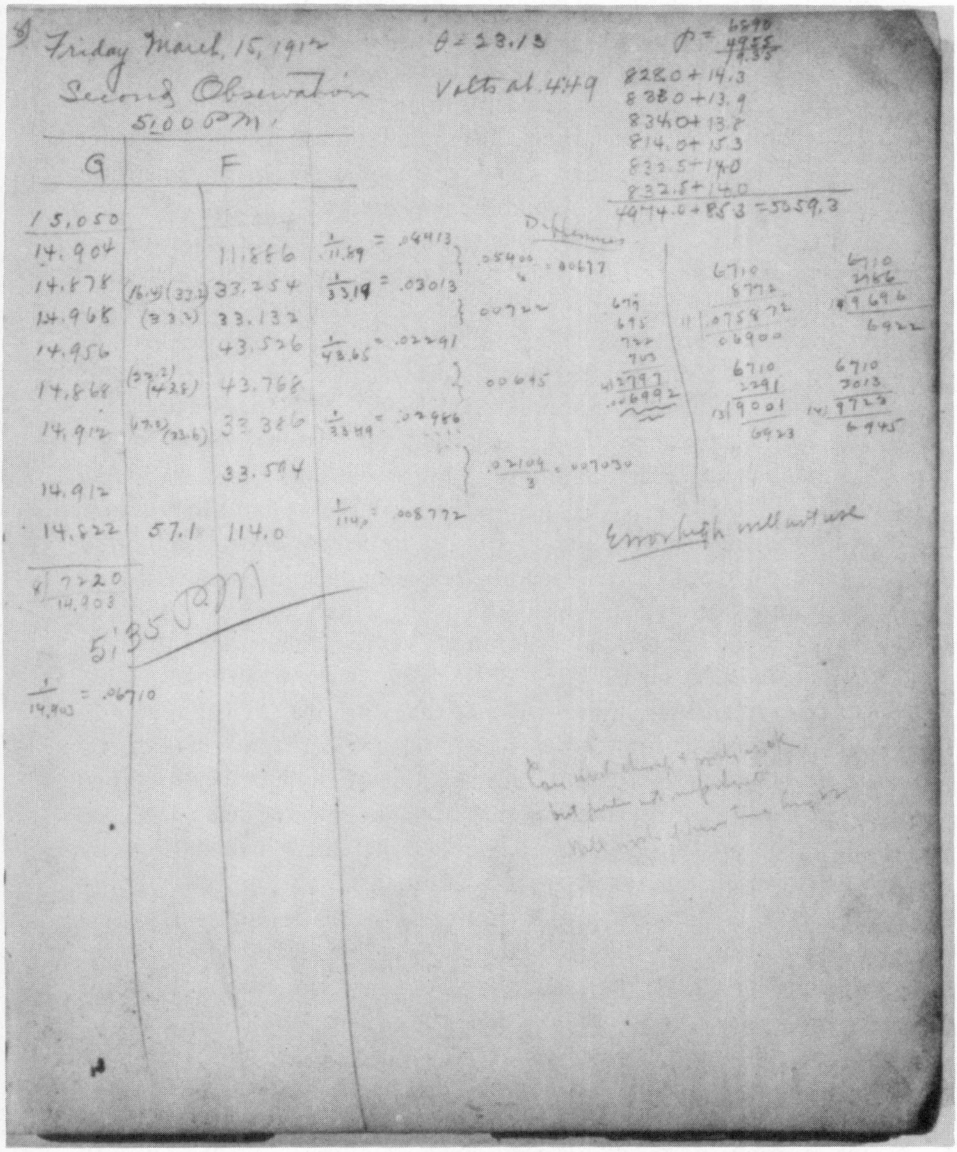

Friday March 15, 1912 θ = 23.13

Second Observation Volts at 4449

5.00 P.M.

G		F
15.050		
14.904		11.886
14.878	(16.4)(22.2)	33.254
14.968	(33.2)	33.132
14.956		43.526
14.868	(22.2)(43.8)	43.768
14.912	(47.0)(33.6)	33.386
		33.584
14.912		
14.822	57.1	114.0

8)7220

14.903

5.35 P.M.

1/14.903 = .06710

1/11.89 = .08413

1/33.19 = .03013

1/43.65 = .02291

1/33.49 = .02986

1/114.0 = .008772

Differences

.05400/4 = .00677

.00722

.00645

.02104/3 = .007030

8280 + 14.3

8380 + 13.9

8340 + 13.8

814.0 + 15.3

832.5 + 14.0

832.5 + 14.0

4974.0 + 85.3 = 5059.3

Error high will not use

Friday, March 15, 1912. Courtesy, California Institute of Technology Archives.

experiment for about five years. The techniques used in 1911–12 are not explorations in unfamiliar territory.

As we trace details of the analysis, we encounter fine details and ingenious decisions, hour by hour. (One remembers Henry A. Murray's definition, quoted in the introduction: ". . . science is the creative product of an engagement between the scientist's psyche and the event to which he is attentive.")

Figure 2.4 refers to data taken at an advanced stage of the work, in the second notebook, and only a month before the end of the series that yielded the 1913 paper. The left-hand page is a representative example, chosen here because it is the raw protocol from which one of the published tables of data (in this case, Drop No. 41 as later renumbered) was drawn. Thus we are looking at the experiment on one of the fifty-eight drops upon which Millikan's final calculation of e was based in the 1913 paper ($4.774 \pm 0.009 \times 10^{-10}$esu) – a value Millikan could stay with for a dozen years, despite all further improvement of technique.

Every part of the page can be coordinated fairly quickly with the corresponding published version which we have seen in Figure 2.3. Thus the first column (G) is equivalent to t_g; the next column is t_F, both for the full distance of the drop's descent of 10.21 mm, and sometimes for half that distance. At the top right are the readings of temperature, pressure, and potential differences (apparently for different sections of the battery, modified by a calibration correction that appears to have been later revised, before publication). The detailed hand calculations by logarithms, in the lower right quadrant, can also be followed up to the determination of e_1. Modifications made during the final computation prior to publication, sometimes several in different inks and pencil on the same page, appear in the notebooks, some pages carrying indications that the recomputations occurred during the summer of 1912.

A key point in Millikan's work is his comparison between two sets of figures for each run. In the first run in Figure 2.4, one set is given under "Differences" (seven entries, starting with [0.00]933). The other is just to the right of it (eight entries, with the computed average of [0.00]9301). Each entry in the first of these two columns is a calculation of a quantity proportional to the elementary ionic charge, e_{ionic}, that is, of

$$\frac{1}{n'}\left(\frac{1}{t'_F} - \frac{1}{t_F}\right)$$

obtained exactly as in the table reproduced in Figure 2.3. For example, 0.009257 is one-fourth of the difference between the reciprocal times of successive ascents

$$\left(\frac{1}{t'_F} = \frac{1}{23.84 \text{ sec}}, \frac{1}{t_F} = \frac{1}{203.2 \text{ sec}}\right)$$

on the assumption that $n' = 4$, that four integral charges were picked up between the measurement of t_F and t'_F.

In the same way, the entries in the other column refer to the calculation of a quantity proportional to the elementary frictional charge, $e_{\text{frict.}}$ that is, of

$$\frac{1}{n}\left(\frac{1}{t_g} + \frac{1}{t_F}\right)$$

again as in Figure 2.3. For example, $1/t_g$ is here 1/24.01 sec (or 0.04166 sec $^{-1}$) for all parts of the experiment. To this are added successive values of $1/t_F$, as previously calculated – but again the assumption is made for each entry that some integral multiple (8, 6, 7, 8, . . .) of charge is present.

Both assumptions become plausible when the scatter of data in each of the two columns turns out to be small, and when the mean values obtained in each of the two columns, so differently based, are nevertheless closely equal. This is just what happens here: 0.009311 and 0.009301 are only about 0.1 percent apart. (Figure 2.3 indicates that recalculations prior to publication changed the first of these values to 0.009314; but it is still a good agreement.)

Millikan expresses his pleasure in the lower left corner. He writes: "Beauty. *Publish* this surely, *beautiful!*"

Taking the readings occupied about a half-hour. Thirteen minutes after that Millikan was ready for another run, as indicated on the upper right-hand page (Figure 2.4). Considering his energy and long experience, Millikan may have used these minutes between runs to make the first rough calculations from his data (although occasionally with small arithmetical errors), subject to later reexamination.

We can look briefly at the right-hand page to see how the next run went. Evidently not well. This was now a heavier drop (t_g is shorter). It did not change its charge drastically between ascents, and it appears to have been lost sooner than one would like, leaving only four "Differences." Worst of all, the average values indicating e_{ionic} (0.006992) and e_{frict} (about 0.00692) are about 1 percent apart. Millikan notes frankly: "*Error high* will not use," and adds (probably later): "Can work this up & probably is ok but point is [?] not important. Will work if have time Aug. 22." It was a failed run – *or, effectively, no run at all.* Instead of wasting time investigating it further, he simply went on to make another set of readings with a new drop, on the next page of the notebook. Again it was a heavy drop, and for a while it was touch and go whether the data could be considered meaningful. He noted on the margin: "Might omit because discrepancy . . ." but then crossed that note out. Ultimately, these data, christened Drop No. 39, made it into the final published set.

The second set of observations made on March 15, 1912, was by no means the worst drop, nor was the first one (No. 41) the best in the series. But it is clear what Ehrenhaft would have said had he obtained such data or had access to this notebook. Instead of neglecting the second observation, and many others like it in these two notebooks that shared the same fate, he would very likely have used all of these. For example, the entries on the right-hand page make excellent sense *if* one assumes that the smallest charge involved is not *e* but, say, one-tenth *e*. Thus in the top right-hand corner of the page, if the sums given (for $1/t_g + 1/t_F$) such as 0.075872, and 0.09001, and 0.09723 are divided *not* by the integers 11, 13, and 14, but by 10.9, 12.9, and 13.9, a value proportional to e_{frict} results which matches almost exactly the mean of 0.006992, obtained earlier for the ionic charge (under "Differences"). From Ehrenhaft's point of view, it is the assumption of integral multiples of *e* which forces one to assume further, without proof, a high "error" to be present, and thus leads one to the silent dismissal of such readings – instead of using them to support a conceivable claim that at least in this experimental run the quantum of electric charge appears to be 1/10th of *e*.

Support for the conception of subelectrons would not fit with the rest of the physics of the time. From Ehrenhaft's point of view it was for just this reason to be regarded as an exciting opportunity and challenge. In Millikan's terms, on the contrary, such an interpretation of the raw readings would force one to turn one's back on a basic fact of nature – the integral character of *e* – which clearly beckoned. Admittedly, it did not come through in every one of these runs, but that was to be expected. In real life, observations of this sort are beset by a number of difficulties, some more obscure than others; but one feels sure that eventually they can be explained and removed or dealt with by plausibility arguments. To cite some difficulties recorded in Millikan's notebooks, generally against "failed" runs: The battery voltages have dropped; the manometer is air-locked; convection often interferes; the distance must be kept more constant; stopwatch errors occur; the atomizer is out of order.

In the meantime, Millikan had quite enough observational material left – 58 drops out of about 140 – to make a sound case, the more so as the integral value of *e* fit very well with other secure and unchallenged facts such as Rutherford's measurement of the charge of the alpha particle. Indeed, Millikan would have warned Ehrenhaft that using *all* readings equally, just as they come in, would be defensible only in a completely routinized situation where the chances for artifacts entering the "open window" have become negligible. This was by no means the case here. Thus at the end of a long run on December 20, 1911, Millikan was puzzled by the value of *e* far outside the expected limits of error. Aware that occasionally some material such as dust might still intrude in the observation chamber, he calmly explained the discordant result to himself by a marginal note: "$e = 4.98$ which means that this could not have been an oil drop."

This remark illustrates again that the results of Millikan and of Ehrenhaft were quite sensitive to the treatment of data – and, before that, to the decision about what is the relevant or even crucial aspect of the experimental design, which data are discordant or suspicious, and which may be dismissed on plausibility grounds. As is generally true prior to the absorption of research results into canonical knowledge, the selection of the relevant portion of experience from the in principle infinite ground is guided by a hypothesis, one that in turn is stabilized chiefly by success in handling that "relevant" portion, and by the thematic presupposition which helps focus attention on it.[128] One is reminded of two other, historic cases in which the strikingly new and simple results announced turned out, on recent reexamination, to be not easy to correlate with or induce from the data obtainable at the time: John Dalton's rule of simple whole-number ratios of weights in chemical reactions, and Gregor Mendel's simple numerical ratios obtained from his botanical experiments.

Of course, Millikan did not need to worry that Ehrenhaft might use the discordant results in his notebook. They belonged to the realm of private science, with many decisions to be made before the work was fully done. Therefore, he evaluated his data and assigned qualitative indications on their prospective use, guided by both a theory about the nature of the electric charge and a sense of the quality or weight of the particular run. It is exactly what he had done in his first major paper, before he had learned not to assign stars to data in public. This practice is familiar to anyone who has done basic experimental research: In the midst of a run, one does respond to small clues of the extent to which the numbers one is recording do in fact stem from the phenomena being observed.

It appears likely that after almost every run some rough calculations of *e* were privately made on the spot, and often a summary judgment appended. Here are some of Millikan's exclamations as the work proceeds, as recorded in the notebooks next to the data and calculations:

> Very low Something wrong [November 18, 1911]. Very low Something wrong [November 20, 1911]. This is almost exactly right & the best one I ever had!!! [December 20, 1911]. Possibly a double drop [January 26, 1912]. This seems to show clearly that the field is not exactly uniform, being stronger at the ends than in the middle [January 27, 1912]. Good one for very small one [February 3, 1912]. Exactly right [February 3, 1912]. Something the matter . . . [February 13, 1912]. Agreement poor.

Will not work out [February 17, 1912]. Publish this Beautiful one . . . [February 24, 1912]. Beauty one of the very best [February 27, 1912]. Perhaps Publish [February 27, 1912]. Excellent [March 1, 1912]. This drop flickered as tho unsymmetrical [March 2, 1912].

This continues, with beauty appearing more consistently as the work progresses, and ends thus, during the last week or so:

Can't get differences [April 8, 1912]. Beauty. Tem & cond's perfect. no convection. Publish [April 8, 1912]. Publish Beauty [April 10, 1912]. Beauty Publish [crossed out and replaced by] . . . Brownian came in [April 10, 1912]. Perfect Publish [April 11, 1912]. Among the very best [April 12, 1912]. Best one yet for all purposes [April 13, 1912]. Beauty to show agreement between the two methods of getting $v_1 + v_2$ Publish surely [April 15, 1912]. Publish. Fine for showing two methods of getting v . . . No. Something wrong with the therm.

Suspension of disbelief

Two rather contrary tendencies are visible as we watch Millikan at work. One is the standard classical behavior of obtaining information in as depersonalized or objective a manner as possible. As every novice is taught, the graveyard of science is littered with those who did not practice a *suspension of belief* while the data were pouring in. But there is the other side of the coin, a strategy without which work of novelty could not get past those first hurdles whose exact nature can be identified in detail only after the fact. To understand this side of the researcher's behavior I introduce the notion of the *suspension of disbelief;* that is, the ability during the early period of theory construction and theory confirmation to hold in abeyance final judgments concerning the validity of apparent falsifications of a promising hypothesis.[129]

This aspect of the operation of the scientific imagination is one of its key features, and one that does not contradict another notion requiring tests on other grounds, namely, the notion that falsification is the crucial duty of the scientist. In the well-known formulation of Karl Popper, "I arrived . . . at the conclusion that the scientific attitude was the critical attitude which looked not for verification but for crucial tests; tests which could *refute* the theory tested . . ."[130] Whether or not this conception is adequate for the analysis of scientific work in its later stages, when it has become part of a public dialogue, a striking feature of Millikan's work in progress is that it exhibits a mechanism for stabilizing belief in the efficacy of a hypothesis, long enough to help it survive to the later stage of testing in public discussion.

If Millikan's only scientific achievement was the oil drop experiment, he might be open to the charge that he was lucky in guessing at the usable data, or fortunate in his obstinacy. Such a charge would collapse in the face of his next and perhaps most influential work, the resumption of research on the photoelectric effect.[131] Here he found himself working with the wrong presupposition, but he knew how to rid himself of it eventually. Millikan launched into that work with the same energy and obstinacy as into his earlier work on the quantization of the charge of the electron, yet with the opposite assumption. As easy as it had been for him to adopt quantization as a thematic hypothesis for electricity, secure in the belief that it was an ancient and a sensible idea, for a long time he regarded the application of the quantum hypothesis to the energy of light as an unacceptable novelty. Millikan wrote that Einstein's "bold, not to say reckless," hypothesis "seems a violation of the very conception of an electromagnetic disturbance"; it "flies in the face of the thoroughly established facts of interference."[132] On accepting the Nobel Prize, Millikan reported: "After ten years of testing and changing and learning and sometimes blundering . . . this work resulted, contrary to my own expectation, in the first direct experimental proof in 1914 of the exact validity . . . of the Einstein equation . . ."[133]

Toward Ehrenhaft's presuppositions

The ability to exploit and, if necessary, transcend one's presuppositions defines a chief difference between Millikan and Ehrenhaft during the period around 1910. I turn once more to

Ehrenhaft, in order to try to understand *his* presuppositions and motivations. That the notebooks of his laboratory group did not survive impedes the fuller study he deserves; but much can be retrieved from the published materials. Out of the wealth of papers issuing from the Vienna laboratories, one that Ehrenhaft published in the *Physikalische Zeitschrift* in 1910 provides important clues.[134] The key data in this instance again support his contention that if an indivisible atom of electricity existed, "it would seem to have to be smaller than 1×10^{-10} esu" (if it can exist at all). Ehrenhaft presents a set of 1,000 individual measurements on fog droplets, created by blowing moist air over white phosphorus. The measurements were taken from the previous publication of Karl Przibram, who apparently had undertaken these measurements at the request of Ehrenhaft, using a method proposed to him by Ehrenhaft.

Figure 2.5 presents the results. Along the abscissa are the observed charges, in units of 10^{-10} esu; along the ordinate, the number of observed cases. The graph displays the first hundred data as the histogram with the lowest profile. To this, the next hundred data are added to make the second histogram, and so forth. The striking fluctuation of the daily maxima was acknowledged to be mysterious but did not touch the essential point: It is clear that the peaks are not separated by simple integral relations, nor is there any reason why a continuation of this process should not yield charges even smaller than those found. The statement in the title of the article is certainly borne out by the results displayed.[135]

As for many years to come, the audience that came to hear and discuss Ehrenhaft's paper, according to the transcript appended to it, was distinguished, puzzled, and unable to propose definite remedies for what to them seemed wrong. In retrospect, it is clear that at least one methodological difficulty had entered the experiment, and it is significant that the kind of remedy for it would not normally be suggested in a public scientific meeting. The experimenters appear to have used all their assiduously collected readings, good, bad, and indifferent. The kind of discrimination we saw at work in Millikan's private data analysis was lacking. On the contrary, the bias now was in the opposite direction. The "window" was opened, and all "measurements"

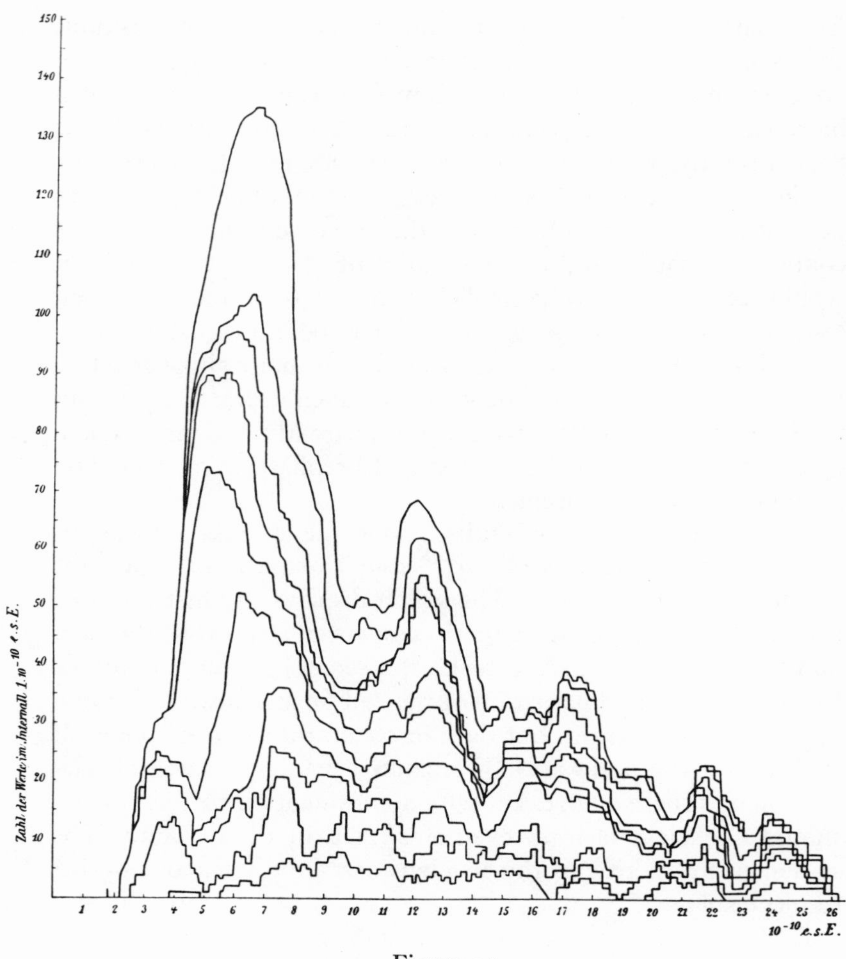

Figure 2.5

were admitted. Ehrenhaft's method was not altogether different from what students do to this day when they repeat a well-established experiment. Figures 2.6 (a) and (b) illustrate this point by showing the widely scattered results in some recent publications of student experiments on the electric charges on oil drops.

Another ironic possibility for explaining Ehrenhaft's results is that the equipment in Vienna was rather more sophisticated than necessary. Millikan's equipment and procedure, at least in the crucial early phase, appear to have been much more primitive than Ehrenhaft's. Millikan's simple apparatus was put together in a rather homespun way. The atomizer was originally a perfume sprayer bought at a drugstore and the telescope was a short focus tube set up 2 feet from the 1.6 centimeter gap in the horizontal (22-cm diameter) air condenser.[136] Ehrenhaft's equipment was far more sophisticated, involving the ultramicroscope (with which Siedentopf and Zsigmondy had caused a sensation in 1902), which permitted observation of objects down to a limit about five hundred times below the resolving power of an ordinary microscope. Ehrenhaft himself had perfected its use in the observation of Brownian movement. The condenser system he used was about an order of magnitude smaller than Millikan's in each dimension, and the range of size of charged objects he could follow was far wider. Thus it permitted measurements on much smaller objects, which fitted with his conception that in looking for the smallest charges one should look at the smallest available objects. As to fears that Stokes's law would break down in that regime, Ehrenhaft had two responses. Insofar as a correction would be needed, it should come by empirical methods rather than, as he perceived Millikan to be doing, by building the conception of a unitary electron into the method of correction. At any rate, corrections to Stokes's law would tend to make the small charges he was finding even smaller.

While Millikan may have appeared to be looking at the world of charged particles through a curiously primitive device, that was just an aspect of Millikan's strength. The particular dimensions of the apparatus he initially chose, and the voltage of the battery available, were

> the element which turned possible failure into success. Indeed, Nature here was very kind. She left only a narrow range of field strengths within which such experiments as these are at all possible. They demand that the droplets be large enough so that the Brownian movements are nearly negligible, that they be round and homogeneous, light and non-evaporable, that the distance be long enough to make the timing accurate, and that the field be strong enough to more than balance gravity by its pull on a drop carrying but one

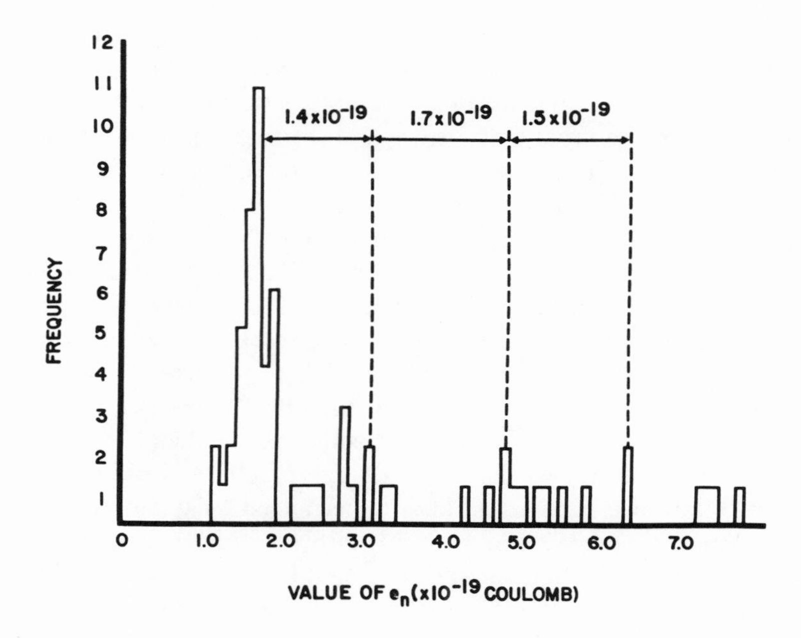

Figure 2.6*a*. "A histogram of 74 drop charges determined by seven student pairs. Redrawn from the original by student R. Williams, Western Michigan University, Fall 1969." From Haym Kruglak, "Another Look at the Pasco-Millikan Oil-Drop Apparatus," *American Journal of Physics* 40 (May 1972): 769.

> or two electrons. Scarcely any other combination of dimensions, field strengths, and materials, could have yielded the results obtained.[137]

Nature is not kind to everyone. Relatively few scientists know how to find or seize upon a "device of choice" that becomes the tool for opening up an area of research. Galileo fastened on the pendulum and the rolling ball as keys to dynamics. Fermi used the slow neutron, and Einstein the thought experiment of a freely falling experimenter noticing the seeming absence of gravitational effects. Ehrenhaft refused to see any resemblance between such cases and Millikan's device. On the contrary, Millikan's work seemed to him unacceptable on epistemological grounds; that Millikan's measurements were restricted to a smaller region of mass, to droplets that are relatively large rather than allowing the use of arbitrarily large and small drop-

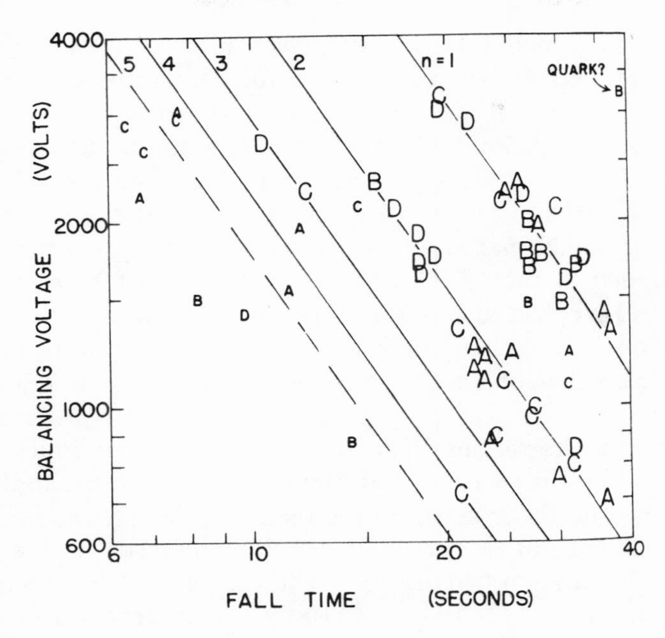

Figure 2.6*b*. "Raw data of balancing voltage and fall time obtained by students in the 1971 class (four laboratory sections of approximately 15 students each are identified by letter symbols). Data points for $n>4$ were discarded as were a few apparent blunders (designated by small symbols)." From Mark A. Heald, "Millikan Oil-Drop Experiment in the Introductory Laboratory," *American Journal of Physics* 42 (March 1974): 244.

lets, was a detrimental feature. Valid findings should exhibit themselves over a large range, rather than within a relatively small sanctuary.

The abandonment of the electron

Ehrenhaft's conversion from his original expressed belief in the elementary quantum of electricity was so rapid and fervent that we can specify the period when it seems to have occurred. His last paper in the tradition of the search for the value of the electron's charge was received by the *Physikalische Zeitschrift* for publication on April 10, 1909.[138] Slightly over a year later, by April 21, 1910, the date of the short note in the *Anzeiger*,[139] he had begun to change his mind: "An indivisible quantum of electricity, which theory presupposes to exist, would have a value considerably below the one accepted hitherto." By May 12, 1910,[140] the atom of electricity had shrunk to below 1×10^{-10} esu, and the question "whether it can exist at all" was proposed as the subject of forthcoming research. When the first full-scale paper appeared in May 1910,[141] the words *elekrische Elementarquantum*, which had been in the titles of the 1909 papers and had slipped to the subtitles of the short notes of April 21 and May 12, 1910, had disappeared from the title entirely.

Although Ehrenhaft made some gestures to connect the new work with that of 1909 (which had used similar experimental equipment), it is clear that by the third week of April 1910 he had at least very serious doubts about the electron of which there were no hints in 1909. By mid-May 1910 he was quite confident about the need for subelectrons that, in principle, might have no lower limit of charge at all. He drew attention to the wide spread or variation between the values reported in the literature for e (from 1 to 6×10^{-10} esu), both by different methods and by different observers using the same method. If one wants to avoid a style of science that piles up "hypotheses and corrections," one is led to the recognition that the apparent variations of charge are grounded in nature.[142] The interpretation of the experiments must be modified correspondingly. A few months later[143] these results had become "certain beyond doubt" for Ehrenhaft; one needed really only to look at what nature herself made directly accessible to the senses of the assiduous experimenter, as one could gather from the data in Figure 2.5.

As Ehrenhaft's publications continued, there was an increasingly epistemological component to the work, that is, the use of these experiments to attack the credibility or necessity of atomism itself. In a long paper of 1914[144] summarizing his work and defending it against his critics, Ehrenhaft still used some of his older arguments. He now believed that quanta of electricity, if they exist, should be at most on the order of 10^{-11}esu. With this he could turn the tables on Millikan, for now the puzzle that needed explanation was why in the experiments of Millikan and others a specific value of e was found again and again. Ehrenhaft hinted at a theory that might explain why his smallest particles exhibited the smallest charges; this was to be expected, he explained, because the smallest quantities of electricity should be on bodies of smallest capacity.

But Ehrenhaft's attention was not chiefly on physical arguments. He deplored the fact that although Ludwig Boltzmann, a few years earlier, still had to argue for the necessity for atomistics in the natural sciences, current views now accepted this conception: "In recent years the atomistic theories of matter, electricity and radiation have gained ground in physics more than ever before."[145] Everyone in physics was convinced of the heuristic value of these theories; but if such a theory was more than pure speculation, it must be solidly based on experiments that could withstand critical examination. Ehrenhaft noted that his study provided such an examination of the foundations of a portion of those hypotheses (the atomistics of electricity) and that his style was to proceed "from the *direct facts*."

Of course, there was never a direct laboratory disproof of Ehrenhaft's claims. In the 1916 edition of *The Theory of the Electron*, H. A. Lorentz still had to remark, "The question cannot be said to be wholly elucidated." In his review of the case, R. Bär noted in 1922[146] that "the experiments [of Ehrenhaft] left, at the very least, an uncomfortable feeling." Like most such controversies, this one also faded into obscurity, without anything as dramatic as a specific, generally agreed-upon falsification taking place at all. Indeed, Ehrenhaft continued to publish on subelectrons into the 1940s, long after everyone else had lost interest in the matter.

"*A Battle of Two Worlds*"

In his Nobel Prize acceptance speech, Millikan had put an end to his side of the debate with a careful review of his work. A year after its publication in 1925, Ehrenhaft also gave a public address which signaled his realization that the controversy had ended for all practical purposes. As it happened, that address was also part of a ceremony, one held in a public park, on a Saturday in Vienna. The occasion was the unveiling of a bust in honor of Ernst Mach, to commemorate the tenth anniversary of Mach's death. Morritz Schlick delivered a eulogy.[147] Another contribution came from Einstein, who had admired Mach and had once specially sought him out during a 1911 visit to Vienna, a meeting apparently arranged by Ehrenhaft. Ehrenhaft's own presentation[148] was brief but revealing. Perhaps for the first time the main pieces of his motivation in the long fight against the atom of electricity came into the open.

Ehrenhaft saw Mach as a lonely fighter. Even the bust of Mach, which the authorities did not want in the arcade of the university building, stood there "alone and isolated." Accepting Mach's own habitual underestimation, Ehrenhaft thought that Mach had "remained not understood, and had so few followers, and those not among physicists . . .":

> . . . I only want to draw attention to this: the great difference between Mach and most physicists arises from the fact that through the further development of physics each of the two opposing views shows itself to be ever more fundamental, ever more contrary and unbridgeable, like two professions of faith. Mach [appears] as an advocate of the much more modest, phenomenological point of view which finds satisfaction merely with the description of the phenomena, and despairs of other possibilities. The others are advocates of views which, through statistical methods and speculative discussions concerning the constitution of matter, are reflected in atomism, and who believe themselves able to get down to the true Being of things.

Ehrenhaft's talk then ended with a Wagnerian crescendo:

> Mach had the courage to set himself with mighty arguments against the current of the atomistic Weltanschauung that was sweeping along almost all others – the very same atomistics which, in the smallest, supposedly indivisible constituents of matter and, recently, also of electricity, believes to have attained the magic keys for opening at last all doors of natural knowledge.
>
> But the world follows a remarkable development. On the one hand, daring researchers storm further into the realm of atomistics, undaunted by such powerful thinkers as Mach; on the other hand, one must admit that the great man whom we celebrate today may be victorious

> in the end. Who dares to render judgment in this Battle of Two Worlds? (*Wer wagt es, in diesem Kampfe zweier Welten das Urteil zu fällen?*)[149]

Ehrenhaft had indeed touched a key point. Whatever else the controversy was about, it was also about two ancient sets of thematically antithetical positions: the concepts of atomism and of the continuum as basic explanatory tools in electrical phenomena, and the use of methodological pragmatism versus an ideological phenomenology.

This is as far as one can safely go on the basis of the documents now available. Some tantalizing questions remain. At some point after his early, striking success in a physics based on atomism, Ehrenhaft evidently had been converted to antiatomism and to "antihypothetical" theorizing, both of which were commonly identified with Mach, although Ehrenhaft was not a Machist in the positive and productive sense of the term. As we saw, the first indications of his change of mind appeared in the papers of late April and May 1910.[150] But to switch from one thema to its opposite is rarely done in science, and we naturally wonder what external influences may have helped Ehrenhaft to reach this new point of view. The rebuff by Millikan (published in February 1910) may well have played a role, although it is not likely to have been the major one.

We do not, and perhaps never will, know the reasons. But there is another unpublished letter in the Mach–Lampa correspondence that concerns Ehrenhaft, and it falls in the critical period when he was making the switch. It may contain a clue. The two-page letter from Anton Lampa in Prague to Mach is dated May 1, 1910, just after Ehrenhaft's first, rather cautious announcement of April 21 and before his more detailed presentations of mid-May.[151]

Lampa first tells Mach about an attack on the philosophy they both shared, an attack coming from Max Planck – the last major physicist who still dared to attack Mach openly, although Mach and his circle saw themselves as a little, beleaguered group. Lampa notes that Planck has published a book "in which he maintains *in extenso* the views of his which you have been fighting against." Planck has embroiled himself hopelessly in contradictions; hence "reading it will give you much pleasure." Although interesting from the point of view of physics, the book is "epistemologically childish."[152]

Then Lampa turns to the results of a recent trip to Vienna. Perhaps this was the occasion announced in his earlier letter of February 9, 1910, in which he had written to Mach: "I look forward with pleasure to be able to greet you personally in a few weeks, and to report to you then on the further developments of the case [the physics appointment, still pending in Prague]."[153] Without preliminaries, as though it were familiar territory to both, Lampa turns to the work of Ehrenhaft:

> If the provisional measurements should be confirmed which Ehrenhaft carried out when I was now in Vienna as part of his continuing research on the charges on colloidal particles, then the electron would be divisible. Even then Ehrenhaft had found particles with half electrons – in the meantime, Lang [Victor von Lang, whose assistant Ehrenhaft had been in 1903] has told me, he appears to have observed some with 1/3, 1/5 electrons.
>
> It would be just too beautiful [*Es wäre doch zu schön*] if now the electron were to undergo the same fate as the atom did as a result of cathode rays. . . .

Indeed, from the viewpoint of the more enthusiastic followers of Mach, that would have been just too beautiful – a long-awaited new hope for a battered cause. All recent events had been discouraging, for example, Perrin's successful crusade on behalf of molecular reality. A particularly severe blow must have been the defection of Wilhelm Ostwald. In the 1908 edition of his text *Allgemeine Chemie*, Ostwald had recanted his antiatomism in the preface, dated November of that year:

> I am now convinced that we have recently become possessed of experimental evidence of the discrete or grained nature of matter, which the atomic hypothesis sought in vain for hundreds and thousands of years. [Experiments such as those of J. J. Thomson and J. Perrin] justify the most cautious scientist in speaking now of the experimental proof of the atomic nature of matter. The atomic

> hypothesis is thus raised to the position of a scientifically well-founded theory.[154]

In this dark period for the Machists, Ehrenhaft must have appeared to them as a bright new star.[155] He, in turn, can hardly have been oblivious to the favorable impression his preliminary new findings were making on them; just in that provisional stage of his work, and just when they were looking for new ideas – and for new men.

3

Dionysians, Apollonians, and the scientific imagination

How do scientists go about obtaining knowledge? How *should* they? Few modern research scientists tend to be introspective about these questions. During apprenticeship, most scientists somehow absorb the necessary pragmatic attitude and then go about their business quite successfully, content to leave it to a small handful to become interested in epistemology when some obstinate difficulty blocks scientific advance.

Outside the walls of the laboratory, however, interest in theories of scientific knowledge runs high among three groups: (1) a small but vigorous set of professional philosophers; (2) students and other laymen who rightly believe that few questions are more practical or urgent today than how knowledge may be reliably gained and where the limits of certainty lie; and (3) critics of culture, including the new Romantics, the remnants of the counterculture movement, and a tiny band of "outsider" scientists and former science students who, disenchanted with the politics and performance of the scientific establishment, are interested in the ideological links among knowledge, power, and values.

With all their differences – some of which, as we shall see, are total, unresolvable, and of great consequence – these three groups have at least one property in common: Certain people in each group have high commitment, eloquence, and visibility and can command the attention of a public wider than that of all scientists put together. Nevertheless, they receive only passing attention from the scientists themselves. This is not surprising, since most scientists see little reason to volunteer for a debate for which they neither claim particular expertise nor expect much reward.[1] Life is short and research is long. In any case, the unwelcome distractions are increasing. Within the laboratory, the accelerating pace and complexity of work make heavier demands every year; at the same time, the external world keeps pressing for new and ever-more-urgent involvement and yet seems to impose ever-larger constraints in obtaining adequate support for scientific research, training, jobs, freedom of inquiry, or even for the public understanding of science itself.

These problems are by no means unrelated. Incorrect apprehensions by nonscientists of how scientific ideas are obtained or tested are at the base of many of the troubles that scientists encounter (or from which they tend to shy away). In selecting for analysis here two quite different views on scientific epistemology, I am keeping in mind their effect on citizens generally. Science policy in a democracy depends on long-range factors such as the popular understanding of science as a cognitive activity, both for its own sake and to permit soundly based participation in the making of science policy. The time scale of change in this public attitude is far longer than the terms of office of congressmen, presidents, or bureaucrats. Don K. Price, in his prophetic address as retiring president of the American Association for the Advancement of Science,[2] during the height of concerns over the downgrading of science in the scheme of national priorities, warned that the short-range difficulties then so evident to scientists should not blind them to the long-range ones. He asked that they look beyond the discomforting "political reaction" of economy-minded politicians and face the "fundamental challenge," which he described as "a rebellion . . . a cosmopolitan, almost worldwide, movement."[3]

> Its mood and temper reflect the ideas of many middle-aged intellectuals who are anything but violent revolutionaries. From the point of view of scientists the most important theme in the rebellion is its hatred of what it sees as an impersonal technological society that dominates the individual and reduces his sense of freedom. In this complex system, science and technology, far from being considered beneficent instruments of progress, are identified as the intellectual processes that are at the roots of the blind forces of oppression.[4]

Price is no Pollyanna; he agreed that "we have not learned how to make our technological skills serve the purposes of humanity, or how to free men from servitude to the purposes of technological bureaucracies" (although he added at once, "But we would do well to think twice before agreeing that these symptoms are caused by reductionism in modern science, or that they would be cured by violence in the name of brotherhood or love"). He called the rebels pessimists but then said:

> I do not think they are even pessimistic enough. To me it seems possible that the new amount of technological power let loose in an overcrowded world may overload any system we might devise for its control; the possibility of a complete and apocalyptic end of civilization cannot be dismissed as a morbid fantasy.[5]

Price pointed out that the intellectual core of the rebellion is not a disagreement over practical proposals to avert catastrophe but a philosophical aversion to the historic establishment of scientific reductionism – "the change from systems of thought that were concrete but complex and disorderly, and that often confused what is with what ought to be, to a system of more simple and general and provable concepts." The relationship the public perceives between science and politics therefore springs out of the popular theory of knowledge: "The way people think about politics is surely influenced by what they implicitly believe about what they know and how they know it – that is, how they acquire knowledge, and why they believe it."

The issue is even more complex, and therefore more interesting, because there exists also another, opposite (one is tempted to say symmetrical) set of forces. If the scientist – whether he takes notice or not – is confronted on one side by writers feeding a rebellion based on popular beliefs concerning scientific reductionism, he is also subject to a barrage from exactly the opposite direction, from a group of philosophers who wish to redefine the allowable limits of scientific rationality. Thus the scientist is caught between a large anvil and a fearful hammer. The one is provided by what I might call "the new Dionysians" – by authors like Theodore Roszak, Charles Reich, R. D. Laing, N. O. Brown, Kurt Vonnegut, Jr., and Lewis Mumford.[6] With all the differences among them, they do agree in their suspicion or contempt of conventional rationality and in their conviction that the consequences flowing from science and technology are preponderantly evil. Methodology is not their first concern; they think of themselves primarily as social and cultural critics. But they would "widen the spectrum" of what is considered useful knowledge as a precondition of other changes they desire. They tend to celebrate elements that they do not see in science – the private, personal, and, in some cases, even the mystical. Their skill is high, and the appeal of their lively prose is large.

If these new Dionysians constitute the anvil, the hammer is wielded by the group I shall call "new Apollonians."[7] They advise us to take precisely the opposite path – to confine ourselves to the logical and mathematical side of science, to concentrate on the final fruits of memorable successes instead of on the turmoil by which those results are achieved, to restrict the meaning of rationality so that it deals chiefly with statements whose objectivity seems guaranteed by the consensus in public science. They would "shrink the spectrum" emphatically, discarding precisely the elements that the other group takes most seriously.

Both groups present their cases with the apocalyptic urgency of rival world views. As in most polarized situations, they inflict the most damage not on each other, but on those caught in middle ground. Indeed, they seem to reinforce each other's position, as cold-war antagonists tend to do. In the face of the enemy, each limits the circle of allowable thought and action: one ritualistically heaping scorn on a caricature it calls rationality, the other on a caricature it calls irrationality. Each is dissatisfied with how science is actually done, and neither hides its distaste.

The new Dionysians

Evidence for the existence of the new Dionysians is not difficult to find.[8] Although their ideas may be fashionable, it would be a mistake to think of them as transient fads. To be sure, the twenty-first century is unlikely to discover a new voice among the present new Dionysian writers, as our century found Nietzsche hidden among nineteenth-century Dionysians; the high level of today's sales of these wares does not seem to be grounded in lasting literary quality or in fresh depths of insight. But even if each of these writers separately lasts only a season, the fact that their message falls on so many believing ears shows that the succession is not likely to wither away soon.

One of their more measured proponents of today, Theodore Roszak, sets the tone of the attack with statements such as these:

> What *is* to blame is the root assumption . . . that culture – if it is to be cleansed of superstition and reclaimed for humanitarian values – must be wholly entrusted to the mindscape of scientific rationality.[9]
>
> I have insisted that there is something radically and systematically wrong with our culture, a flaw that lies deeper than any class or race analysis probes, and which frustrates our best efforts to achieve wholeness. I am convinced that it is our ingrained commitment to the scientific picture of nature that hangs us up.[10]

When Roszak proposes to redefine true knowledge as "gnosis," within which traditional science is only a small part of a larger spectrum (that part which seeks merely to gather "candles of information"), one recalls that almost exactly a hundred years ago DuBois-Reymond's essay on "Die Grenzen des Naturerkennens" led to the controversy that culminated in the slogan "the bankruptcy of science," and that George Santayana wrote in *Reason and Science:*

> Science is a halfway house between private sensation and universal vision . . . a sort of telegraphic wire through which a meager report reaches us of things we would fain observe and live through in their full reality. This report may suffice for approximately fit action; it does not suffice for ideal knowledge of the truth, nor for adequate sympathy with the reality.

Indeed, today's critics of what they take to be the method and pervasiveness of science fit into a long and often brilliant tradition – Thoreau, Shelley, Coleridge, Wordsworth, Blake ("I come in the grandeur of inspiration to abolish ratiocination"), Goethe, Rousseau, Vico, Montaigne, and back to the ancient Greeks. Epicurus is reported to have said, in a letter to Menecaeus, "In fact, it would be better to follow the myths about the gods than be a slave of the physicists' destiny; myths allude to the hope of softening the gods' hearts by honoring them, while destiny implies an inflexible necessity." To me, the exemplar in this case is Dostoyevsky's *Notes from Underground.* I must pause here to recollect how moving the case was when it was still presented with literate passion:

> Nevertheless, there's no doubt in your mind that he [modern man] will learn as soon as he's rid of certain bad old habits and when common sense and science have completely re-educated human nature and directed it along the proper channels. You seem certain that man himself will give up erring *of his own free will* and will stop opposing his will to his interests. You say, moreover, that science itself will teach man (although I say it's a luxury) that he has neither will nor whim – never had, as a matter of fact – that he is something like a piano key or an organ stop; that, on the other hand, there are natural laws in the universe, and whatever happens to him happens outside his will, as it were, by itself, in accordance with the laws of nature. Therefore, all there is left to do is to discover these laws and man will no longer be responsible for his acts. Life will be really easy for him then. All human acts will be listed in something like logarithm tables, say up to the number 108,000, and transferred to a timetable. Or, better still, catalogues will appear, designed to help us in the way our dictionaries and encyclopedias do. They will carry detailed calculations and exact forecasts of everything to come, so that no adventure and no action will remain possible in this world.
>
> Then – it is still you talking – new economic relations will arise, relations ready-made and calculated in advance with mathematical precision, so that all possible questions instantaneously disappear because they receive all the possible answers. Then the utopian crystal palace will be erected; then . . . well, then, those will be the days of bliss.
>
> Of course, you can't guarantee (it's me speaking now) that it won't be deadly boring (for what will there be to

> do when everything is predetermined by timetables)?. . . . Well, chances are that man will then cease to feel desire. Almost surely. What joy will he get out of functioning according to a timetable? Furthermore, he'll change from a man into an organ stop or something like that, for what is man without will, wishes, and desires, if not an organ stop?

As a contemporary version of that tradition, I choose a book that has had a wide popular impact precisely because it did not pretend to have anything so sophisticated as an explicit epistemological message. Long selections from it first appeared in the fall of 1970 in *The New Yorker.* An unprecedented storm of publications followed: It was simultaneously available under six different imprints, one of them going through twelve printings in six months, another through twelve more printings in eight months, some still available on the bookstands. It seemed permanently installed on the best-seller list. Seven articles in rapid succession discussed the phenomenon in *The New York Times.* Its meaning was so widely analyzed that another widely circulated book sprang up devoted entirely to reprints of the reviews. Its message was castigated by many worthies, from then-Vice President Spiro Agnew (to whom it seemed permissive and immoral) to radical activists (who considered it counterrevolutionary); its popular success has never been fully explained. That book is, of course, Charles Reich's *The Greening of America.*

Reich's basic attitude toward nature, science, and rationality is still representative of the new Dionysians. In fact, I find the book more revealing as an example of that world view than more recent ones – somewhat as an art historian interested in the popular understanding of the arts would do well to attend not only to what is hanging in museums but also to some samples of widely liked *Kitsch.* It furnishes direct answers to the question how, in this framework, knowledge about the natural world should be sought.

Reich's is, on the whole, an optimistic book that promises a kind of paradise or utopia for the United States; but it says little about the problems faced by the majority of the world's people. This relatively parochial platform is a hint of the fundamental solipsism that pervades the book. Indeed, the first rule of the new attitude Reich espouses, which he calls Consciousness III, is that it "starts with self . . . The individual self is the only true reality. Thus, it returns to the earlier America: 'Myself I sing.' "[11]

With this inward turning to the idiosyncratic individual self, Reich juxtaposes no antithetical command that would allow the self to be transcended. The direct result is a Ptolemaic, homocentric conception of the world order. It may allow intense, moving and satisfying experiences; but doing or understanding science is not among them, nor is any field of scholarship in which the warrant of validity stems not from private enthusiasm but from some form of community consensus, for those activities require the recognition that the individual self is not the "only true reality." Rather, precisely the contrary is the case. The task in science is to achieve results that, to the greatest extent feasible, allow one to "describe a reality in space and time which is independent of ourselves".[12]

This is a point on which almost all scientists will agree, from beginners to sages. In its extreme form, this view has been stated perhaps most eloquently by Max Planck and Albert Einstein. Einstein stated a "basic axiom" in his own thinking, namely, that "It is the postulation of a 'real world' which, so-to-speak, liberates the 'world' from the thinking and experiencing subject,"[13] and repeatedly insisted that "physics is an attempt to grasp reality as it is thought independently of its being observed."[14] In the essay "Religion and Science," Einstein reiterated the tension between the two contrary drives in these words: "The individual feels the futility of human desires and aims, and the sublimity and marvelous order which reveal themselves both in nature and in the world of thought." Einstein thinks of this sympathetically as "the beginnings of cosmic religious feeling," which, together with the "deep conviction of the rationality of the universe," he recognizes as "the strongest and noblest motive for scientific research."[15]

In the constant struggle to go beyond what he called the futility of individual human desires and aims, Einstein came to agree fully with Planck that the final aim of science is the very opposite of its necessary initial stage of private, even heroic, struggle. That final aim of Public Science is the search for a world picture that is "real" insofar as it is covariant with respect to differences in individual observers.

The starkness of that vision – and although not all scientists would explicitly follow Einstein and Planck so far, they share it operationally to a large degree – may be one reason why the new Dionysians seem inevitably tempted to reach for the word "dehumanizing" when discussing the methods of science. Yet these methods, to yield testable truths, must go beyond Private Science, even though they cannot get started without going through that stage first. Nor do they contradict the fact that human concerns can (and should) remain central in those activities that have direct societal impact. Thus Einstein said, "Concern for the person must always constitute the chief objective of all technological effort." Moreover, the path from the "merely personal" through the projection of a rational world order does, after all, eventually lead back to the solution of complex and pressing human problems – physical, biomedical, psychological, social. Indeed, as I shall soon note, it is the only known method for finding such solutions.

But let us return to Reich, and through him to the whole movement of which we take him to be an indicator. Another "Commandment" of Consciousness III, Reich tells us, is that it is open "to any and all experience. [Elsewhere he calls experience "the most precious of commodities."] It is always in a state of becoming. It is just the opposite of Consciousness II, which tries to force all new experience into a pre-existing system, and to assimilate all new knowledge to principles already established."[16] With this premise, Reich announces an important thema, one that characterizes the movement: the primacy of *direct* experience – nonreductionistic, unanalyzed, unreconstructed, unordered. This is the guiding attitude, on the one hand, toward music ("the older music was essentially intellectual; it was located in the mind . . . ; the new music rocks the whole body and penetrates the soul"[17]) and, on the other, toward nature itself.

The new Dionysians are, of course, all *for* nature and the experience of nature, but in a specific way. In one of his most revealing passages, Reich explains that the Consciousness III person "takes 'trips' out into nature; he might lie for two hours and simply stare up at the arching branches of a tree . . . He might cultivate visual sensitivity, and the ability to meditate, by staring for hours at a globe lamp."[18] (He might also find at that point that "one of the most important means for restoring dulled consciousness is psychedelic drugs." Although Reich does not stridently advocate the use of drugs, he holds that "they make possible a higher range of experience, extending outward toward self-knowledge, to the religious." Incidentally, this seems to be the only reference to religion in the book.)

Nature, thus, is what one takes "trips" out into. By nature, Reich explains, he means "the beach, the woods, and the mountains,"[19] which he claims are "perhaps the deepest source of consciousness . . . Nature is not some foreign element that requires equipment. Nature is them."[20]

This homocentric epistemology, in which man and nature overlap in a total experience of natural phenomena – an act of imagination without criticism – rules out the very possibility of rational understanding of natural phenomena. And it is meant to do so: "Consciousness III . . . does not try to reduce or simplify man's complexity, or the complexity of nature . . . It says that what is meaningful, what endures, is no more nor less than the total experience of life."[21]

Even the scientists who are farthest from the usual rationalistic stereotype would have to disagree vigorously. To a mystic like Kepler, experience had a very different function. It triggered a puzzle in the mind, and it was through the working out of such puzzles that, in the view of Kepler and other neo-Platonists, persons could feel that they were communicating directly with the Deity. Newton, at the end of the *Opticks*, expressed an analogous hope for moral benefits to be derived from the study of nature:

> If Natural Philosophy in all its Parts, by pursuing this Method, shall at length be perfected, the Bounds of Moral Philosophy will also be enlarged. For so far as we can know by Natural Philosophy what is the first Cause, what Powers he has over us, and what Benefits we receive from him, so far our Duty towards him, as well as towards one another, will appears to us in the Light of Nature.

Even Goethe, though more secular in his expectation, intended that his holistic, noninstrumental approach to nature study would improve the state of science, by opening up new subjects for study – "Optical illusion is optical truth" – and by recruiting from a large, previously untapped reservoir that could yield new types of working contributors to science. He writes on the last page of the *Farbenlehre:* "All who are endowed only with habits of attention – women, children – are capable of communicating striking and true observations . . . *Multi pertransibunt et augebitur scientia.*"

None of these or similar ambitions are, however, reflected in the holism of the neo-Dionysians. What counts there is experience freed from analysis, from questions, even from the perception of complexity itself. The method is a direct shortcut through complexity. But the "method" that Newton had in mind in the preceding quotation consisted of two steps: "As in Mathematics, so in Natural Philosophy, the Investigation of difficult Things by the Method of Analysis ought ever to precede the Method of Composition." It is so in scientific work to our day: first reduction, then synthesis. Einstein, too, in the essay "Motiv des Forschens"[22] wrote with some regret that we must be satisfied with first "portraying the simplest occurrences which can be made accessible to our experience." More complex occurrences cannot be constructed with the necessary degree of accuracy and logical perfection. He acknowledged that one must choose: "supreme purity, clarity, and certainty, *at the cost of completeness.*" But, he thought, this should be considered only the halfway house. He noted that the reality of human limitations restricts the efficacy of logic in two ways; it would be foolish to hide it or deny it and self-defeating to define the permissible use of reason in science to narrow ground. On the one hand, from the general laws on which the structure of theoretical physics rests (he wrote in 1918),

> it should be possible to obtain by pure deduction the description, that is to say the theory, of natural processes, including those of life – if such a process of deduction were not far beyond the capacity of human thinking. The physicist's renunciation of completeness for his cosmos is therefore not a matter of fundamental principle.

> [In addition, there is another limitation:] To these elementary laws there leads no logical path, but only intuition supported by being sympathetically in touch with experience [*Einfühlung in die Erfahrung*] . . . There is no logical bridge from experience to the basic principles of theory . . . Physicists accuse many an epistemologist of not giving sufficient weight to this circumstance.[23]

Thus, Einstein cautions, after a preliminary world image has been constructed by the method of reduction and simplification, one can hope that, as science matures, it will turn out to apply to natural phenomena as they offer themselves to us in all their complexity and completeness. The history of science provides a wealth of examples that attest to this truth. The effort to encompass the totality of experience is in principle achievable in physical science – not at the beginning but at the end of the two-step process. But Einstein has also introduced at this very point a warning that he was to repeat frequently, one that, for quite different reasons, must appear as surprising to the new Romantics as to the new positivists: The supreme task is to arrive at universal elementary laws from which the cosmos can be built up by pure deduction – but there is no "logical bridge" to the laws.

Obviously, I have chosen Einstein because of the clarity, honesty, and independence of his methodological remarks. The process he describes is one most scientists will recognize as applicable to really fundamental work (although the use of the word "intuition" is bound to embarrass some of them). Moreover, almost by definition, the methods an Einstein used cannot reasonably be denied the label "rational," no matter how different they are from the models for rationality set up as strawmen by the new Dionysians or icons by the new Apollonians. It might therefore be fruitful to clear a little area in the battlefield between them and look in more detail at the credo of Einstein[24] concerning the use of the scientific imagination.

Einstein on the use of the scientific imagination

Einstein discussed his view on the nature of scientific discovery, and of theory construction in particular, in a generally consistent way on many occasions, notably in his essays "On the Method of Theoretical Physics" (1933), "Physics and Reality" (1936), and "Autobiographical Notes" (completed in 1946). He gave what was perhaps his clearest and most succinct presentation of his thoughts on the act of scientific reasoning in a letter (written on May 7, 1952) to his old friend Maurice Solovine. He began this portion of the letter by explaining that Solovine had misunderstood certain of Einstein's previous statements concerning epistemology. Einstein apologized and then asserted: "I probably expressed myself badly. I view such matters schematically thus . . ."[25]

There followed a diagram (Figure 3.1) – not entirely surprising. (As we know from Einstein's autobiographical writings and other evidences, he preferred to think visually.) Einstein went on to explain:

> (1) The *E* (experiences) are given to us [represented by the horizontal line along the bottom of the figure].
>
> (2) *A* are the axioms, from which we draw consequences. Psychologically the *A* rest upon the *E*. There exists, however, no logical path from the *E* to the *A*, but only an intuitive (psychological) connection, which is always "subject to revocation" [disavowal].

This point is one of the most persistent methodological remarks of Einstein, from 1918 on, when he was still doing his best consciously to toe the positivistic line. He wrote in the Spencer Lecture (1933)[26] that the axioms are "free inventions of the human intellect," and similarly in many of his letters. To return to the letter to Solovine:

> (3) From the *A*, by a logical route, are deduced the particular assertions *S*, which deductions may lay claim to being correct. [As he had said in the Spencer Lecture: "The *structure* of the system is the work of reason."]
>
> (4) The *S* are referred [or related] to the *E* (test against experience). This procedure, to be exact, also belongs to the extra-logical (intuitive) sphere, because the relation between concepts that appear in *S* and the experiences *E* are not of a logical nature. [In his "Reply to Criticisms" (1949)[27] Einstein elaborated on this point: The distinction between sense impressions or experience

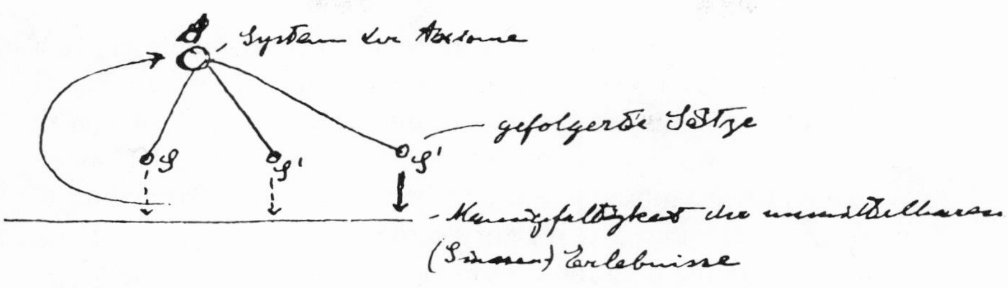

Figure 3.1. From a letter of A. Einstein to M. Solovine, May 7, 1952. Courtesy of the Estate of Albert Einstein.

> on one hand and ideas or concepts on the other is a necessary distinction, regardless of the reproach that using it makes one "guilty of the metaphysical 'original sin.' "]
>
> These relations of the *S* to the *E*, however, are (pragmatically) much less uncertain than the relations of the *A* to the *E*. (For example, the notion "dog" and the corresponding experiences.) If such correspondence were not obtainable with great certainty (even if not logically graspable), the logical machinery would be without any value for the comprehension of reality (example, theology).
>
> The quintessence is the externally problematic connection between the world of ideas and that of experience . . .

Now let us suppose that a connection can be made between the prediction *S* and the experiences *E* that are at hand. *Does this constitute an adequate test of the theory being examined?* Einstein had discussed this question in his "Autobiographical Notes."[28] At least in the case of the grand theories of greatest interest to him – those whose "object is the *totality* of all physical appearances" – he asserted that comparing the predictions of a theory with experiment is but one of two criteria according to which one can "criticize physical theories at all."[29]

The first criterion is that of "external confirmation." This is the easier one to meet, since one can often ("perhaps even always") make an adequate connection by suitable *ad hoc*, "artificial additional assumptions." Moreover, Einstein phrased this criterion in a remarkably generous way: "The theory must not contradict empirical facts." This principle of disconfirmation or falsification is, of course, quite different from the much stronger injunction that was usually associated with scientific "confirmation" by empirical test. Just how effectively he followed this first criterion was shown repeatedly, for example, in his steadfast and unswerving adherence to his ideas when, from time to time, evidence came that purported to show that his predictions, though not in unambiguous flat *contradiction* to the "facts of experience," at the very least were not being *supported* by experimental test. Moreover, though unwilling to accept the possibility of confirmation of a theory by "verification" of its prediction, Einstein in practice also held to the falsification principle only skeptically (or weakly) when the theory being purportedly falsified by experimental test had in his views certain other merits compared with its rivals. (See, for example, his refusal to accept Walter Kaufmann's experimental "falsification" of 1906 of Einstein's newly published special theory of relativity. The limited, ad hoc character of the rival theories that seemed to be borne out by Kaufmann's experiments signaled to Einstein that those theories "have a rather small probability." It turned out that he was right; the experiment, as is so often the case, was far less decisive or "crucial" than others had thought.[30])

Going on to the second criterion, Einstein explains that it "is concerned not with the relation to the material of observation, but with the premises of the theory itself, with what may briefly but vaguely be characterized as the 'naturalness' or 'logical simplicity' of the premises (of the basic concepts and of the relations between these which are taken as a basis.)" Clearly, here is a place for individual aesthetic or other preferences – although as soon as this is confessed, Einstein feels he, too, must apologize for it. "The meager precision of the assertions contained in the last two paragraphs I shall not attempt to excuse by lack of sufficient printing space at my disposal, but confess herewith that I am not, without more ado, and perhaps not at all, capable to replace these hints by more precise definitions."

We now need an important clarification. At the heart of the method of scientific discovery shown schematically in Figure 3.1, there was the leap up from the plane of experience E to the premises A. That leap, as Einstein stressed, is logically discontinuous; but it cannot be entirely "free" after all, if the premises later are to pass the tests of naturalness and simplicity (and the like) in order to meet the second criterion for a good theory.

In fact, the leap *is* channeled and guided. One such guide, at least for Einstein himself, was given by the fact that he attained the concepts for use at the *A* level by a form of mental play with visual materials "to a considerable degree unconsciously" – by a powerful iconographic rationality which he added to the more conventional semantic and quantitative ones. Another guide in the leap from *E* to *S* is one shared by all scientists engaged in a major work on novel ground: the guidance provided by explicit or, more usually, implicit preferences, preconceptions, presuppositions.

Einstein himself saw this and commented on it repeatedly. A good statement occurs in his essay "Induction and Deduction in Physics":

> The simplest conception [model] one might make oneself of the origin of a natural science is that according to the inductive method. Separate facts are so chosen and grouped that the lawful connection between them asserts itself clearly . . . But a quick look at the actual development teaches us that the great steps forward in scientific knowledge originated only to a small degree in this manner. For if the researcher went about his work without any preconceived opinion, how should he be able at all to select out those facts from the immense abundance of the most complex experience, and just those which are simple enough to permit lawful connections to become evident?[31]

It has always been this way in major scientific work. As shown, for example, in the cases treated in the first chapters of this book and also in *Thematic Origins of Scientific Thought*, we recognize the existence of (and even the necessity, at certain stages in scientific thinking, of postulating and using) precisely such unverifiable, unfalsifiable, and yet not arbitrary conceptions or hypotheses – a class to which I have referred as thematic presuppositions – as necessary for scientific work as the empirical and analytical content. In Einstein's own scientific papers we can watch him stating his presuppositions boldly, as, for example, when he first announces his two basic postulates of relativity, almost brusquely declaring them to be hunches (*Vermutungen*) that he decides to elevate to the status of postulates – without even bothering to connect them plausibly with the experimental material on the *E* level.

There is, of course, another side to this thematic origin of scientific thought. Dedicating oneself to some presuppositions or themata means one is likely to exclude others, as Einstein indeed did when he refused to accept the themata that were so basic in the work of the Copenhagen school on quantum mechanics. Here again one sees that the "leap" to *A*, the "system of axioms" in Figure 3.1, is not entirely "free" but "guided."

We can now see that much of the fight of the priests of the counterculture against what they attack as overly rationalistic science is a sham: It is largely a fight against strawmen of their own making. They conceive of scientific rationality as limited to strictly quantitative and semantic-logic processes, but that applies at most, and only to a degree, to Public Science, that is, to science as a pedagogical or as a consensus-seeking activity. What they attack is, however, only a poor caricature of Private Science, the process by which reasoning men and women make discoveries. There the discontinuous and thematic characteristics cannot be overlooked. To call them "irrational" is at best playing with words and denying rationality to some of our best thinkers. On the other hand, to let oneself be frightened into doubting the validity of thematic choices during the play of the scientific imagination would endanger the very process of scientific discovery itself.

If the new Dionysians have noticed the failure of scientists like Einstein to conform to such models, they do not let on. They feel that nature should be studied by neither induction nor the analytical-synthetic method, not even if it allows a speculative leap where human limitation makes it necessary and human ingenuity makes it possible. Rather, authors like Reich advocate that one coast through total, unselected experience with one's hands off the wheel and one's rational gearbox in neutral.

The true enemy in books such as *The Greening of America* is, in fact, not science, not the Corporate State, not the Department of Defense, not even the regrettable failures of science – the cases in which scientists or technologists allowed themselves to be used knowingly for destructive purposes. The real enemy is rationality itself, of which science is seen to be a preeminent exemplar. Thus we read that the Corporate State has "only one value, the value of technology-organization-efficiency-growth-progress. The state is perfectly rational and logical. It is based upon principle."[32] It would appear that the vision of Saint-Simon had really triumphed in our day.

What, then, is wrong with rationality? Reich gives the answer on the second page of his book, where we read that the rationality of the modern state must be "measured against the insanity of existing 'reason' – reason that makes impoverishment, dehumanization, and even war appear to be logical and necessary." Among the evils of rational thought, discussed at greater length later, are not merely its failures to prevent the recent wars, but the intellectual justifications that were given for those wars.[33] Thus we arrive at the remedy – a recipe for escape from rationality: "One of the most important means employed by the new generation in seeking to transcend technology is . . . to pay heed to the instincts, to obey the rhythms and music of nature, to be guided by the irrational, by folklore and the spiritual, and by the imagination."[34] "Accepted patterns of thought must be broken; what is considered 'rational thought' must be opposed by 'non-rational thought' – drug-thought, mysticism, impulses."[35]

Technically, one could analyze Reich's many conceptual difficulties in more detail. As Charles Frankel has accurately noted in a critical article:

> The Irrationalist's theory of human nature is steeped in the tradition of the dualistic psychology it condemns. It talks about "reason" as though it were a department of human nature in conflict with "emotions." But "reason," considered as a psychological process, is not a special faculty, and it is not separate from the emotions; it is simply the process of reorganizing the emotions.[36]

Precisely because such flaws are simple to expose, the chief puzzle about the new Dionysians is, and will remain, the large extent of their popular appeal. And here it may be significant to notice an ironic asymmetry. Many scientists throughout history have written about their motivation for turning to their work as if it were an intellectual and emotional turning away from the turmoil within and all around.[37] Reich also wrote at a turbulent time, in the Vietnam war year of 1970 – at the height of a reign which *his* audience, at any rate, seemed to recognize as tragic and stupid even without the subsequent evidence of the secret bombing of Cambodia, the Pentagon papers, the sale of the public trust, the conspiracy to abridge civil rights, the arrogance that led to the Watergate crisis, not to speak of the continuation of a senseless arms race and the widening of the world's poverty. Reich, however, charges the horrors of his time to the sovereign rule of reason and urges his readers to turn inward, thereby abandoning their chief weapon for organizing and validating any realistic attack on the ills he deplores. Indeed one key to the wide appeal of the Dionysians may be that they release their followers from all responsibility for effective action. Furthermore, at a time when so many feel they can only sit by in helpless disbelief to watch the unrolling of an absurd tragedy, the new Dionysians, in their attacks on scientific thought, furnish a convenient, safely passive target for expressing intellectual distaste.

The new Apollonians

Now to the hammer. The philosophers who have taken it on themselves to protect rationality in the narrowest sense of the word are also members of a long tradition. Some of their genes can be traced back to the logical positivists of the pre–World War II period, who are themselves descended from a long line of warriors against the blatant obscurantism and metaphysical fantasies that haunted and thwarted science in the nineteenth and early twentieth centuries. Rereading today Otto Neurath's influential essay "Sociology and Physicalism" (1931–32), one can glimpse the fierce doctrine that helped this school to achieve its victories:

> The Vienna Circle . . . seeks to create a climate which will be free from metaphysics in order to promote scientific studies in all fields by means of logical analysis . . . All the representatives of the Circle are in agreement that "philosophy" does not exist as a

> discipline, along side of science, with propositions of its own. *The body of scientific propositions exhausts the sum of all meaningful statements* . . . They wish to construct a "science which is free from any world view."

But the line of descent goes back much further, all the way to Lucretius, to Democritus, to all who undertook the antimetaphysical mission of liberating mankind from the enchantment and terror of superstition. Thus a modern Lucretius, Bertrand Russell, proclaimed that "all these things, if not quite beyond dispute, are yet so nearly certain that no philosophy which rejects them can hope to stand":

> That Man is the product of causes which had no prevision of the end they were achieving; that his origin, his growth, his hopes and fears, his loves and beliefs, are but the outcome of accidental collocations of atoms; that no fire, no heroism, no intensity of thought and feeling, can preserve an individual life beyond the grave; that all the labours of the ages, all the devotion, all the inspiration, all the noonday brightness of human genius, are destined to extinction in the vast death of the solar system, and that the whole temple of Man's achievement must inevitably be buried beneath the debris of a universe in ruin.[38]

Although it is no longer fashionable to force the rationalists' message upon a fearful populace with quite so much glee, the ancient division between thematically incompatible world views continues to exist and is not likely to disappear.[39]

Some of today's most eloquent defenders of rationality have been associated with the school of Karl Popper, who himself was influenced, at an early point, by the prewar positivist movement. Out of Popper's many contributions over the decades, I shall refer here only to one small portion that happens to have relevance to this particular study. He considers that the rationality of science presupposes a common language and a common set of assumptions which themselves are subject to conventional rational criticism. There does exist a contrary opinion, namely, that there may be cases of individual scientific work that have not been and perhaps never can be subjected fully to such a critique. To Popper this is quite intolerable; he writes that this so-called Myth of the Framework is "in our time, the essential bulwark of irrationalism."[40]

In his view, progressing from one valid stage of scientific theory to another cannot involve breaking the thread of continuous, rational, progressive development. "In science, and only in science, can we say that we have made genuine progress: that we know more than we did before."[41] To be sure, "an intellectual revolution often looks like a religious conversion." But a critical and rational evaluation of our former views must remain possible in the light of the new ones. If it were not possible, what guarantee would we have that science was indeed accruing a content of truth? What guarantee that the changes in science are indeed a progressive sequence of steps toward objective knowledge, and not merely a sequence of conversion experiences from one unfounded set of beliefs to another?

A critical discussion of this position is, however, made difficult by a set of self-inflicted taboos. Popper writes:

> I cannot conclude without pointing out that to me the idea of turning for enlightenment concerning the aims of science, and its possible progress, to sociology or to psychology, or to the history of science, is surprising and disappointing. In fact, compared with physics, sociology and psychology are riddled with fashions and with uncontrolled dogma. The suggestion that we can find anything here like "objective, pure description" is clearly mistaken. Besides, how can the regress to these often spurious sciences help us in this particular difficulty? . . . No, this is not the way, as mere logic can show.[42]

What, exactly, is at stake here? On one level, it is the definition of where the philosopher of science should look for valid problems and tools. Popper rules out, as of no interest, the context of discovery, and hence the actual working out of a problem by an actual person.

> The initial stage, the act of conceiving or inventing a theory seems to me neither to call for logical analysis, nor to be susceptible to it. The question of how it happens . . . may be of great interest to empirical psychology; but it is irrelevant to the logical analysis of

> scientific knowledge. The latter is concerned only . . . with questions of justification or validity.[43]

Fair enough as a statement of preference – although one may not personally subscribe to it, particularly if one's own fascination is precisely with a historical study of the "personal struggle." One may even regret that Popper shares with many scientists – and, for that matter, with Reich and the new Dionysians – a complete lack of interest in studying the creative act of scientists (even, one must assume, of Einstein, who, Popper claimed, was "perhaps the most important" influence on Popper's own thinking[44]), thereby denying the possibility of a critique of the scientific imagination.

But if not to real cases, where is one to turn for data to examine Popper's logic of discovery and to test out his hypotheses? It is at this point that some modern philosophers of science have recently evolved a technique of criticism that tries to force the understanding of scientific work as far to the right as the new Dionysians wish to force it to the left. Instead of looking at actual case studies in their historic setting (a technique of what they call the "spurious sciences"), they look at a "rational reconstruction" of the events.

Popper himself proposed the technique in a rather gentle way:

> Admittedly, no creative action can ever be fully explained. Nevertheless, we can try, conjecturally, to give an idealized reconstruction of the problem situation in which the agent found himself, and to that extent make the action "understandable" or "rationally understandable," that is to say, adequate to the situation as he saw it. This method of situation analysis may be described as an application of the rationality principle.[45]

This proposal was taken up by others and clothed more dogmatically, most vigorously by Imre Lakatos, Popper's former student and successor to his chair at the London School of Economics. In the influential work of Lakatos the opinion of what constitutes a valid study of a historical case is laid down in words such as these:

> In writing a historical case study, one should, I think, adopt the following procedure: (1) one gives a rational

> reconstruction; (2) one tries to compare this rational reconstruction with actual history and to criticize both one's rational reconstruction for lack of historicity and the actual history for lack of rationality.[46]

Lakatos then gives examples of what happens to a historical case study when done in this style, including his own reconstruction[47] of "Bohr's plan . . . to work out first the theory of the hydrogen atom [1912–13]." "His first model was to be based on a fixed proton-nucleus with an electron in a circular orbit . . . ; after this he thought of taking the possible spin of the electron into account . . .[48] All this was planned right at the start." As it happens, Bohr's early work has been very carefully studied by historians of science; but this version, produced by "rational reconstruction," is an ahistorical parody that makes one's hair stand on end.[49] Otto Neurath's dictum that "Philosophy does not exist as a discipline, alongside of science, with propositions of its own" has been stood on its head: The study of the actual work of scientists does not exist as a discipline, alongside of philosophy, with propositions of its own.

The resulting rationalization of actual historic cases, although not without technical interest in philosophy itself, is so risky an idea and so unacceptable to most historians of science[50] that one is forced to speculate it may be motivated by higher stakes than appear on the surface. In the writings of the more extreme new Apollonians, one senses that their philosophical position is not being developed simply for its own sake, or for the sake of its potential evaluation in the crucible of rational critique, but that their ambitions are much larger. They seem to hope to save scientists from the threat of the irrational, suspecting that scientists will be unable to do a good job without expert help in deciding which of their theories are truly scientific and which are merely pseudoscientific. Thus Lakatos confessed sadly:

> If we look at the history of science, if we try to see how some of the most celebrated falsifications [of hypotheses] happened, we have to come to the conclusion that either some of them are plainly irrational, or that they rest on rationality principles radically different from the ones just discussed.[51]

Hence rational reconstruction; hence the effort to replace the "naïve" version of methodological falsification actually followed by scientists when left to their own devices with a "*sophisticated* version . . . and thereby rescue methodology and the idea of scientific *progress*. This is Popper's way," Lakatos tells us, "and the one I intend to follow."[52]

Hanging over the whole stage is the shadow of David Hume, with his repugnant message, as Popper puts it, that "not only is man an irrational animal, but that part of us which we thought rational – *human knowledge*, including practical knowledge – is utterly irrational."[53] The new Apollonians dedicate a major effort to the disproof of this specter, with particular attention to scientific reasoning.

But their ambitions, and the perceived threat, are even larger than that: Mankind must be saved – from obscurantism, astrology, and revolution. Lakatos writes that a recent theory of scientific progress – which allows the role of changing exemplars rather than of logical proof alone – makes "scientific change a kind of religious change."[54] Such a theory, he says, not only poses a threat to technical epistemology, but "concerns our central intellectual values," hence affecting "social sciences . . . moral and political philosophy." Moreover, it "would vindicate, no doubt unintentionally, the basic political credo of contemporary religious maniacs ('student revolutionaries')." Elsewhere, Lakatos is led so far as to speculate on the possibly sinister personal influence of the author of such a theory of scientific change: "I am afraid this might be one clue to the unintended popularity of his theory among the New Left busily preparing the 1984 'revolution.' "[55]

Now we recognize what is really at stake: civilization itself. These philosophers of rationalism see themselves as the soldiers at the gates, fending off a horde of barbarians. Popper himself has, of course, made no secret of his mission. Long before the new Dionysians were as prominent as they are now, he said that the conflict with advocates of irrationalism "has become the most important intellectual, and perhaps even moral, issue of our time."[56] Irrational attitudes and the flagging of the critical habit, he warns, could well open the way for demagogues which promise political miracles. One must preserve what has been gained, with all its shortcomings, for "our present free world, our Atlantic Community . . . ruled by the interplay of our individual consciences . . . is the best society that has ever existed."[57] Lakatos, for his part, warned that a work on the nature of scientific change, with which he disagreed, is "a matter for mob psychology," "vulgar Marxism," and "psychologism," and has even triggered "the new wave of sceptical irrationalism and anarchism."[58]

Thus, each of the opposing Dionysian and Apollonian groups is imbued with a sense of urgency to save the Republic. Each thinks that following a proper process for gaining valid knowledge is a key for salvation and proposes to clarify the understanding of that process, but in fact does not look at the way the scientific imagination works in action. One side condemns the scientists for being too rational; the other chides them for being too irrational. Caught in between, scientists, virtually without exception, pay no attention to either side, not even to defend themselves against grotesque distortions of what it is that they really do.[59] In effect, scientists hand the public platform over to the propagation of two sets of quite different but equally erroneous answers to questions such as those posed at the beginning of this chapter: How do scientists actually go about gaining knowledge, and how *should* they?

Postscript

This is not the place, nor is it my intention, to build a prescription for a cure on an analysis of the symptoms. Deeper involvement of research scientists in discussions concerning their methods would surely improve the understanding of science – including their own. Certainly, those four phases of scientific work which rest on rationality, by any definition of the word, could benefit from more modern analysis: rationality in the deductive portions of private theorizing; rationality in the structure of a theory once it has been worked out moderately well; rationality in the process of communication and validation among scientists operating in the area of public science; and the perception, at least among our more exalted spirits, of an underlying rationality and uniqueness in the world order seen through science – perhaps the only order open to human perception which is not a Rashomon story, inherently different for each observer.

In addition, sound pedagogical materials are needed to show that there are processes at work in the making of science which, while they are acts of reason, cannot be forced into the logical-analytical framework. Entering into such processes are the ways by which new ideas arise and are handled during the nascent moment; the sources of individual thematic choices, and the reasons for cleaving to them; the connection between the elementary concepts, of both science and everyday thinking, and the complexes of sense experience; and the eternally surprising fact that we so often find the logically simple suitable for building a theory of nature's phenomena.

As Peter Medawar has observed, the hypothesis of the interaction of essential dual components may still be the most fruitful one.

> Scientific reasoning is an exploratory dialogue that can always be resolved into two voices or episodes of thought, imaginative and critical, which alternate and interact . . . The process by which we come to form a hypothesis is not illogical but non-logical, i.e., outside logic. But once we have formed an opinion we can expose it to criticism, usually by experimentation.[60]

This is not compromising between rationality and irrationality. On the contrary, it is widening the claim of rationality, and widening also the scope of much-needed research on the nature of scientific rationality in practice. In opposition to the two groups I have analyzed, Medawar holds that

> the analysis of creativity in all its forms is beyond the competence of any one accepted discipline. It requires a consortium of the talents: psychologists, biologists, philosophers, computer scientists, artists and poets would all expect to have their say. That "creativity" is beyond analysis is a romantic illusion we must now outgrow.[61]

Whether for pedagogic purposes or as a field of research, whether as a part of philosophical analysis or as a key to a study of politically significant intellectual rebellions and reactions, the methods by which humans use their scientific imagination are themselves much in need of more thorough scientific study. Possibly the worst service the new Dionysians and the new Apollonians render is that their antithetical attacks continue to discredit the accommodation of the classically rationalistic with the sensualistic components of knowledge. We should, rather, strive to acquire a clearer notion of how actual mortal beings, with all their frailties, have managed to use both these faculties to grasp the outlines of a unique and fundamentally simple universe, characterized by necessity and harmony. Such knowledge, we may hope, can be of practical use at a time when our species seems to depend on tapping all the resources of reason for the generation of new ideas that are both imaginative and effective.

4

Analysis and Synthesis as methodological themata

The terms "analysis" and "synthesis" bring to mind, on the one hand, certain methodological practices in the works of Plato, Descartes, Newton, Kant, Hegel, and others and, on the other hand, techniques in fields as disparate as chemistry and logic, mathematics and psychology. The width of this spectrum of associations alerts us to the realization that at the base of these two related terms there lies a specific methodological thema-antithema ($\Theta\bar{\Theta}$) pair. Indeed, it is one of the most pervasive and fundamental ones, in science and outside.[1] This chapter attempts to uncover and identify this thematic content, to clarify the meanings and uses of the terms "analysis" and "synthesis," and especially to distinguish among four general meanings: (1) Analysis and Synthesis, and particularly synthesis, used in the grand, *cultural* sense, (2) Analysis and Synthesis used in the *reconstitutional* sense (e.g., where an analysis, followed by a synthesis, re-establishes the original condition), (3) Analysis and Synthesis used in the *transformational* sense (e.g., where the application of Analysis and Synthesis advances one to a qualitatively new level), and (4) Analysis and Synthesis used in the *judgmental* sense (as in the Kantian categories and their modern critiques).

Analysis and Synthesis in the "cultural" sense

High on the list of achievements our culture has traditionally defined as best are grand, synoptic, and unifying works usually characterized as "syntheses" of the thinking of a period or a field. Examples are the philosophical treatises of Aristotle and Aquinas, Spinoza and Kant; the scientific syntheses of Euclid,

Descartes (*Principles*), Newton (*Principia*), Darwin, Maxwell, Mendeleyev, Freud, Einstein; and, in our day, the groups responsible for the unification of biochemistry and genetics (e.g., Watson and Crick) and of evolutionary biology (e.g., Dobzhansky and Mayr). Many significant literary works also have this unifying character and intention, for instance, the Greek epics, the works of Dante, Milton, Goethe, and Tolstoy. Although the latter deplored in *War and Peace* (Book V) that now a "science of the whole" was no longer possible, he felt that at least in the arts a synoptic view of man's life and worth could exist.

With all their differences in both intent and method, these cultural products share a property that helps to explain the power they have over the human imagination: The synthesis provides a framework of interpretation and analysis of particulars that helps to propel thought and feeling to important truths. In the Aristotelian cosmos, a stone, falling from a height to its "natural" place, is understood to follow not any arbitrary but a necessary scenario within the total setting; in the same sense, the young reader, caught up in the world view exemplified by the work of an Aquinas or a Goethe, can thereby construct persuasive interpretation (whether profound or not) of the complex or anguishing details of his own experience.

Certain high cultural products are considered specifically *analytical* rather than synthetic in intent – for example, the mathematical analyses of Descartes and Fourier or the philosophical works of G. E. Moore and Bertrand Russell; sometimes an analytic portion may be embedded in the work of synthesis. However, although it is not difficult to enumerate synthetic works in a list of cultural achievements, and although synthetic works pervade the training and consciousness of educated persons from an early age, works of explicit analysis rarely achieve such high status. (The poets Wordsworth and Coleridge, for example, pronounced the analytical activity of experimental scientists as the work of inferior minds.) Yet when we examine the Analysis and Synthesis conception from the point of view of praxis rather than "high culture" (e.g., Analysis and Synthesis as expressed in professional, scientific, and scholarly work), the positions of analysis and synthesis are entirely reversed, and the former is more prominent.

A working intellectual will rarely claim to be concerned with "synthesis," and if he does so, he is more likely to be referring to a small range of specific activities, such as the chemical synthesis of materials, whether ammonia or hormones, fibers or resins, or to a test of a chemical structure in which analysis is confirmed by synthesis.[2] "Analysis," on the other hand, appears prominently in a large number of intellectual activities. The dictionary definitions of analysis, for example, tend to center on the reductionistic or pragmatic core of meaning: breaking or resolving the complex into simple elements, and the determination of these elements (as in chemistry, resolving into simpler molecules; in optics, finding spectral compositions; in grammar, finding elements of which phrases or sentences are composed; or in thematic studies of folklore, music, literature, or science, determining the basic themes on which the structure of the work depends).

The professional manifestations of work concerned with analysis stretch from psychology to linguistics, from economics and business practice to chemistry, from engineering to medicine. For a philosopher, the task of analysis refers to the process of reaching conceptual clarification. Mathematics makes the most widespread and diverse use of the conception. One is apt to encounter the word "analytic" first in school in courses on analytic (or coordinate) geometry, dating back chiefly to René Descartes (1637), who succeeded in showing that each point on a geometric figure, or in space, can be reduced to an ordered set of numbers, later called coordinates.[3]

In view of their praxis-oriented purposes, major encyclopedias not surprisingly tend to have many and elaborate discussions of analysis but very little on synthesis (neither the *Encyclopaedia Britannica*, subtitled "Knowledge in Depth," nor the *Dictionary of the History of Ideas* has an entry for "synthesis") – a situation contrary to the relative places assigned to analysis and synthesis as earmarked of the high cultural achievements of the past. Therefore, it is the more important for us to seek out the relations *within* the Analysis and Synthesis couple in order to understand the full power of each of the components, rather than be misled by the asymmetrical valuations of them in contemporary theory and practice – possibly the result of the preponderance (and success) of reductionistic thought in our time. It is appropriate, of course, to give credit to those successful examples of synthesis-seeking endeavor that do exist today, for example, the kind of "global" thinking of certain environmentalists; the quest in fields such as particle physics toward the unification of all fundamental forces; and, in the field of education, a number of interesting experiments under titles such as transdisciplinary, interdisciplinary, or general education programs. At the same time, we should remember that some obfuscationist movements also often grow under the banner of "synthesis," "unification," or "holism."[4]

An exemplar of Analysis and Synthesis: the "Newtonian Synthesis"

To specify the properties of a "synthesis" in operational terms, no historic case is more inviting than that exemplar of all successful scientific syntheses, the so-called "Newtonian Synthesis," the historic unification of celestial and terrestrial physics. We can discern in it the strengths and weaknesses of synthesis as an intellectual strategy, and the interaction between analysis and synthesis as parts of a method that allows the production of a cultural object named a "synthesis."

The roots in the history of physical sciences. The significance of the Newtonian contribution is clearer if it is first positioned with respect to the history of physical science itself, and with respect to the history of the method of scientific discovery. For the Newtonian synthesis, these roots reach back to antiquity, to the two earliest grand syntheses in natural philosophy, associated with the names of Thales and Pythagoras, respectively. The former was essentially positivistic and materialistic, with a certain resemblance to modern empiricism, whereas the latter was metaphysical and formalistic, with a certain resemblance to rationalism. One typically used the observed fact of the three states of an observed material (water) as a key for understanding the problem of persistence and change in the material world; the other typically based itself on the properties of numbers and geometrical figures, and its methods were closely associated with religious ritual. It is significant that these two systems, each coming into Western culture at about the same time and impelled by the persisting drive to find basic unity underlying the diversity of all experience, nevertheless were diametrically opposite in assumption and mutually exclusive in content.

From each of these two schools, a separate chain of distinguished followers emerged over the next centuries. Aristotle's pivotal position derives from the fact that he is the first major thinker who is not chiefly a follower of either of the two main trends, but who made a powerful attempt to adapt elements from both of the antithetical systems in a new synthesis (although at the cost of persisting internal divisions, e.g., among physics, mathematics, and metaphysics). Nothing even faintly analogous was done successfully in natural philosophy until Kepler's and Galileo's joining of neo-Platonic and materialistic conceptions, and then only on a much less ambitious scale. They are the necessary forerunners of Newton, whose synthesis must be understood as the last grand bridging of the materialistic-positivistic and the formalistic-metaphysical traditions in natural philosophy.

From then on, other such attempts were made in increasingly narrow fields within the pure sciences. For example, Faraday's central theme, in his research on relations between gravity and electricity, was what he called "the long and standing persuasion that all the forces of nature are mutually dependent, having one origin, or rather being different manifestations of one fundamental power." (Even though he failed to find the connection between gravity and electricity, as did Einstein with analogous ambitions, Faraday opened the door to Maxwell's work when he found the direct relation and dependence between light and the magnetic and electric forces.) Other achievements that followed in the same tradition were J. R. Mayer's view of natural phenomena as the playing out of a law of conservation of energy; Maxwell's joining of the phenomena of electricity, magnetism, optics, and radiant heat in one theory of electromagnetism; and the work of Einstein, which resolved the clash between the electromagnetic world view and the mechanistic world view in the first part of this century and which found connections between previously separated conceptions such as space and time, and mass and energy.

The roots in the history of methods of scientific discovery. The second prerequisite for appreciating the impact of the Newtonian synthesis on modern thought concerns the history of methods of scientific discovery. Here it is significant that to this day, most scientists who understand Newton's view will, by and large, still agree that it also applies to their own work, and will recognize the chief elements in what is now often called the hypothetico-deductive method, at the center of which analysis and synthesis are located.

As used in scientific works since the beginning of the seventeenth century, analysis and synthesis refer to parts of a transformational procedure of reasoning much indebted to Plato's discussion (e.g., in the *Republic*, 509–511 and 533–534), and subsequently named analysis and synthesis in Greek, resolution and composition in Latin.[5] Plato warned that merely descending from the high ground of axiom, or from an unexamined hypothesis whose truth is not assured, may suffice to erect self-consistent systems (as in "geometry and the sister arts") or even systems that work well enough on a technical level, but it does not lead to science (*episteme*). Using modern terms, this warning amounts to saying that attempting synthesis without a previous analysis does not lead to truths. To gain true knowledge, one must proceed by going first up and then down, as if on an arch made of steps – that is, at the beginning one puts forward a hypothesis (literally, a placing under). As Plato proposed:

> Using the hypotheses not as first principles, but only as hypotheses – that is to say, as steps and points of departure into a world which is above hypotheses, in order that she [reason] may soar beyond them to the first principle of the whole; and clinging to this . . . by successive steps she descends again without the aid of any sensible object, from ideas, through ideas, and in ideas she ends.

In terms of later commentary, reason here is seen, in the first half of the process, to proceed "upwards" by induction to the perception of "first principles." Then reason follows the arch downward in the second half of the process, to produce a demonstration: Postulating those truths as first principles, the searcher for knowledge descends to a conclusion by deduction in a series of steps. The second part of the cycle, corresponding to synthesis, can succeed because it was properly preceded by analysis in the first part.

With all their differences, Descartes and Newton (at least in their exhortations) agreed with this Platonic ordering: Descartes wrote, "It is certain that, in order to discover truth, we should always begin with particular notions in order to reach general notions afterwards, though reciprocally, after having discovered the general notions, we may deduce from them others which are particular." Newton put it similarly in a famous passage (*Opticks*, Book IV, Query 31):

> The investigation of difficult things by the method of analysis ought ever to precede the method of composition. This analysis consists in making experiments and observations [these are of course Newton's crucial addition to the process!], and in drawing general conclusions from them by induction, and of admitting no objections against the conclusions but such as are taken from experiments or other certain truths . . . By this way of analysis we may proceed from . . . effects to their causes . . . And the synthesis consists in assuming the causes discovered and established as principles, and by them explaining the phenomena proceeding from them, and proving the explanations.

The sequence of analysis/synthesis or ascending/descending was referred to by Newton in, among other places, his preface to the *Principia:* "The whole burden of philosophy seems to consist in this: from the phenomena of motion to investigate the forces of nature, and then from these forces [e.g., the postulated universal gravitation] to demonstrate the other phenomena." Whereas Descartes's process of identifying postulates gave a large role to clear and undisbelievable ideas and the role of intuition, Newton relied on observation and experiment to anchor the first principles in experience, at the top of the arch. It is this difference more than any other which causes modern scientists to trace their philosophical roots to Newton rather than to

Descartes. What neither Descartes nor Newton, nor any major scientist up to Einstein's time, fully faced was that the "hypotheses" can never be altogether purged of their origin in the fallible human imagination. Einstein held not only that the axioms from which testable consequences must be deduced are "free inventions of the human intellect," but that there are elements in both halves of the arch which "belong to the extra-logical (intuitive) sphere."[6]

In what respect was Newton's work a synthesis? We can distinguish seven contributing aspects by which the achievement of Newton (and that of his followers who built on his findings and had some of the same ambitions) may be regarded as an exemplar of synthesis in the cultural sense.

1. The starting point was the initial identification (and further analysis when necessary) of individual and seemingly disparate *elements*,[7] chiefly the various separate classes of objects that encompass an infinity of individual cases. These cases ranged from the motion on earth of projectiles, including the falling apple, to the precision of the equinox, and the complex perturbations in the moon's motion, the tides, the motion of comets, and the motion of planets.

Newton also was often conscious of *excluding* candidate-elements from the eventual synthesis. This refers not only to "occult qualities" that were no longer desired but to light and its propagation, chemical reactions, much of fluid mechanics and the theory of elasticity, sensations in the human body, and the properties of the ether (which he confessed he considered a necessary substructure of space, needed for the propagation of light, of gravity, probably of sensation, and probably also as the manifestation of the sensorium of God). Also sacrificed were the commonsense methods of intuitive "reasons" (e.g., of our muscles) which had still sufficed to make plausible the Cartesian solution for the solar system, representing it as a huge vortex of motion in the ether.

2. The key act in the synthesis was Newton's induction and postulation as a first principle of the law of universal gravitation, applicable to any two objects regardless of kind, size, distance, material intervening between them, or whatever. It was im-

portant to the success of the synthesis that scientists and other intellectuals could immediately see the law to be endowed with a startling simplicity and universality, since it applied to all material objects throughout the cosmos. Moreover, the law evoked the synthesizing image of mutual binding forces literally pulling the fragments of the far-flung cosmos together.

3. The law of universal gravitation, together with Newton's own laws of motion and the mathematical apparatus for handling problems in kinematics and dynamics, allowed Newton to show systematically that each of the previously mentioned fragments could be deduced and gathered together as special cases of the motion of real (ponderable) bodies. This unification could accommodate not only individual observations (such as those concerning the motion of the moon) going back to Babylonian times but also previously found laws (e.g., the three empirical laws of planetary motion of Kepler), which were thereby "explained."

4. One must not underestimate the philosophical effect on Newton's contemporaries of his demonstrations that causal and quite "ordinary" actions were at work in producing complex or frightening effects (e.g., tides or comets) and that the world of obvious change was explainable by the persistence of a few simple laws that any schoolboy could memorize. By extending the reign of familiar terrestrial processes and showing them to be at work throughout the knowable world, a single, almost hypnotic and seductive image could suggest itself, that of the universe as a majestic clockwork.

5. Three related aspects helped make the synthesis *scientific* in the modern sense:

(a) Newton showed how to treat the complex phenomenic world by means of mathematical method (some of which Newton had to invent for the purpose).

(b) He introduced the evidence of observation or simple experiments at crucial points in his work, if only occasionally as *Gedanken* experiments.

(c) His system allowed predictions that later could be and were successfully checked (e.g., the calculation that the comet of 1682 would have a period of approximately seventy-five years, being merely an object moving on a Keplerian ellipse and subject to the Newtonian force of gravity toward the sun; e.g., the determination on dynamical grounds of the shape of revolving planets such as Earth and Jupiter; e.g., the discovery of previously unsuspected planets such as Neptune and Pluto, by deducing their positions from the perturbing gravitational effect they exert on visible planets).

6. What helped make the synthesis so *lasting* was its vast extent over individual classes of cases, so that the "unpacking" of the cases kept physical scientists busy for over two centuries. We need only mention two examples of this power: The law of universal gravitation suggested that electric forces obey the same kind of inverse-square law; and the motion of the whole spiraling galaxy in which we now know our solar system to be embedded shows that the parts are under the mutual actions of the Newtonian force of gravitation.

7. What helped make the Newtonian synthesis a *powerful exemplar* was that it not only modified chief parts of natural philosophy but also changed civilization. These changes were not only through technological consequences (for to begin industrializing a society, one must first learn Newtonian physics) but also through effects on the imagination in biology, psychology, economics, sociology, theology, and the arts. As Fontenelle expressed it, if written by the hand of someone knowledgeable in the mathematical sciences, "a work of morals, of politics, of criticism, perhaps even of eloquence, will be finer."

This same interplay between analysis and synthesis, induction and deduction, was thought to be applicable in all of the natural and social sciences. The same hope to capitalize on both the experimental and the mathematical part of the "scientific method" stood before them all. Nature, society, religion, and the human mind were equally open to the promise of its success. It seemed that the problems in all of these fields might yet be reduced to the mathematical treatment of quasi-mechanical interactions of parts that obeyed specific laws under the general reign of "the Laws of Nature and of Nature's God," to cite a powerfully motivating phrase in the American Declaration of Independence.

Limitations of the Newtonian synthesis. The use of a real case allows us to stress that syntheses, by their very nature, can be successful only within limits; beyond that, they fail.

The "failure" of the Newtonian synthesis was not merely that, to Newton's own dismay, it did not encompass fields such as contemporary chemistry (or, as we now would put it, any of the four forces of nature other than gravitation), but that in the long run it could not account adequately for the ever-widening range of phenomena, such as those in cosmology or in the realm of the very small (atomic and nuclear physics). Newtonian science is now linked at one end with relativity theory, which is particularly important for bodies with very great mass or moving at very high speed; and, at the other, it approaches quantum mechanics, for particles of extremely small mass and size. For the vast range of problems between these extremes, Newtonian theory gives accurate results and is far simpler to use; moreover without Newtonian mechanics, relativity theory and quantum mechanics could not have emerged in the first place.

An additional failure developed from the initial boundary conditions. A synthesis necessarily excludes its antithetical alternatives, and any one of these may develop a cultural force of its own. This is at the base of the rebellion of the romantic movement, and the modern counterculture movements espouse many of the same attitudes and arguments, as noted in Chapter 3. The case for that point of view has been put most succinctly and clearly by the historian and philosopher of science E. A. Burtt:

> . . . the great Newton's authority was squarely behind that view of the cosmos which saw in man a puny, irrelevant spectator (so far as being wholly imprisoned in a dark room can be called such) of the vast mathematical system whose regular motions according to mechanical principles constituted the world of nature. The gloriously romantic universe of Dante and Milton, that set no bounds to the imagination of man as it played over space and time, had now been swept away. Space was identified with the realm of geometry, time with the continuity of number. The world that people had thought themselves living in – a world rich with color and sound, redolent with fragrance, filled with gladness, love and beauty, speaking everywhere of purposive harmony and creative ideals – was crowded now into minute corners in the brains of scattered organic beings. The

> really important world outside was a world hard, cold, colorless, silent, and dead; a world of quantity, a world of mathematically computable motions in mechanical regularity. The world of qualities as immediately perceived by man became just a curious and quite minor effect of that infinite machine beyond.[8]

A clarifying digression is in order: The phraseology I have used in introducing this quotation should not be taken as a declaration of adherence to a Hegelian theory. The theory of Dialectic, especially as Hegel developed it, claimed that human thought through history developed in stages characterized by the dialectic triad, thesis, antithesis, and synthesis. The last is seen as the resolution of a necessary struggle between the two others, going beyond each – although this synthesis too may subsequently function as a thesis that calls forth a new antithesis and hence a new struggle. One essential feature of the theory is that the thesis "produces" its antithesis. It exists apart from, and is imposed upon, the individual thinkers, whose discourses are thus merely the audible expressions of the rule of irresistible higher forces; Hegel claimed in his *Encyclopedia*, Part I, Chapter 4, that Dialectic is "the universal and irresistible power before which nothing can stay."

Another essential feature of the theory of Dialectic is that the resolution of contradictions between thesis and antithesis is not achieved by finding some parts of one or the other in error, or by issuing a modification of both that allows a new accommodation, but by accepting contradictions within the synthesis, thereby "negating" or canceling them.

The difficulties with this theory, logical and otherwise, are many;[9] but the chief difficulty comes when we examine actual cases to see whether such scenarios really do occur. It turns out that they do not, provided one takes seriously the methods and findings of modern history of science.

In their disenchantment with the limits of the Newtonian world view that in fact developed, the romantic opponents might have had an ally in Newton himself, to some degree. While most others were captivated by his demonstration in the *Principia* that the raw materials of the world are forces, matter, motion, and mathematics, Newton himself was not. As we now know from the analysis of Newton's previously unpublished papers (e.g., in Frank E. Manuel's book *The Religion of Isaac Newton*, discussed in Chapter 9), Newton saw himself as not only a scientist but also a historical scholar who had a duty to study the scriptures as a form of objective historic record. Although the study of the causes of natural phenomena, Newton admitted, does not bring one directly to the First Cause and Creator, it does perfect Natural Philosophy, which in turn enlarges "the Bounds of Moral Philosophy."

In treatises such as that on "The Revelation According to Daniel," Newton showed the fusion of his religious and natural-philosophical concerns. He believed that to find the design behind the obscure prophecies in the Apocalypse was as important as to find the cause of the motion of the moon and of the planets. Moreover, he hoped to find one great, unifying structure within which all details, physical or not, are parts of one coherent cosmos. Indeed, the properties of and necessity for God are built into Newton's very physics, into his conceptions of absolute space and time, the ether, gravitation, and sense perception. Whereas to us the *Principia* and the *Opticks* are breathtaking synthetic works in science, to Newton they must have appeared as preliminary way stations to a much grander synthesis that eluded him, one by which he had hoped to attain knowledge of the Creator of both the book of nature and the book of the scriptures. How far he had to go must have been oppressively obvious to him, as indicated by the fact that he published almost none of his voluminous theological writings, on which he spent a large fraction of his time even during his most productive scientific period.

One must therefore distinguish more carefully than is often done between the *Newtonian* Synthesis and *Newton's* synthesis; the former refers to the successful development of physical science from the late seventeenth century, the latter to Newton's own achievement – of which his magnificient *Principia* was only the first, incomplete stage of a much more ambitious quest.

There are three points to add to this examination of the building of a grand synthesis in the cultural sense. (1) In the sciences, such a cultural synthesis is achieved by a constant interplay of analysis and synthesis in the transformational sense. (Thus

Newton based himself on previous analytical triumphs such as those of Galileo, e.g., the method of resolution and composition of vector quantities). (2) Any synthesis must fit into a set of boundary conditions, of which the choice of initial "elements" often is the clearest expression. Thereby it leaves open the potential for the rise, sooner or later, of an antithetical attempt at synthesis centering on the omitted elements. Any synthesis fashioned by the human mind is incomplete, often even by its own standards. (3) Nevertheless, the persistence of synthetic attempts and the high place we give them show that a thematic drive toward synthesis exists and is essentially unavoidable. William James referred to one of its manifestations in the remark that "the ideal which this philosophy strives after is a mathematical world-formula" – that is, in some form or other we are heirs to the Laplacian vision, a view of the future as the comprehensible extrapolation of present, measured states – in principle, to arbitrarily fine degrees of inclusiveness, detail, and accuracy. Helmholtz, too, hoped in his way to achieve (in principle) "the complete comprehensibility of nature." As we shall note, this drive has the earmarks of the ancient hope for the transcendental knowledge of the "One."

Sociobiology: a new synthesis?

It is appropriate to turn from our oldest grand scientific synthesis and apply the same apparatus to the most recent attempt of the same sort – that of current sociobiology. For this purpose I shall base my remarks largely on Edward O. Wilson's book.[10] What interests us here are the overall claims sociobiology makes, how these claims and aims fit into the history of ideas, and whether sociobiology has the earmarks of being indeed the beginning of a major synthesis.

The overarching aim of sociobiology – the task that Wilson says is to be completed within the next 20 or 30 (p. 5) or perhaps 100 years (p. 575) – is of course signaled first in the subtitle of his book, *The New Synthesis*. We find it again in the title of Chapter 1, "The Morality of the Gene," and in statements throughout, particularly on the early pages.

A basic axiom is that the "individual organism is only [the

genes's] vehicle, part of an elaborate device to preserve and spread them," and, as a corollary, that to "explain ethics and ethical philosophers" one need understand "the role of evolution in shaping the whole device." Our "hypothalamic-limbic complex" floods our consciousness with all the emotions – hate, love, guilt, fear, and others – and has been "programmed to perform as if it knows that its underlying genes will be proliferated maximally only if it orchestrates behavioral responses that bring into play an efficient mixture of personal survival, reproduction, and altruism." Wilson quotes Richard Lewontin approvingly: "Natural selection of the character states themselves is the essence of Darwinism. All else is molecular biology." In the same sweeping way, Wilson asserts later: "In the microscopic view the humanities and social sciences shrink to specialized branches of biology. History, biography, and fiction are the research protocols of human ethology; and anthropology and sociology together constitute the social biology of a single primate species."

It is an ambitious vision. To name only entities to which Wilson himself refers: Elements in the intended synthesis include evolutionary biology, genetics, biochemistry, ethology, anthropology, psychology, sociology, the humanities, and ethics. One can almost glimpse a mutual accommodation of conceptions such as bonding, sex, division of labor, communication, territoriality, patriotism, warfare, learning, aggression, fear, altruism, and the structure of DNA. Indeed, if one thinks of what has been left out of this projected synthesis, one comes up with a very short list. But significantly at the head of such a list stand the notions of the transcendental and of free will.

Let me inject here, for what it is worth, that regardless of the ultimate merit and success of this great program, I find the aim useful. I have a fourfold set of reasons for this judgment. First, science needs more such wide-ranging, intellectually stimulating efforts than we get, more than our usual fare of small additions to the pale sandheap of individual analytical results. Second, even if it fails eventually, the challenge that sociobiology has thrown down before the neighboring disciplines is bound to have a strong, perhaps transforming effect on some of them, even if not along the lines envisaged by its proponents. That is how progress is made. Third, I view Wilson's book as a significant cultural artifact in its own right because it represents rather accurately, and eloquently, one typical, current world view characterizing this part of the twentieth century – for example, in its plea for a sophisticated form of flexible, almost stochastic predeterminism and materialism, in its apparently dispassionate concern with a secularized ethic, and in its accent on rationality and its underemphasis on symbolic forms. In short, with all its limitations it exemplifies what is widely considered to be some of the best thinking today. Fourth, and lastly, the discussions of the work among scientists and others present them with an opportunity for the difficult and hence often neglected task of assessing the possible ethical and human-value impacts of their own scientific work.

Some precursors. To understand the aims and claims, the powers and limits of sociobiology, it is essential to realize that this field of research, and the motivating spirit behind it, is part of a long evolutionary development. Sociobiology too has its phylogeny, and was already well established in the middle of the nineteenth century, at the time when the mechanists and vitalists were doing battle.[11] In 1845 a group of young physiologists, among them Helmholtz and DuBois-Reymond, swore an oath "to account for all bodily processes in physical-chemical terms." They did not prohibit all metaphysical discussions of that science, but merely declared, in DuBois-Reymond's famous phrase, "ignorabimus," that is, that we shall never know the great world riddles, other than those portions that reveal themselves within mechanistic science.

This group was distinguishable from a parallel but more extreme group of experimental biologists and medical materialists who may be called the "nothing-but" school. To those, all things were to be reduced to a homogeneous mechanistic scheme, including the world riddles despaired of by the others. This naturally led them to attack the established order, the alliance between church and state, and all the other impedimenta to radical progress, in science and without. Not surprisingly, many of them were socialists and visionary fighters for social justice. For example, Rudolf Virchow, one of the sympathizers, supported the German Revolution of 1848 and became the chief of the liberation opposition to Bismarck. It is both ironic and significant that from the present perspective, the medical materialists and the Helmholtz group were far closer to each other than to any of their common enemies; they were, for example, united in being antitranscendentalists.

To me, the most interesting figure among all these was the splendid biologist and German Darwinist, Ernst Haeckel. A fiery materialist, monist, and sociologist, he scoffed at all myth-mongers and offered a complete science-based world view, one that would solve all puzzles. His turbulent book of 1899, written toward the end of his career but at the height of his fame, was in fact titled simply *The Riddle of the Universe.* It swept over Europe like a crusade against mystification, against what he regarded as "the untruth foisted on the people by their spiritual and economic masters." Science was to triumph over theology by spreading the gospel of evolution infused with a modicum of pan-psychism. Haeckel's chief point was that there was a monism or unity of the inorganic and the organic world, grounded in the laws of the conservation of matter and energy (what he called "the law of substance").

It was indeed a replay, complete in many details, of an ancient message. Here is, first, Lucretius, introducing the world view of the first Greek atomist:

> I will essay to discourse to you of the most high system of heaven and the gods, and will open up the first-beginnings of things, out of which nature gives birth to all things and increase and nourishment . . .
>
> When human life to view lay foully prostrate upon earth crushed, down under the weight of religion who shewed her head from the quarters of heaven with hideous aspect lowering upon mortals, a man of Greece ventured first to lift up his mortal eyes to her face and first to withstand her to her face . . . On he passed far beyond the flaming walls of the world and traversed throughout in mind and spirit the immeasurable universe; whence he returns a conqueror to tell us what can, what cannot come into being, in short on what principle each thing has its powers defined, its deep-set boundary mark . . .

> This terror then and darkness of mind must be dispelled not by the rays of the sun and glittering shafts of day, but by the aspect and the law of nature; whose first principle we shall begin by thus stating: Nothing is ever gotten out of nothing by divine power. Fear in sooth takes such a hold of all mortals because they see many operations go on in earth and heaven, the causes of which they can in no way understand, believing them therefore to be done by divine power. For these reasons, when we shall have seen that nothing can be produced from nothing, we shall then more correctly ascertain that which we are pursuing, both the elements out of which everything can be produced and the manner in which all things are done without the hand of the gods.[12]

In Haeckel's battle against notions such as personal immortality, the conventional belief in a creating God, or in the belief in a mind or a purpose behind evolution, Haeckel did not have to refer explicitly to Lucretius or to the distant predecessors, Leucippus and Democritus. His sentences had their own grand, Teutonic sweep:

> All the particular advances of physics and chemistry yield in theoretical importance to the discovery of the great law which brings them to one common focus, the law of substance. This fundamental cosmic law establishes the eternal persistence of matter and force, the unvarying constancy throughout the entire universe. It has become the pole-star that guides our monistic philosophy through the mighty labyrinth to a solution of the world-problem.

The promise of eternal persistence and of a guiding pole-star was vivid in the colorful and reassuring chapters in Haeckel's book: "The History of Our Species," "The Phylogeny of the Soul," "Consciousness," "Immortality," "The Evolution of the World," "The Unity of Nature," "Our Monistic Ethics," and, finally, "The Solution of the World-Problems." In comparison, Wilson's book is an exercise in understatement and scientific objectivity. I doubt that it is able to arouse a small fraction of the hopes and fears that Haeckel's book did for about a half-century.

Another grand precursor of Wilson is Jacques Loeb, the author of *The Mechanistic Conception of Life* (1912). He was born in 1859, the very year of Darwin's *Origin*. A scientist in the old style of philosopher and social innovator, he too was certain that scientific findings might lead directly to political and social-development consequences. Influenced by Schopenhauer as were so many others of his generation, he seems to have turned to biology in order to find evidence against the conception of the freedom of will. Perhaps his best work was on animal tropism, the involuntary movements imposed by environmental conditions such as light upon organisms; he considered it a model for understanding behavior in terms that avoid the use of the noxious conception of "will." The accomplishment for which he is most famous, artificial pathogenesis by physical-chemical means, fell in the same category of scientific research findings with anti-transcendental and antimetaphysical implications.

From 1911 on, cheered by the proof of the existence of molecules by Perrin and others as the "final vindication of mechanistic philosophy," he spoke and wrote on "the mechanistic conception of life," and published his book of that title in 1912. As Donald Fleming puts it, in it he reduced life to a "physical-chemical phenomenon, free will to an illusion generated by tropistic causes, and religious faith to an absurdity. He proclaimed the total validity of mechanistic principles and derived from them a system of human ethics based on instincts whose unobstructed expression would rejuvenate world society." As early as 1911, "he said that the main task for students of heredity was to determine 'the chemical substance in the chromosomes which are responsible for the hereditary transmission of a quality.' "

In his book, Loeb asks whether "the wishes and hopes, efforts and struggles" – man's inner life – should be "amenable to a physical-chemical analysis." And he answered yes, even if the proof would have to come from much research that still waited to be done: "For some of these instincts, the chemical basis is at least sufficiently indicated to arouse the hope that the analysis from the mechanistic point of view is only a question of time."

In the last pages of Loeb's book, just as in Haeckel's and in Wilson's, Loeb has a section entitled "Ethics." Here is a passage:

> We eat, drink and reproduce not because mankind has reached an agreement that this is desirable, but because machine-like, we are compelled to do so. We are active because we are compelled to be so by processes in our central nervous system . . . The mother loves and cares for her children, not because metaphysicians had the idea that this was desirable, but because the instinct of taking care of the young is inherited just as distinctly as the morphological characters of the female body . . . Not only is the mechanistic conception of life compatible with ethics: it seems the only conception of life which can lead to an understanding of the source of ethics.

In comparison, Wilson's is a soberer, more scientifically grounded and modest effort. Ironically, just for this reason, it will not have the same popularity that those predecessors had.

Evaluating the potential for synthesis. We can now return to our earlier discussion of the properties of a synthesis in operational terms, to test against the earlier model the claims inherent in sociobiology of producing a major synthesis in our time. We shall make due allowance for the obvious fact that the Newtonian synthesis, while a natural standard of comparison, is also by far the most distinguished case that we have; even while referring to it to compare the half-dozen major structural elements of synthesis, we thereby calibrate, so to speak, the top reading on that kind of thermometer.

1. Almost by definition, a synthesis does not spring out of nowhere. It has roots in the history of the fields within which it produces coherence. We noted that for the Newtonian synthesis, these roots reach back to antiquity, in one part to the grand scheme of Thales. We indicated that Wilson's work, too, has a distinguished phylogeny. Today's sociobiology is the current terminal point on a trajectory or line of system builders issuing primarily from the same materialistic-mechanistic and antimetaphysical school of Thales of Miletus and his followers – Anaximander, Anaximenes, Heraclitus, Leucippus, Democritus, Anaxagoras, and Lucretius. The more recent stations on this trajectory of physiologues, teaching "disenchanted" or "positive" explanations of nature's phenomena, are some aspects of Newton

and Laplace, D'Alembert and Condorcet, Comte, Virchow, Helmholtz, DuBois-Reymond, Herbert Spencer, T. H. Huxley, Haeckel, Loeb, Mach, Julian Huxley, Haldane, the early Lysenko, and Schrödinger (in "What Is Life?").

If we look beyond their many differences, they all share fundamental ambitions, approaches, and themata. For example, the matrix of social values and the moral base are taken not as a priori but as susceptible of explanation within a materialistic world view. These natural philosophers tend to opt for continuity instead of uniqueness, for unity rather than discreteness. In modern sociobiology, the old thema of classical physical casuality exists, although modified and recast in terms of tendencies and "potentials" – an even older yet still current thema. Among the moderns, many are social innovators, and opt for an essentially optimistic and liberal, that is, evolutionary, political stance.

Thus looking at this aspect of syntheses in general – synthesis as the climactic achievement of a long-term trajectory – the ambition of sociobiology is entirely recognizable.

2. *Inclusion and exclusion of elements*. The raw materials from which a synthesis must be fashioned are individual, seemingly disparate elements or separate classes of entities. Thus the Newtonian laws govern the motions of objects having from atomic to galactic size. Yet Newton also specifically excluded large sets of elements from his synthesis.

Looking at sociobiology from the outside, as I must do, it seems that the field is in some danger of not knowing how to exclude explicitly some tempting candidate-elements. One has the impression that the range of behaviors, traits, events, and so on, clamoring for inclusion is enormously large. To be sure, history reminds us that exclusion very frequently is not, and perhaps cannot be, an a priori conscious decision, but can only come at the end of a long series of unsuccessful attempts at inclusion. That is, exclusions are the result of the discovery of "impotency principles." And to find those, one needs time.

3. *A first principle*. After Newton, nothing so basic as the intuition of the universal law of gravitation as the one first principle on which to build a system will perhaps ever be granted to another synthesist. But sociobiology does make several basic,

fundamental postulates, for example, the central theorem of modern evolution that animals behave so as to optimize their inclusive fitness; that there is some molecular base of behavior, in other words, that the genes "program the potentials"; that for all phenotypes, including behavior, there is selection by interaction of genes and environment; that there is a continuity of mammalian traits in humans. For an eventual success in the large sense, it would seem to me necessary to postulate explicitly the smallest number of independent statements, and insofar as possible to exhibit the role of parsimony and necessity among those postulates that do remain.

4. *Cohesion of a general system.* Again, since Newton (with the possible exception of General Relativity), nothing so magnificent can be expected to arise again in our time by way of a system based on one or a few principles – least of all in a theory that is still under construction before our very eyes. Still, there are beginnings here, such as understanding the size of families and colonies, and diffusion speeds. More such victories, and a good cataloging of them, will be needed to make this synthesis widely persuasive. No doubt that is the growing edge of the whole effort.

I recognize that those working in other sciences may well be depressed (or overly impressed) by the success under this heading which the 300-year effort of modern physics has had at its culmination, for that is a model that physics willy-nilly puts before the other sciences. The power of the deductive network produced in physics has been illustrated in a delightful article by Victor F. Weisskopf.[13] He begins by taking the magnitudes of six physical constants known by measurement: the mass of the proton, the mass and electric charge of the electron, the light velocity, Newton's gravitational constant, and the quantum of action of Planck. He adds three or four fundamental laws (e.g., deBroglie's relations connecting particle momentum and particle energy with the wavelength and frequency, and the Pauli exclusion principle), and shows that one can then derive a host of different, apparently quite unconnected, facts that happen to be known to us by observation separately: for example, the size and energies of nuclei, the mass and hardness of solids such as rocks, the height of mountains on earth, and the size of our sun and of similar stars. This is indeed fulfilling Newton's program, triumphantly. The program of Lucretius, and the related one of Wilson, must, however, not expect to reach this level soon.

5. *Demystification and central "image."* As Newton demystified comets, sociobiology in its current version holds out a promise of "explaining" complex or disturbing effects in the processes of human society, from homosexuality to warfare. Even if there is only a quite partial delivery on that promise, the effects upon the world view of our society will be enormous. There is missing, however, a central visual image, analogous to the clockwork conjured up by the Newtonian synthesis. There is not even a complex one such as Darwin's "tangled bank." The voluminous and painstaking work chronicled in Wison's book and similar sources may never lend itself to such a feat of fruitful oversimplification.

6. *Prediction.* Predictive capability is usually regarded as the ultimate test of how "scientific" a synthesis is – the prouder and the more confident, the "harder" the science. In this respect, sociobiology seems to be in only an early stage of development. However, we may be touching here on an essential difference between the biological as against the physical sciences, rather than on a *sine qua non* of scientific synthesis as such.

7. *Cultural reach.* The claim of the Newtonian synthesis as a powerful exemplar of a cultural synthesis that changed civilization has been amply documented. (If one were allowed only a single example, an analysis of the sources of the imagination in Thomas Jefferson's draft of the American Declaration of Independence might suffice.)

By this measure, of course, the strategy with respect to sociobiology is, once more, patience. It does appear that both the proponents and the more vociferous and politically oriented opponents of sociobiology are united in the expectation that the New Synthesis of which Wilson speaks will be one that changes our culture, that is, that it will be a cultural synthesis. If it does not, the synthesis will still be one in the "transformational" sense.

In sum, we may conclude that on many, perhaps most, counts the Wilsonian synthesis has a fair chance eventually to bear out its claims, particularly if the crucial elements now lacking are supplied in the ongoing research, notably with respect to cohesion and so-called prediction. All this has nothing to do with whether one likes the New Synthesis, or even whether the ideas of sociobiology can be as severely abused as were, on occasion, Darwin's and Einstein's. The issue of *scientific validity* will be decided in the laboratories and in field research. If the evidence in favor of sociobiology brought in by experimenters does turn out to be voluminous, varied, and positive, the fathers of the field will be installed in the Pantheon. But it is a curious question whether Haeckel and Loeb and the others who will be waiting for them there will in fact approve of this new version of the ancient quest. We can imagine they will at least raise an eyebrow that in our time the offsprings of Lucretius no longer find any theological opponents to engage head-on, but that they address themselves to the modern equivalent of the ancient seat of moral force, that is, to society and social science.

It may take a long time to find out whether the promised synthesis is in fact possible, whether the social behavior of animals can generally be linked, across all species, to the mechanism of natural selection. I am inclined to think that even if the New Synthesis will not have all the features the most optimistic sociobiologists would like to see, biology, sociology and many neighboring fields will have been shaken up in a fruitful way. We may not get in one jump to the new New Jerusalem, but we can still hope to end up much wiser about the behavior of people and of other animals. One must remember that even in making a synthesis, between complete success and failure there is a large and useful middle ground.

Distinctions and dangers

Having examined uses of the terms "analysis" and "synthesis," and the characteristics of syntheses in two examples in science, our tasks are now to identify dangers and even pathologies in the Analysis and Synthesis process; to distinguish more adequately between analysis and synthesis; and also to differentiate Analysis and Synthesis from other activities and concepts. We shall want to keep in mind that often the lines cannot be drawn in a natural way and are thus imposed either too sharply or not sharply enough.

Assumption of unitary structure, and of elements for the Analysis and Synthesis process. From the beginning, the process of Analysis and Synthesis forces us to assume a priori the existence of some unity in nature and in knowing, a unity that is penetrable, fragmentable, or made otherwise manageable by the analysis of cognate pieces, and reestablishable by synthesis. The natural phenomena themselves, however, do not show these properties in any direct or easy way. Unlike, say, "analogy," a concept that may force itself early on the alert intellect by reason of the existence of formal, simple symmetries in the materials of observation, few if any natural processes by themselves persuasively impose the concept of Analysis and Synthesis. Thus in observing a natural process of decay or growth directly, it would be a mistake to believe that one discerns processes of analysis or synthesis "by themselves" rather than, say, the transformation of qualities. To be sure, we know that at every cell division the unwound DNA helix replicates the original double helix by synthesizing a structure from the materials of its environment in proper order; similarly, a physiological psychologist has found that brain-damaged animals (including man) frequently recover by going through a rebuilding in a sequence which exhibits recapitulation of the organism's growth from its early stages. But all such examples are the results of structuring the observations in the frame of a sophisticated model theory.

The guidelines for discerning the fissures along which to break a complex into separate "fragments" for analysis, or for identifying the pieces that serve as raw material for the synthesis, are usually imposed on us by the *tradition* in the particular scholarly field. A modern example is offered by the treatment of the universe of the Zinacanteco Indians, as well described by the anthropologist Evon Z. Vogt.[14] Acknowledging that any effort to delineate separable features in the total way of life of these Mexican Indians leaves one with elements that are still intimately interrelated, Vogt says he has "quite arbitrarily divided them into a series of chapters – beliefs about the universe, the organization into a ceremonial center and outlying hamlets, social life, economics, the life cycle, and ritual." These chapters, each within traditionally established lines of demarcation, help us to dissect, analyze and understand better the culture in terms of these traditional components. Conversely, he hopes, the reader "will be able to synthesize the diverse elements superimposing one description onto another to form a picture of a highly integrated system, and thus to approach more clearly the Zinacanteco view of their own way of life." It is clear that each reader's synthesis will be an individual product with a particular orientation, depending on the cultural and professional framework of the individual reader, for there are of course no absolute criteria for the process of either analysis or synthesis.

Further effects of the cultural framework. The cultural framework rules one's conception of the Analysis and Synthesis couple as a method as well as ruling the distinction within the couple. The type of synthesis in a given field will depend not merely on whether the *Zeitgeist* is dominated by Plato and Euclid or, say, by Sartre and Gödel. Even within a fixed time period, large cultural differences can assert themselves. A simple way to illustrate that is to point to differences today in dealing with disease within different ethnographic contexts. The symptoms perceived by the patient as well as by the healer, the presumed cause, the analysis of likely stages of the disease, and of course the cure prescribed – all depend heavily on the specific societal context. The Western patient will be concerned with physiological and chemical changes; a Mexican Indian may typically be concerned with redirecting his lost "inner soul" into the corral of his gods. The sense of unity and holism in the latter case (which has its price, too) stands in sharp contrast with an intellectual system which tends to partition man *ab initio* into separable spheres such as the bodily, mental, and spiritual. The expressions of disease among "primitive" groups, even within an industrial society, tend to be influenced by a more holistic orientation; consequently, the link that members of such groups tend to draw between "functional" and "organic" pathology will also be quite different.

Problematics of Analysis and Synthesis in use. If Analysis and Synthesis are relativistic with respect to the cultural framework, they can also differ for different fields within the same culture; as John Locke held, modes of reasoning can differ, for example, between mathematics and morals. Moreover, advances in knowledge change what are considered to be the necessary parts of a complete and coherent picture of the universe. Even at a given time in a given field, cognitive modes are different among intellectual workers; the processes of synthesis or analysis that suggest themselves will necessarily differ for those who tend to think visually and those who tend to think semantically.

For such reasons, it may be *in principle* irresolvably disputable what the specific elements are toward which analysis should proceed and what the shape is of the desired structure to which the process of synthesis strives. Thus when Descartes said, "Divide each problem that you examine into as many parts as you can and as you need to solve them more easily" (*Discourse on Method*), Leibniz objected that such a method provides no guide that would prevent one from dividing the problem into "unsuitable parts" and so "increase the difficulty."

This very objection has frequently been raised against systems analysis (as the technique is used, for example, to design capabilities for meeting energy needs of a nation for some period in the future). There, typically the "problem" is broken into a few subunits within which quasi autonomy and quasi equilibrium are believed to exist, and where higher-order interactions are, at least for the moment, not important. Judgments of this sort can be seriously open to error, the more so where actions taken on the basis of a first-stage systems analysis are difficult to modify when later stages of reexamination reveal faulty presuppositions. Similar objections can often be leveled against "cost-benefit" analyses.

An even simpler case is the fallacious method of selecting the "elements" in accord with the ease of making gross measures for them. Some examples are given in Chapter 6 on making science indicators; to give another example, pollution is measured in the United States (under the Clean Air Act of 1970) by the relatively simple-to-obtain total weight of each of six basic contaminants – particulates, sulfur oxides, carbon monoxide, photochemical oxidants, nitrogen oxides, and hydrocarbons. Now the Council on Environmental Quality, a federal agency that advises the President and Congress, has found this yardstick fails to distinguish between the relatively harmless large particles and the small ones that are deleterious to the human lungs. Moreover, it provides no gauge of toxic substances such as acid sulfates, nitrates, trace metals, and organic compounds, nor of the possibly hazardous interaction among two or more of these substances.

Analysis versus synthesis. Opposing thematic commitments of professionals within a particular group at a given time may cause them to accentuate either analysis or synthesis. For example, the author of a recent review of anthropological kinship studies notes that after a hundred years of discussion,

> despite the apparent diversity of perspectives, two camps may be sorted out: the splitters and the synthesizers. To the splitters, kinship is divisible, for analytic purposes, into a series of discrete aspects – kinship nomenclature, alliance (marriage) and dissent systems, household and family organization. To the synthesizers, it is a kind of social idiom, a way of understanding social life that is inseparable from other ways of talking about life in society – in terms of economics, politics, or ritual. The splitters are, in most instances, formalists or ethnoscientists of one variety or another, whereas the synthesizers are often advocates of positions, such as symbolic anthropology or structuralism. Both camps are concerned with the descriptive foundations of social anthropology . . . Kinship is a kind of crucible for theories of description and analysis.[15]

Generally only the more exalted spirits among scholars and scientists are able to avoid falling into the camp of either the splitters or the synthesizers, and keep in view the coherence of Analysis and Synthesis. Freud saw this when he wrote, in his letter of April 7, 1907, to Carl Jung, "It seems downright deceitful to conceal the fact that psychosynthesis is the same thing as psychoanalysis. After all, if we try by analysis to find the repressed fragments, it is only in order to put them together again." Similarly, Descartes knew that his use of algebra in developing analytical geometry was meant to bring together the various parts of mathematics and, moreover, to make one generally applicable method of analysis; Descartes thought his universal mathematics more powerful than any other instrument of knowledge, "being the source of all others."

We noted that it may be a mark of our time – one that tends to espouse pluralism and to frown on monism – that there is an asymmetry between analysis and synthesis, for example, that text entries under analysis are far more frequent than those under synthesis. But the monists are by no means routed, nor ever can be if they are responding to a thematic drive. In William James's *Pragmatism* (1907), Lecture 4, he considered the difference between pluralists and monists as nothing less than "the most central of all philosophic problems," and he indicated that the dispute is not a resolvable one:

> I wish to turn . . . upon the ancient problem of "the one and the many" . . . I myself have come, by long brooding on it, to consider it the most central of all philosophic problems, central because so pregnant. I mean by this that if you know whether a man is a decided monist or a decided pluralist you perhaps know more about the rest of his opinions than if you give him any other name ending in *ist.* To believe in the one or in the many, that is the classification with the maximum number of consequences.

There is little doubt that the division between synthesizers (or "lumpers") and splitters can be found in practice, and that the synthesizer belongs among the monists, as the splitter belongs among the pluralists. Whereas the one reaches out for the universal and whole – recall Thales' "all is water," Pythagoras's "all is number," Democritus's "all is atom and void," and John Wheeler's "There is nothing in the world except empty curved space"[16] – the other finds himself at home among the particular parts. Each takes courage for his breathtaking simplifications by a thematic commitment, in one case to break through the barriers, in the other to neglect, in reductionistic analyses, all second- and higher-order interactions among the pieces. To be sure, each of these two approaches by itself can produce very successful work and may correspond to a fundamental intellectual ability and skill, separately testable and separately listed in the taxonomy of educational objectives.[17] Yet Plato's warning, cited earlier, remains appropriate. Analysis alone or synthesis alone leaves the work incomplete; reason at its best first ascends through analysis and then descends through synthesis.

The best example of a major spirit of science in our century refusing to favor one polar position or the other was Niels Bohr. In launching the *International Encyclopedia of Unified Science*,[18] he put it into a single, powerful paragraph:

> Notwithstanding the admittedly practical necessity for most scientists to concentrate their efforts in special fields of research, science is, according to its aim of enlarging human understanding, essentially a unity. Although periods of fruitful exploration of new domains of experience may often naturally be accompanied by a temporary renunciation of the comprehension of our situation, history of science teaches us again and again how the extension of our knowledge may lead to the recognition of relations between formerly unconnected groups of phenomena, the harmonious synthesis of which demands a renewed revision of the presuppositions for the unambiguous application of even our most elementary concepts. This circumstance reminds us not only of the unity of all sciences aiming at a description of the external world, but above all, of the inseparability of epistemological and psychological analysis. It is just in the emphasis on this last point, which recent development in the most different fields of science has brought to the foreground, that the program of the present great undertaking distinguishes itself from that of previous encyclopedic enterprises, in which stress was essentially laid on the completeness of the account of the actual state of knowledge rather than on the elucidation of scientific methodology. It is therefore to be hoped that the forthcoming *Encyclopedia* will have a deep influence on the whole attitude of our generation which, in spite of the ever increasing specialization in science as well as in technology, has a growing feeling of the mutual dependency of all human activities. Above all, it may help us to realize that even in science any arbitrary restriction implies the danger of prejudices, and that our only way of avoiding the extremes of materialism and

mysticism is the never-ending endeavor to balance analysis and synthesis.

Possible physiological aspects. Evidence is accruing that a physiological component may help explain the existence of the analysis/synthesis dichotomy in human thinking, as well as the preference for one or the other in particular cases. Recent research has found that the right and left halves of the human brain appear to be capable of distinct modalities of thought. For right-handed persons, the left half appears to be more concerned with language processing, linguistic structures, and "logical" sequences. The right half appears to deal preferentially with holistic images and is also identified with spatial orientation, the production of music, and pictorial representation. There is also some evidence that the right half may be the seat of the more intuitive activities and may act to a large degree independently of and simultaneously with the left half.

A number of consequences may follow. One is the possibility of a difference of development in one or the other of these modalities, corresponding to individual physiological differences. Another is that problem solving that involves both synthesis and analysis may be divided in time sequences between the two halves of the brain, starting, for example, with ideas that have relaxed standards of logicality in the right-hand side, then preliminarily focused by the logical processes of the left side, and then handed over again to the right side, where sometimes a sudden and totally unexpected solution may be recognized. It should, however, be emphasized that these interesting suggestions are still at an early stage of testing.

Extreme monism. Analysis and synthesis should be distinguished from other forms of human knowledge that are frequently not considered part of ordinary human reasoning. To Plotinus, as to Aquinas, the direct perception of the One was transcendent, divine knowledge. The sensorium of God was undifferentiated and without process, whereas man's intellectual activity was dualistic (subject/object dualism) and discursive. The closest human knowledge could come to that state was through the process of intuition; as Descartes also understood and used this conception, intuition arrived at knowledge without stages, directly and with certainty.

At the end of our earlier discussion of the Newtonian synthesis, we made reference to a core feature of the concept of synthesis. In its heightened form, the occasional occurrence of a more or less "direct" perception at the end of the process of induction can open up a prospect of all-pervasive and seamless quality which, in Plato, evidently led him to accept the first principles so produced. In Plotinus, this vision surfaced in the conception of the One.[19] However, in its most extreme form it can also be characteristic of the state of mind in certain mental disorders.

Different forms of Analysis and Synthesis? Many social scientists have been concerned with the extent of the habit to model the methods (including that of analysis and synthesis) used in other sciences too closely on the physical sciences. In fact, we may suspect that historical accidents favored the physical sciences in a way that makes them too "simple" to be altogether adequate models.

Thus the exemplar for physical science work has indelibly been its early success in astronomy, for example, by focusing on the wandering pinpoints of light representing the planets and by creating a model of the solar system from observations of their relative motions against the background of the fixed stars. The profusion of complexities and data had to yield a simple system eventually, for the planets *were* independent of one another; a single force on each from the sun is a close enough approximation for understanding their motions. The planets formed therefore a "pure" sample, possibly the simplest of all the samples with which science has had to deal ever since, and hence the success of early systems makers of astronomy was a success of a philosophy of oversimplification hardly acceptable now, even to the physical sciences. It was possible to make cumulative observations over very long time spans on "the same things"; the samples were stable – the most stable in existence. The data were therefore reproducible, allowing the formation of an international community of scientists who could learn to gain consensus by reference to repeatable observations made within their individual horizons. Moreover, the history of physical science allowed the parallel development of careful observers like Tycho Brahe and daring speculators like Kepler who could be quite divided on all essential presuppositions, and yet collaborate.

When we turn to the social studies, the situation appears to be strikingly different. For one thing, the "world" of samples is rarely clean and stable enough to allow easily reproducible measures by other groups or by other techniques, particularly at the microlevel or with fine-structure. Nor is it likely that the distance and decoupling between the observer and the observed can be achieved as successfully in the social studies as it has been in most of the physical sciences. In short, one should not be surprised if the forms Analysis and Synthesis can take in these different fields are markedly different.

Kant and Einstein on synthetic judgments a priori

One of the major, if doomed, attempts to explore the power of the Analysis and Synthesis conception was Kant's distinction between analytic (or explicative) judgments – a particular class of "necessary" truths – and synthetic judgments. To illustrate, he said that the statement "All bodies are extended" does not amplify in the least one's concept of body but has only invited an analysis of the concept. On the contrary, the judgment "All bodies have weight," he held, contains in its predicate something not actually incorporated in the general concept of body; it therefore amplifies this knowledge and must be called synthetical.

Kant further divided judgments and propositions into a posteriori (empirically based truths) or a priori (not so grounded, therefore transcendental, shown by "pure reason"), a division that gives rise to the two corresponding types of knowledge. Furthermore the dichotomies *analytic* and *synthetic*, on the one hand, and a posteriori and a priori, on the other, invited the concept of four pairs of possible subdivisions of judgments and propositions.

Whether this device, so evidently seductive, could yield operationally meaningful categories was another matter. It is now rather generally granted that the conception of the synthetic a priori, for example, is a labile notion, much battered,[20] and certainly not workable for mathematics as Kant thought. (Many hold now that mathematical propositions are analytic a priori only.) Any notion that claims to be a priori is under the constant threat of challenge by new data (e.g., from neurophysiology or cognitive psychology) which has given an entirely new view of the complex relations between "experience" and prior genetic potential. As noted previously, the change of world view from a Euclidean and Newtonian universe, in which Kant lived, to the one dominated by Einsteinian and other recent notions of science, changes the ground completely, as in the very meaning of what one may consider "empirical."

Thus Einstein himself, in his own work and letters, documented a pilgrimage that had begun at an early point in a philosophy of science close to Mach's sensationism and operationalism, and that in the end led him to a rationalism that accepted the concept of an objective, "real" world behind the phenomena to which our senses are exposed. As Einstein said in one of his letters (January 24, 1938, to C. Lanczos), "Coming from skeptical empiricism of somewhat the kind of Mach's, I was made, by the problem of gravitation, into a believing rationalist, that is, one who seeks only the trustworthy source of truth in mathematical simplicity."

The problem he had encountered, as he explained in his essay "Physics and Reality" (1936), was that the aim of the general theory of relativity had been to connect it in "as simple a manner as possible" with "directly observed facts"; but the aim proved to be unachievable. The inclusion of nonlinear transformation, as the principle of equivalence demanded, was inevitably fatal to the simple physical interpretation of the coordinate; that is, it could no longer be required that coordinate differences should signify direct results of measurement with (ideal) measuring sticks and clocks. The solution of the dilemma, from 1912 on, was to attach physical significance not to the differentials of the coordinates but only to the Riemannian metric corresponding to them.[21]

As Einstein put it in his Spencer Lecture of 1933:

> Experience may suggest the appropriate mathematical concepts, but they most certainly cannot be deduced

> from it. Experience remains, of course, the sole criterion of physical utility of a mathematical construction. But the creative principle resides in mathematics. In a certain sense, therefore, I hold it true that pure thought can grasp reality, as the ancients dreamed.[22]

Indeed, in his lecture "Physics and Reality,"[23] Einstein explicitly stated that Mach's theory of knowledge is not sufficient because of the closeness it supposes between experience and the concepts. Einstein then advocated going beyond this "phenomenological physics" to achieve a theory whose basis may be further removed from direct experience, but which in return has more "unity in the foundations."

Coming to write about Kant from this point of view later in life in his "Autobiographical Notes,"[24] Einstein felt that the differentiation Kant made between the indispensability of certain concepts taken as necessary premises for every kind of thinking, and of concepts of empirical origin cannot be defended. "I am convinced, however, that this differentiation is erroneous, i.e., that it does not do justice to the problem in a natural way. All concepts, even those which are closest to experience, are from the point of view of logic freely chosen conventions . . ."

Einstein had just recounted his delight in discovering at an earlier stage that the mental "objects" with which geometry deals seemed to him to be of no different type than the objects of sensory perception, "which can be seen and touched":

> This primitive idea, which probably also lies at the bottom of the well-known Kantian problematic concerning the possibility of "synthetic judgments *a priori*" rests obviously upon the fact that the relation of geometrical concepts to objects of direct experience (rigid rod, finite time interval, etc.) was unconsciously present. If thus it appeared that it was possible to get certain knowledge of the objects of experience by means of pure thinking, this wonder rested upon an error. Nevertheless, for anyone who experiences it for the first time, it is marvelous enough that man is capable at all to reach such a degree of certainty and purity in pure thinking as the Greeks showed us for the first time to be possible in geometry.[25]

At the end of the same volume, Einstein returned once more

to Kant and what he called those "who today still adhere to the errors of 'synthetic judgments *a priori*.'" He felt that his own theoretical attitude was

> distinct from that of Kant only by the fact that we conceive of the categories not as unalterable (conditioned by the nature of the understanding) but as (in the logical sense) *free conventions*. They appear to be *a priori* only insofar as thinking without the positing of categories and concepts in general would be as impossible as is breathing in a vacuum.[26]

This Einsteinian perception of the *necessity of positing categories* was one of the few survivors of the battle that surrounded synthetic judgments a priori. Indeed, *the Analysis and Synthesis dichotomy itself is a reflection of the existence of a fundamental thematic couple.* Some of its manifestations appear under names such as the many (manifold) and the one; parts and whole; disaggregation and aggregation; reductionism and holism; dismemberment (dichotomization, categorization, reduction, etc.) and unification; fragmentation and wholeness; and so on. The single best terminological characterization for the Analysis and Synthesis couple may, however, be the opposition *Differentiation and Integration.*

Analysis/Synthesis as thematic Differentiation/Integration

Earlier, we gave a number of examples of the thematic role of Differentiation and Integration, separately and jointly. We noted the search for categories to disaggregate and order different types of "elementary particles" in modern physics and, conversely, the attempts to find a "unified theory" for them all (see Chapter 1), and we noted that the method of dealing with complex entities by Differentiation (resolution, reduction, etc.) found its use in science very early (resolution of motion into components in Greek astronomy and in Galilean ballistics). Many other examples could be added to illustrate the pervasiveness, in science and beyond, of Differentiation and Integration, but we shall limit ourselves to instances of the two chief types of Differentiation, each of which accounts for part of the immense range and power of the methodological thema.

Taxonomic systems of categories, levels, or hierarchies. The use of taxonomic systems has been illustrated in the discussion on current elementary-particle physics (Chapter 1), for example, in the division of forces into four categories and of particles into families and superfamilies. Other instances suggest themselves readily, from the systems of Linnaeus and Mendeleyev (taxonomy without then-known developmental relationship of parts) to the "stages" in the life-cycle theory of Erik Erikson or in the genetic epistemology of Jean Piaget (progress by going through sequential and causally connected steps).

These are, however, the exceptions. In every field of scholarship, *successful* taxonomic systems are a minute fraction of the ever-growing total. At their best, hierarchical schemata of this kind convey the conviction of causal relationships and allow us to make confirming experiments and insightful extensions. At their worst, taxonomic schemes produce the very opposite effect; in the absence of some "first principle" to which to rise and from which to derive an ordering matrix, they may simply fall back on the production of pigeonholes with plausible labels which invite one to disaggregate the incoherent vastness of possible observables. Since those labels were obtained in the first place through some intuitive perception of portions within the grand aggregate, the danger of a vicious circle is obvious. One can easily end up masking perplexity under the guise of parceling out portions of it into the various corners of some lofty but arbitrary construction.

A special case of differentiation via taxonomy is *Differentiation by spectroscopy*, that is, by laying out next to one another presumably *non*interacting parts of a whole, whether they are separate axioms, separate "traits," or separate wavelengths. In his influential book *Analysis of Sensations*,[27] Ernst Mach expressed his hope to combine reductionism and sensationism and to find six separate "elements" in preconception and physiology: "If the seemingly limitless multiplicity of color-sensations is susceptible of being reduced, by psychological analysis (self-observation), to six elements (fundamental sensations), a like simplification may be expected for the system of nerve processes." (In fact, he believed that the six fundamental sensations were white, black, red, yellow, green, and blue.)

Differentiation by spectroscopy is so basic a conceptual tool that it seems (and probably is) inescapable for the human mind. A few well-known examples remind us how constantly we are assailed by the products of this strategy: The Stoics divided the sciences into physics, ethics, and logic (or dialectic); Aristotle's work abounds in differentiations – causes are either efficient, material, formal, or final; dialectical reasoning is of two forms (syllogistic and inductive); the principal sciences are three in number (medicine, mechanics, and morals); the units of language allow handling it grammatically versus logically;′ and statements are necessary or contingent.

The medieval distinctions between the parts of the trivium and quadrivium (alive today in the practice of splitting education into the natural sciences, social studies, and the humanities) show the same crutch of the imagination at work. Immanuel Kant abounds in such bifurcations and divisions: He holds that the methods of investigating nature and mind are either rational or empirical; philosophy is divided into speculative and practical uses of pure reason; the faculties of the soul are three (of knowledge, of pleasure and pain, and of desire); and so on.

John Locke lists five types of primary qualities – "solidity, extension, figure, motion and arrest, and number." Similarly Isaac Newton had listed the "universal qualities of all bodies whatsoever" – "extension, hardness, impenetrability, mobility, and inertia." Schools of scholarship similarly differentiate themselves into two, three, or more groups: those looking for "revolutions" in science versus those stressing continuities and evolution; those explaining life phenomena essentially on mechanistic-materialist grounds (E. Haeckel, Jacques Loeb, E. O. Wilson) versus rational (including primarily environmental, or Marxist) explanations or versus nonmaterialistic (transcendental, theological, or emergent) explanations. It is practically impossible to read a page of scholarship in any field without encountering this type of differentiation.

Differentiation by dichotomization. The simplest, most frequent, most powerful case of Differentiation is, however, *dichotomization.* It is also the most dangerous. Wolfgang Pauli considered the dichotomous divisions forced on our thinking by Aristotelian logic as "an attribute of the devil." The list of dichotomous categories is nearly endless in every field. Wilhelm Ostwald distinguished between classical and romantic scientists, A. Szent-Gyorgyi between Apollonian and Dionysian scientists, and Carl Jung between sensation-based scientists and intuitive scientists (the first associated with reductionistic thinking, the second with holistic thinking). Psychologists have found (Chapter 7) that young persons early make a choice either for a "people"-oriented career or for a "thing"-oriented one. A random listing of the enormous number of antitheticals in constant use would contain matter/antimatter, animate/inanimate, subjective/objective, observer/observed, order/chaos, Parmenidean unchanging structure versus Heraclitean flux, a priori/a posteriori, either/or, plus/minus, yes/no, induction/deduction, internal/external, macro/micro, formal/functional, classic/romantic, synchronic/diachronic, Being/Becoming, male/female, lumpers/splitters, sacred/vulgar, friend/enemy, good/evil, as well as raw/cooked, day/night, sun/moon, and the host of dyads by which not only in structural anthropology but in general, meaning is given to each of the two opposing members by virtue of the particular opposition within other dyads to which each member may also be related.

Our very language seems so constructed; when Peter Mark Roget, physician and Secretary of the Royal Society, had categorized 997 basic ideas into 6 great Classes and 24 Sections of his *Thesaurus*, the chief order that emerged was the division of synonyms and antonyms – yet another reminder that the binary approach is built into our semantic structures. Themata–antithemata groups, too, are usually dichotomous (although sometimes trichotomous).

The pervasiveness of dichotomization as the prime form of Differentiation or analysis (sometimes to the point of oppressiveness) can lull one into failing to see that dichotomization may sometimes do little justice to the complexity of the case. "Either/or" may be not a solution but a new problem, to be solved by "both" (as in modern physics since complementarity was introduced), allowing for the *superposition* of alternatives instead of forcing a choice between them.[28] Yet the almost automatic tendency to bifurcation makes one suspect that it has its roots in the ontogeny or phylogeny of ideas, or both. We are told by child psychologists that one of the first and most important cognitive accomplishments of the newborn is to dichotomize the total experience into self and other and then to dichotomize further among these entities (as well as up/down, right/left, ahead/behind, etc.). Along the same lines, Ernst Gombrich has argued that the learning of children proceeds essentially by differentiation, by constantly delimiting further the original "undifferentiated mass."[29]

It is probably no accident that the cosmogonic ideas of antiquity, those earliest works of cultural synthesis, are also full of accounts of the developmental breakup into antitheticals. The first thirty-one paragraphs of Genesis are again, from the thematic point of view, an account of proliferation by repeated dichotomization. Thus on the first day occurs the separation of heaven and earth, and of light and darkness (day and night, evening and morning). On the second day, heaven and the waters are separated. On the third we encounter the dichotomy of seas and earth, as well as the animate versus the inanimate (the earth bringing forth plants). On the fourth appear the greater and the lesser of the lights in the firmament of heaven; on the fifth, two kinds of animals, in water and in air; and on the sixth, earth animals and man. The whole sequence exemplifies the dictionary meaning of dichotomy – literally, a cutting asunder, a separation of a class into two subclasses that differ in some quality or attribute.

Similarly, the ancient Milesian cosmogenic myth held that at the beginning there emerged, out of a primal unity, parts of "opposite things" which later interacted or reunited in meteoric phenomena or in the production of individual living things. The persistence of fundamental themata is nicely illustrated by the fact that quite analogous basic ideas infuse the cosmogony of the evolutionist camp of modern cosmologists. According to one of their recent proposals, the universe originated in a single "particle" called universon. It immediately divided into two such particles, a cosmon and anticosmon, which separated into our cosmos and, elsewhere, an anticosmos, each condensing into stars and galaxies (or into matter and antimatter, respectively). Within each of these worlds, the initial mixture of radiation and neutrons, through subsequent stages of differentiation by expansion and neutron decay, finally leads to the building up of heavier elements by thermonuclear fusion processes, preparing the ground for the later formation of molecules, including those on which life is based. It is not too daring to predict that at every stage in the future, no matter how rapidly and diverse the progress of science may be, there will be some school of cosmogony which continues the use of the same thematic tools of analysis and synthesis.

We began this essay by noting the existence of works that act as cultural syntheses. Like other syntheses, they are the results of the action of both members of the Analysis and Synthesis couple, themselves a form of the general and basic thematic couple Differentiation and Integration. In their power and limitation lie the power and limitations of the cultural syntheses. Works as different as those of, say, Newton and Einstein, therefore can turn out to have fundamental functional similarities. In this example, both scientists tried to analyze the phenomena of light by heuristic models and also tried to unify the phenomena of light and matter. Moreover, both were concerned about the limitations of the syntheses they had achieved. If we study not only their published scientific work but also their philosophical essays and personal correspondence, we find that both men were not satisfied with any primarily "scientific" world construction but yearned for a still grander synthesis. Their unrivaled successes within the limits set by the problems themselves, and their yearning to go beyond them, help us to discern both the upper boundaries of human ability and the spirit that motivates it.

Part II: Studies in recent science

5

Fermi's group and the recapture of Italy's place in physics

The turning points of modern history have sometimes the character of mythological events. Such an event took place in Rome one morning in October 1934, in an upstairs room at Via Panisperna 89A, an old physics laboratory of the University of Rome. There Enrico Fermi and his young collaborators, to their surprise, came upon a key observation from which one may well date the effective beginning of the nuclear age. It was the discovery of the powerful effect which a beam of neutrons can have in initiating the instability of nuclei in a target, provided the incident neutrons are first slowed down or "filtered" by passing through a moderator, such as hydrogen-containing paraffin. Fermi himself later described the nascent moment in a conversation with the astrophysicist Professor S. C. Chandrasekhar:

> I will tell you how I came to make the discovery which I suppose is the most important one I have made. We were working very hard on the neutron-induced radioactivity, and the results we were obtaining made no sense. One day, as I came to the laboratory, it occurred to me that I should examine the effect of placing a piece of lead before the incident neutrons. And instead of my usual custom, I took great pains to have the piece of lead precisely machined. I was clearly dissatisfied with something: I tried every "excuse" to postpone putting the piece of lead in its place. When finally, with some reluctance, I was going to put it in its place, I said to myself: "No, I do not want this piece of lead here; what I want is a piece of paraffin." It was just like that: with no advanced

> warning, no conscious, prior, reasoning. I immediately took some odd piece of paraffin . . . and placed it where the piece of lead was to have been.[1]

Fermi was the least Dionysian, the most Apollonian, of physicists. The impulsive filtration of the rays was, methodologically, perhaps unique in his career. But so were what Segrè calls "the miraculous effects of the filtration by paraffin,"[2] in terms of producing a vastly enhanced radioactivity in the silver cylinder which served as a target for the slowed – "moderated" – neutron beam.

Sergè and others of that group report that within hours Fermi had puzzled out how and why it happened. With this, a chain reaction of unforeseeable but almost inevitable results was initiated. The most immediate effect was on the work of Fermi's own group and with it on the transformation of physics in Italy and beyond. It led ultimately to the award of the Nobel Prize to Fermi on December 10, 1938, as the official proclamation reads, "for his demonstration of the existence of elements produced by neutron irradiation and for his related discovery of nuclear reactions brought about by slow neutrons." Less than two weeks after that ceremony, Otto Hahn and Fritz Strassmann sent to *Die Naturwissenschaften* their paper which indicated the presence of barium in a uranium sample which had been irradiated by neutrons slowed down by Fermi's technique.[3] Otto Frisch and Lise Meitner quickly understood that here was evidence for uranium fission which could no longer be overlooked. And almost exactly four years later, the first nuclear reactor, built in Chicago under Fermi's personal direction and again with the use of the same moderating principle, produced the first self-sustaining nuclear chain reaction under the control of man.

Against the background of these memorable events, one may be surprised to learn that physics as pursued in Italy about a half-century ago, when the physicists of Fermi's generation began their careers, was on the whole and with few exceptions far below first rank. During the 1920s and 1930s in Italy, important work was done in physics at centers such as Florence, Naples, Padua, Palermo, Pisa, Torino, and elsewhere.[4] But it cannot be doubted that the events of 1934 in the Rome laboratories of the Fermi group were crucial to the position physics achieved in modern Italy.

Hence it seemed worth gathering all relevant information, whether new or previously available in the widely scattered literature, in order to try to discern in this complex of scientific, institutional and personal connections the main components, and so to see the confluence of forces which led to the energizing events. Some aspects of this case will undoubtedly be unique to Fermi's particular group – Occhialini calls it "absolutely the perfect group"[5] – one that included Fermi's young collaborators, Edoardo Amaldi, Oscar D'Agostino, Franco Rasetti, Bruno Pontecorvo, and Emilio Segrè, as well as associated members such as Orso Mario Corbino, Ettore Majorana, and G. C. Trabacchi. Other aspects of this case, however, will be unique to Italy itself; and others still may be common to other countries also.

Fermi's mastery as physicist and teacher

At the head of any list of factors we certainly must place Fermi's own insight into physics. His considerable experimental skill was of course a major part of it, but, as has often been noted, his style of experimentation avoided complex and difficult equipment in the manner of his contemporary E. O. Lawrence. Moreover, it would be quite false to see Fermi as working chiefly in experimental physics; in fact, until 1934 he seems to have thought of himself only occasionally as an experimentalist. He was indeed a master of theoretical physics, including its mathematical aspects. This shows up from the beginning; by 1922, among the first of the 270 papers he ultimately published, he had shown his ability to handle probability theory and general relativity in a way which attracted considerable attention to the twenty-one-year-old newcomer to the field.

In addition, his attention to the various aspects of physics made him practically encyclopedic. Segrè called him "the last universal physicist in the tradition of the great men of the 19th century,"[6] and "the last person who knew all of the physics of his day."[7] This wide-ranging interest opened Fermi's mind to new ideas long before others in Italy took notice of them. Thus all who knew him agreed that he was probably the first physicist in Italy who took quantum physics seriously and undertook to learn it. Similarly, Rasetti recalls:

> I remember then that there came one day an issue of *Nature* with a review article by Rutherford on the nucleus [1920, on the experiments on the first artificial disintegrations of nuclei. G.H.] I knew nothing about it, next to nothing about it. Perhaps I had just heard of the existence of the nucleus, but I knew nothing then of the structure of the nucleus. And Fermi was extremely excited. Nobody in the physics department would have noticed this article in *Nature*. Fermi immediately noticed it, and I remember he gave a talk [to Rasetti and Nello Carrara, another fellow student. G.H.].[8]

Superimposed on Fermi's well-nigh encyclopedic interests was, however, an ordering principle which prevented the large range of new, incoming ideas from becoming dispersed and self-destructive. All biographies assert that Fermi ordered the overwhelming and vast amount of knowledge into a set of very few principles and "cases," which allowed him to understand almost any new problem as an example of one of about seven primitive or primary physical situations. Fermi would return throughout his career to a listing or digest of the chief ideas in physics which he had made when he first organized the field for himself as a young student.

This perception of nature's basic parsimony appears to have played a major role in the events of 1934. The best witness here is Amaldi.[9] Amaldi and Segrè had been working on the effects of pressure on the high terms of the spectra of the alkalis and observed an experimental effect for which Fermi subsequently produced and published the theory.

> We were working with electrons in such excited states, they were bound with approximately (1/100)th of electron volts, extremely weakly bound, and moving almost free, bound so weakly that they were almost free and with very long wavelengths. And in that paper, in order to explain the effect that we had found, he made the theory of a collision of a very slow electron against an atom. And this is exactly the same theory that was used

> one and a half years later for slow neutrons against nuclei . . .[10]

Although it would be too much to say that the theory of slow electrons suggested directly the theory of the slow neutrons, the incident is one of several illustrations of Fermi's re-use of theoretical results in novel and apparently unconnected situations.

A key to understanding Fermi's mastery, however, lies neither in his experimental skills, nor in his theoretical insight, nor in his encyclopedic knowledge, nor in his striving for fundamental simplicity and parsimony, but rather in the subtle balance of all these. Although it is commonly agreed that physicists more or less of the class of Bohr, Rutherford, and Einstein in the early and mid-twentieth century may each have had an advantage with respect either to large new theoretical ideas or to purely experimental intuition, nobody combined these capabilities – which are often so separate – as effectively as Fermi did. The commonly accepted ideal for the style of doing physics is today much closer to Fermi's than was the case even in the 1930s. Of that period, O. R. Frisch wrote:

> Madame [Joliot-] Curie had little respect for theory. Once when one of her students suggested an experiment, adding that the theoretical physicists next door thought it hopeful, she replied, "Well, we might try it all the same." Their disregard of theory may have cost them the discovery of the neutron.[11]

Very similar stories of the disdain of theorists for the work of experimentalists are also common. Nothing of this sort, however, could be charged against the Fermi group.

Another important aspect of Fermi's position was that by preference and necessity he was self-taught. His first contact with physics came when he bought a second-hand book on the elements of physics as a fourteen-year-old boy in the market of Campo dei Fiori. At about that time also he met a colleague of his father, Adolfo Amidei, who lent him a series of books on physics and mathematics which Fermi successfully studied on his own. By the age of seventeen he appears to have taught himself the essentials of general relativity. One may take it from the testimony of his colleagues that his seminal papers – for example, those on Fermi statistics and on beta decay – to a considerable extent were written by Fermi as exercises to apply to and thereby check on conceptions and theories which he had taught himself in connection with other problems.[12]

Again, virtually all his colleagues and former students agree that Fermi was unmatched as a teacher, and that his teaching reflected the characteristics which have been discussed previously. For example, his introductory course was enormously wide ranging in the subject matter covered.[13] More important, however, is the fact that much of his teaching was carried out by thinking aloud before the class – in a rational, organized way – about the subjects on which he was then engaged in firsthand study himself. The joy that teaching gave Fermi makes it likely that it was one of his most intense human relationships, as well as being yet another opportunity for self-education. This fact, together with the characteristics cited – and of course his capacity for prolific output – made Fermi the natural center of a lively and productive group, not only in Rome in the 1930s, but also in his later career in the United States.[14]

On Fermi's scientific style

Possibly even more important than the scientific excellence and pedagogic capabilities of the central figure in the Rome group was the particular methodology – conservative and pragmatic at the same time – for the choice of problems and the conduct of research. Here we encounter a distinctive style which was crucial to the success of the Italian group as a whole. In a sense, it represents a fusion of Fermi's personal characteristics and his scientific experience on the one hand, and the needs and opportunities for physics research on the other, within the setting given by existing scientific institutions and the state of science in the world at the time.

We have some glimpses of this methodology, even from the very first reports we have of Fermi as a young student in Pisa, where – in the hot and smelly chemistry laboratory – Fermi was given the task of analyzing a sample chemically. With a group of students whom he had organized for this purpose, Fermi took the sample into a room in the physics department and put it under a microscope. Franco Rasetti recounted:

> We determined first, from the shape of the crystals, how many ingredients there were, and we finally separated the 3 or 4 or 5 ingredients . . . For each ingredient we looked at the crystal shape; that already tells a lot, if it's a cubic crystal, if it is a monoclinic crystal . . . then we looked at the color of each sample . . . if it, let us say, was potassium permanganate, it was almost black, very dark purple; if it was copper sulfate, it was light blue, and so on. Then we said, well, they don't waste expensive elements. Nobody is going to put there gallium chloride . . . so they must be salts of common elements. So by all these various reasonings, we guessed what every one of these ingredients was . . . At the end, an assistant who was in charge of this lab, said: "This is really marvelous, how you analyze this, not following any of the rules."[15]

A great deal of Fermi's enterprise – and that of his later colleague in Rome, Rasetti – is already visible here: his ability to organize, to distinguish between essentials and incidentals, to make very reasonable,[16] simple assumptions and shortcuts contrary to current "rules," to make a commonsense, qualitative approximation before any detailed quantitative solution, to improvise – and to succeed. Neither aesthetic nor other philosophical or quasi-metaphysical principles – except that of simplicity – could preoccupy or delay Fermi.[17]

The pragmatic approach to soluble problems which would lead to a reasonably quick payoff differed fundamentally from the method in some other centers. For example, Chadwick testified to the fact that he had looked for neutrons on and off for a period of about twelve years after joining Rutherford in 1920. The research grant of the Cavendish being about £2,000 a year for all the work which had to be supported, Chadwick felt that he could not obtain or afford the necessary equipment for trying out certain of his ideas. He persisted, nevertheless, as best he could. In recounting the abortive experiments, he rather proudly added, "I wasted my time – but no money."[18] Lengthy investigations with admittedly inadequate apparatus would have been much less likely in a laboratory run along the lines of the Italian group.

Another style which Fermi's group eschewed was that associated with Niels Bohr, who in many ways was completely different from Fermi, and who in turn tended to regard Fermi's solution as too simple to be profoundly important. As one observer has said: "Bohr is such a bad authority on these [Fermi] papers because Bohr really had it in his mind that there was some profound problem with neutrinos and energy and so on, and didn't want to have it solved except in a mystical and deep way. It was solved by Fermi in 'too elementary' a way."[19] Bohr's favorite procedure was to drive contradictions patiently to their ultimate extreme, and to ponder the ensuing conceptual conflict as a necessary preparation for its ultimate resolution.[20] From the beginning, Fermi quite consciously and explicitly rejected as somewhat mystical and too philosophical the approach of Bohr, and indeed of others whose theoretical work dominated the scene. Fermi would say, though smiling and not with doctrinaire belief, "We proceed according to the rules of Bacon . . . The facts. We will make our experiments and then the experiments will tell what it is."[21] To the same effect, Enrico Persico writes:

> From his adolescence onward, Fermi had a quite definite, positivistic view of the world, although it is doubtful that he would have accepted this or any other conventional label for his philosophy. He had not been raised in a religious environment, and so did not have to pass through a religious crisis, as many Italians do when they reach the age of autonomous thinking. As a matter of fact, philosophical discussions did not interest him very much, and even the development of scientific philosophy that occurred during the years of his maturity, through the activity of the Vienna circle and other groups, seems to have left him rather indifferent. This was perhaps because many of the fundamental ideas of logical positivism were already deeply rooted in his mind as self-evident truths, and because philosophical subtleties and polemics did not appeal to his taste.[22]

Possibly the most important part, however, of this pragmatic style was Fermi's ability to choose the right moment, the exact time when the state of knowledge and experimental capabilities matched the opportunity as it opened up.[23] Fermi's ability to perceive, before anyone around him could tell him, that relativity and quantum theory were the correct areas for future work in physics, marked the very beginning of his career. His ability to perceive the opening of a major opportunity in a new field, to detect and shift to the advancing frontier – rather than to wrestle with some "crisis" in existing conceptual structures – characterized his career throughout. Segrè recalled:

> But already at Los Alamos [during World II] Fermi had the feeling that his next phase of activity would not be in neutrons but in something new, and he reminded me that just as [in the early 1930s] he discarded all his investments in spectroscopy to go to nuclear work, so now he would leave the slow neutron in order to proceed to new conquests in the field of high-energy physics. In a half-joking mode he quoted Mussolini: "Rinnovarsi o perire – to renew oneself or to perish."[24]

This brings us directly to the circumstances under which the Fermi group in Rome, after its formation, turned its attention to the work in nuclear physics which ultimately gave it its preeminence and which led to the events of 1934.

Choice of problem and method of research

From the time when he was brought to the University of Rome in 1926 until the end of the 1920s, the work of Fermi and his growing group was largely in spectroscopy and atomic physics.[25] But soon after Fermi had been appointed professor of theoretical physics at the University of Rome in 1926, he began to see that the ongoing revolution in quantum mechanics would necessitate deep study – both in order to do justice to the intellectual excitement aroused by papers such as those of Schrödinger and Dirac, but also from a practical point of view – to see what the advent of quantum mechanics meant in terms of the research program of Fermi and his students. A cool assessment of the situation showed him by the end of the 1920s that the vigorous development of quantum mechanics "signaled the completion of atomic physics." Segrè writes:

> These ideas suggested a radical change in the research projects of the Institute because our experimental tradition in Rome went back to the spectroscopic work

> initiated by Rasetti under Puccianti at Pisa. All our successes in experimental physics up to that time had been on spectroscopic subjects; our equipment was spectroscopic; and our knowledge was mainly in the field of atomic physics.[26]

Amaldi recalls the feeling at the institute at the time: "You had to move to something where the phenomena, the facts, were still unknown – it was very clear that this was urgent."[27]

In a field advancing as rapidly as physics, it is now not at all unusual for a group of collaborating scientists to note that they may be overtaken by events, and to decide to explore a more promising field. But rarely if ever in the history of modern physics was an entry into a new field made in a way so pragmatic, unsentimental, well-phased, and ultimately successful.[28] The decision of Fermi and his collaborators to change to nuclear physics out of all possible new fields was full of risks and would require great labor from everyone, since there was no experimental or theoretical background available to them – except in Fermi's own case, for example, by his study of Rutherford's work on the artificial disintegration of nuclei, referred to earlier, and of the influence of the magnetic moment of the nucleus on the hyperfine structure of spectra, published in 1930.

There were fewer than five years between the time when Fermi's group decided to enter the field of nuclear physics as novices and the time when this decision was carried out and results were achieved. Within this period, the major creative work took place during a few feverish months in 1934, and the group began to break up late in 1935.[29] During the period from late in 1929 to late in 1934, the sequence of events proceeded in a series of eight steps, all of which were, in retrospect, evidently necessary for achieving the goal. Of course, the Fermi group did not consciously proceed according to a rationally constructed grand plan.

First, a decision had to be made concerning the direction and magnitude of the effort, and it had to be carried to the highest level of government to obtain financial and administrative support. This is the chief significance of Orso Mario Corbino's speech on September 21, 1929, at the Società Italiana per il Progresso delle Scienze, on "The New Goals of Experimental

Physics."[30] Corbino – a senator of the Kingdom of Italy as well as a professor of experimental physics, and the director of the Physics Institute at the University of Rome within which Fermi and his group worked – carried to the public, the scientists, and the Senate the message that research in physics in Italy must change in the direction of research into nuclear physics.

Corbino named Fermi as the person to play a dominant role. The talk was "in fact written certainly with intimate collaboration with Fermi."[31] It carefully stated reasons why, by choosing one among all the fields of physics then existing, an opportunity could be created for contributing at the highest, internationally recognizable level, so that "Italy will regain with honor its lost eminence"[32] in physics. To this end, ". . . the only possibility of great new discoveries lies in the chance that one might be able to identify the internal nucleus of the atom. This will be the worthy task of physics of the future."[33] Traditionally, "the highest category in physics research" had been "the discovery of new phenomena,"[34] with x-rays and radioactivity being memorable examples of startling findings beyond the horizon of then current theory. But this style of research was exhausted, for "modern physics already possesses the basic knowledge of the possible phenomena which may develop or be produced experimentally on earth."[35] There was only one exception: "the field of artificial modifications of the atomic nucleus."[36] Moreover, new branches of physics would not arise until one obtained possible artificial modifications of the atomic nuclei.[37] Rutherford had "given us the only possibility of artificial transmutation,"[38] but by his method the effect occurred very infrequently.

"Will it be possible to attack the atom in some other way?"[39] It seemed to Corbino that accelerators would have to be built in Italy, as they were then in construction elsewhere. But the financial and technical difficulties were evident. At any rate, "One can therefore conclude that while great progress in experimental physics in its ordinary domain is unlikely, many possibilities are open in attacking the atomic nucleus. This is the most attractive field for future physicists."[40]

With a kind of national goal for physics research put before them – although, it is important to notice, only in general terms, not as a specific mission to do a specific experiment – the second step was for Fermi's group to engage in a process of basic, general self-education in the field of nuclear physics. For this purpose Amaldi was charged, in the autumn of 1931, to give a seminar in which the new edition of the basic work of Rutherford, Chadwick, and Ellis was read and studied.[41] The membership of the seminar consisted – in addition to Amaldi – of Fermi, Rasetti, Majorana, and Segrè.[42]

Third, the group had to remain scientifically alive and visible, and hence to publish, even during the transition period while shifting from atomic and molecular physics to nuclear physics. Thus, as late as 1933 Segrè and Amaldi still published largely on spectroscopy, though on subjects that in part were transitions toward nuclear physics.[43] Fermi, too, continued to publish prolifically – from early 1930 to his first experimental nuclear physics paper in 1933 – more than twenty papers in a large variety of physics subfields, usually alone, but sometimes with one collaborator. But the general direction of his motion, from the winter of 1930–31 onward, was toward nuclear physics, while waiting for the laboratory and the group to get ready – and for the opportunity to strike. Thus, as he was turning from his earlier predominantly theoretical work toward a period in which he would be predominantly engaged in experimentation, he studied the hyperfine structure of spectral lines caused by nuclear spin; it may be regarded as one of the "bridges" from the old to the new.[44]

During the same period, Fermi began his self-education in constructing instruments for research in nuclear physics. With the help of Amaldi, he tried to construct and operate a cloud chamber. To circumvent the poorly equipped and inefficiently staffed machine shop, Fermi had to

> make use of the "do it yourself" methods that were characteristic of him, both in theoretical and in experimental work. In order to minimize shop work and build a cloud chamber with his own hands, aided only by the most elementary tools, he first inspected several hardware stores and bought assorted kitchenware and gas plumbing.[45]

But by the spring of 1931, Fermi had to give up the project, and go back to theoretical work. It must have been a sobering experience, and may have contributed to the decision that members of the group would have to go abroad to learn techniques at established centers of research in nuclear physics.

The fourth stage saw a great, but temporary, dispersal. It was in fact the second phase of a two-phase set of expeditions. Rasetti had earlier gone to Millikan's laboratory in Pasadena to work on the Raman effect. Segrè had gone to visit Zeeman in Amsterdam to work on the Zeeman effect of quadrupole radiation. That was before the decision to go into nuclear physics. The second phase of the "expeditions" started in 1931, with Rasetti going to Lise Meitner's laboratory at the Kaiser Wilhelm Institut at Berlin-Dahlem and learning how to make a cloud chamber, prepare polonium samples and neutron sources, and make counters.

Segrè went to Hamburg to work with Otto Stern, and Amaldi to Debye's laboratory in Leipzig. Their purpose was not directly to learn nuclear work, but was somewhat more general. The intention was

> that we all would go to a place where you learn a new experimental technique, and bring them all back . . . with an eye to enlarging our field. We were even considering at a certain moment building a cyclotron . . . We knew that one had to learn vacuum technique; we couldn't make a vacuum of our own between all of us put together . . . And so then we would have more variety, more freedom.[46]

In the summer of 1931, Amaldi recalls, he met Rasetti and Segrè in Norway, and then they talked further with Fermi. They were getting ready for some important work. Precisely what was not yet any clearer than it had been in Corbino's speech, two years earlier. But the poor state of the current nuclear theory – the neutron had not yet been discovered – and the confidence derived from learning that the experimental techniques were quite manageable, combined to enable them to guess correctly where the bull's-eye of the target would be, once a specific target began to come over the horizon. As Amaldi put it: "Then we said that it was better that we concentrate our efforts in something that is in so primitive a state that there is a lot to be done. That was quite clear."[47]

The next step may have seemed a rather brazen one. Before

any significant work had yet been done by the Rome group in the field of nuclear physics, an International Congress on Nuclear Physics, organized by Fermi and sponsored by the Reale Accademia d'Italia, was held in Rome in October 1931. It was the first full-scale international conference on the subject. Most of the active experimental and theoretical contributors to nuclear physics came. The list included Blackett, Bothe, Bohr, Marie Curie, Ehrenfest, Ellis, Fermi, Geiger, Goudsmit, Heisenberg, Meitner, Pauli, Sommerfeld, and others.[48] This conference is usually regarded as the announcement by the Rome group of the inauguration of its work in nuclear physics.

By the autumn of 1932, Fermi and Rasetti were ready to pursue in earnest "a joint program of research in nuclear physics in Rome."[49] The research budget of the department had been raised to between $2,000 and $3,000 per year: "a fabulous wealth when one considers that the average for physics departments in Italian universities was about one tenth of that amount."[50] A large and excellent cloud chamber was designed, and constructed by a private firm of mechanics, as was a gamma-ray crystal spectrometer. The group tried its hand at making Geiger-Müller counters, using techniques brought back to Italy by Rossi after his visit to Bothe's laboratory. Rasetti prepared "a neutron source comparable to the most powerful ones then in use elsewhere."[51] Rasetti and Fermi, with their gamma-ray spectrometer, observed the gamma rays of mesothorium of 2.6Mev. Amaldi comments: "That was the only piece of work that had been done in nuclear physics [in the Rome laboratory]."[52]

In other laboratories, the situation was not a great deal better. That is to say, experimental equipment for nuclear physics research was generally still very much in an uncertain and often homemade state. Until 1932, the chief sources of particles for studying nuclear disintegration were emitters of natural alpha particles, either short-lived decay products of radium with heavy gamma radiation background, or the long-lived polonium, "which was difficult to come by (in fact, one practically had to go to Paris)."[53] Similarly, detectors for measuring nuclear radiation were quite unreliable. Decent counters were just being developed, and what was available was "still too noisy to be of much use."[54]

As the experimental facilities at Rome were being built up, the expectations for the right phenomenon coming along heightened. It was as if the group were poised for the moment when, in the words of Segrè, they could say: "Well, now we are all equal, because on this one, we have the source, and we start."[55]

The *annus mirabilis* in nuclear experimental physics was 1932, and for the physicists of Rome it must have been both exhilarating and disturbing. They were still on the sidelines. The existence of the neutron, Chadwick's great discovery which had been missed by the Joliot-Curies, had been recognized by E. Majorana in Rome in the Joliot-Curie experiments; despite urging, he had not published nor taken seriously enough these ideas. In the same year, the cyclotron of E. O. Lawrence and M. Stanley Livingston began to operate, ushering in the era of nuclear reactions by means of particle beams under the experimenter's control. Cockcroft and Walton's accelerator at the Cavendish Laboratory also went into action, with striking results (disintegration of lithium nuclei).

These successes at established centers might have discouraged other nuclear "beginners," and the continuation of the confused state of the theory of the nucleus during the early 1930s might have reinforced a feeling of discouragement in an ordinary group.[56] But Fermi at that very point demonstrated his mastery of theory with the completion in 1933 of his paper on the theory of beta decay. V. Weisskopf later called it "a fantastic paper, which I think stands out as a monument to Fermi's intuition . . . Beta decay, with Fermi's idea, stands apart from the rest of nuclear physics because it is the creation of particles."[57]

The approach and results of Fermi's beta-decay theory were sufficiently novel for the manuscript to be rejected by the editor of the journal *Nature* as containing abstract speculations which were too far from physical reality. Fermi obtained quick publication by sending it to *Ricerca Scientifica*, the weekly journal of the National Research Council (C.N.R.) of Italy, where it was published at the end of 1933. This journal had previously been used only rarely by Fermi and his group, but it was soon to become one of their chief outlets for quick publication of important results. Professor I. I. Rabi commented later that this "made the *Ricerca Scientifica* one of the most prized journals in physics, and caused all physicists interested in nuclei to learn enough Italian to follow these fascinating researches."[58]

And now, the moment for which the whole Rome group had been waiting – without knowing clearly what it would be like – that moment came: the announcement, at the January 15, 1934, session of the Académie des sciences in Paris, of the discovery by the Joliot-Curies that bombarding boron or aluminum with alpha particles produced in the targets new radioactive isotopes (of nitrogen and phosphorus respectively), with the emission of positrons. It was entirely unexpected; the discoverers had stumbled on it, and nearly missed it.[59]

For the first time, elements had been made radioactive in the laboratory "artificially." A whole new horizon opened up to nuclear physics – and essentially in the way the prophetic Corbino had envisaged in his speech of 1929.

The Joliot-Curies' publication reached Rome in about February 1934. The excitement seems to have been very great. G. C. Wick recalls the meeting of all members of the Rome group in Fermi's office to discuss the opportunities for research which were suddenly evident.[60] Suddenly, all the pieces could be fitted together after the long period of preparation.[61] It is striking that almost all the expertise needed came from within the group of collaborators who, for some years, had been together in the Rome group – with the exception of the chemist D'Agostino, who was brought in during this period and became an excellent collaborator.

The idea for exploiting the new effect in a new way was, as one might expect, Fermi's. The Joliot-Curies had produced their artificial radioactivity by using alpha particles as projectiles. Fermi however said "that obviously should be done with neutrons. Neutrons should be much better."[62] The means were already on their bench. Rasetti had prepared a neutron source of polonium-beryllium, drawing on his experience in Lise Meitner's laboratory. When it quickly became obvious that it was necessary, Segrè and Fermi were able to make much stronger radon-beryllium sources.[63]

It is essential to remember that Fermi's suggestion would elsewhere have been dismissed as unlikely at best and absurd at worst. O. R. Frisch has said:

> I remember that my reaction and probably that of many others was that Fermi's was really a silly experiment because neutrons were much fewer than alpha particles. What that simple argument overlooked of course was that they are very much more effective. Neutrons are not slowed down by electrons, and they are not repelled by the Coulomb field of nuclei.[64]

Amaldi recounts that they knew that they would get something interesting only if they could organize the work quickly and in an effective way, and made a systematic search for the effect.[65] With an additional grant of $1,000 from the C.N.R., apparently obtained easily and without any strings attached, they immediately set out systematically to try the effect of neutron bombardments on all elements. Thus started a period of activity that, for a time,[66] made the Rome group the leading one in the field, and exceptional in its internal organization.

The organization of the Rome group now differed from that of its own earlier phase and from the usual procedure in, for example, the Cavendish Laboratory. At the latter, the work was more dispersed, with one, two, or at the most three investigators, above the level of technical support, working together on one of the many rather different problems within the area of the laboratory's general field of preoccupation. As director, Rutherford would serve as a more or less occasional participant and adviser, once the main direction of research had been set for a given person or small group. Fermi's laboratory, on the other hand, now organized itself to focus all forces on one project and to concentrate all its expertise in one major effort, once that was identified as the area which would bring the maximum result for the laboratory as a whole.[67] The change here was from a wide-ranging group of subgroups to a group acting as a "team." It was an institutional innovation.

The effectiveness of the group was enhanced by an act of almost intuitive choice on the part of Fermi in assembling his group. Specifically, he had avoided having members of the group either overlap so closely in their competences and interests that they would begin to assert their own territorial imperatives, or to be so far apart that they would not overlap sufficiently. This balance has since been recognized as a key element in the assembly of a successful group of collaborators. As the chief theorist, Fermi himself was intellectually at the center of the group. This avoided the kind of problems which from time to time characterized work at the Cavendish Laboratory, where, it has been said, the experimentalists were rarely in touch with the theorists.

In the long run, the effect of this style of doing physics was felt profoundly everywhere, and not least in Italy itself. One can see the same style motivating and infusing the work during the period bridging the decay of Fermi's own group in Rome and the renaissance of physics in Italy after the war. U. Fano believed the "establishment was established" as a model for the period after 1934, and Ageno claimed he had learned from Fermi not only physics, but a whole style of laboratory work, including the organization and conduct of research.[68] Amaldi shows how he, Bernardini, Gian Carlo Wick and others, assembled after Fermi's departure and determined, during the war, to concentrate their group effort in Rome upon a carefully chosen, manageable and important problem, that is, cosmic rays.[69] Others participating were M. Conversi, E. Pancini, and O. Piccioni. The effectiveness and concentration of the group were essential for physics in Italy to emerge in a productive state at the end of World War II, when the vigorous high-energy physics research was rooted firmly in the cosmic-ray research which had been carried out just before and during the war.[70]

Early in 1934, the first fruit of the Italian attack on nuclei with neutrons was the discovery of artificial radioactivity in fluorine and aluminium. This was published in March 1934 under Fermi's name.[71] In rapid succession, forty of the sixty elements the group irradiated revealed the existence of at least one new radioactive isotope. A new one was found every few days; the members of the group looked back later to this as the most glorious and satisfying part of their lives.

Publication after publication followed rapidly. Almost casually, two institutional "inventions" were made at that point. One was multiple authorship: after the first two papers by Fermi alone, he drew in the rest of the group, and as many as six authors appeared under the title of the publications – an unusually high number at the time.

Another "invention" was the preparation and mailing out – to a list of about forty prominent physicists around the world – of what now are called "preprints" of articles in press at *Ricerca Scientifica.*[72] In this way – improving even on the journal's remarkably fast publication process of about two weeks – Fermi's group could make its discoveries known in printed form within days of the work being finished, and could send the preprints to active nuclear physicists who might not have access in their own libraries to the journal itself. Amaldi remarks: "This procedure was facilitated by the fact that my wife Ginestra was working at that time at the *Ricerca Scientifica.*"[73]

The comments from scientists abroad showed quick acceptance of and interest in the results from Rome.[74] In Rutherford's letter to Fermi, the line "You seem to have struck a good line to start with" may seem to have the somewhat patronizing overtone of welcoming a newcomer to experimental nuclear physics; but that was undoubtedly not intended. Rutherford was certainly right in one implication: After a long period of preparation for the transforming event which he could only have dimly foreseen, Fermi had indeed "struck gold."[75]

The discovery of the existence and effectiveness of *slow* neutrons follows in about six months. In a sense, it was only a matter of time before the Italian group came upon it, but between their discovery of the effectiveness of neutrons in causing radioactivity in fluorine as the first instance, and the determination that slowed-down, that is, moderated, neutrons can produce vastly increased and quite different activities, everyone seems to have thought that the more energetic the neutrons, the greater would be their effectiveness, that slow neutrons would have only a small capture cross section. All excitation curves known at the time for reactions produced by protons, deuterons, and alpha particles showed a rapid decrease with decreasing velocity of the particle inducing the reaction.[76] This was a consequence not so much of orthodoxy as of the deficient state of an incomplete theory; the effect of photomagnetic capture on the total capture cross section for neutrons was not yet known.

And yet, partly by his own work one and one-half years earlier on the paper which dealt with the effect of slow electrons, Fermi may well have been pondering the possibilities for slow neutrons, and at the very least was sensitive to clues on this point. When the "mythological event" did take place in October 1934, it took him very little time to develop the correct beginnings of a theory for the role slow neutrons play in activation. He and his group thereby entered into what has been called a heroic pioneering period of startling results, which served, within a period of a few months, to reinforce and vastly strengthen the position this group had already achieved in physics. Not only was physics changed thereby. Scientifically, Italy had indeed regained "with honour its lost eminence," as Senator Corbino had hoped. And world history itself had been turned in a new direction.

Fitting into a tradition

Fermi's mastery of physics and his style of research may not have been enough in themselves to produce the transformation in physics in Italy. With these characteristics alone he still might have had the same fate as Amedeo Avogadro, the first holder of a chair in theoretical physics in Italy (*fisica sublime*), whose contributions to chemistry are, of course, very well known, but who neither received adequate recognition in his time nor formed a school. From the beginning Fermi understood the tradition within which science was carried on in Italy and took good advantage of the system which existed for the institutionalization and professionalization of science. "Fermi was fully aware of the system and was eager to reach the top as fast as possible."[77] So when, on returning from Leiden, Fermi "looked eagerly for a job," he undertook to

> advance his career through a substantial number of publications. Thus he wrote (in Italian) as many papers as he could. Although he kept his standards high, it is clear that he counted them carefully and felt satisfaction in seeing the pile of his reprints mount ever higher. He wanted to reach the next step in the academic career, the *libera docenza*, as rapidly as possible, and he believed that the sheer number of publications was important – especially if the judges should be too lazy, or unable, to assess the value of his contributions.[78]

An interesting aspect of Fermi's sensitivity to the tradition and to the existing opportunities within the tradition was his ability, in an early and crucial stage of his career, to seek out and respond to mathematicians, and to obtain support and patronage from that group. Fermi had studied mathematics, even with passion, but chiefly because he considered it necessary to the study of physics, "to which I want to dedicate myself exclusively," as he explained to his first mentor, Adolfo Amidei, in 1918.[79]

But physics, and particularly theoretical physics, had fallen into a poor state in Italy by 1920. The last "great" Italian physicist had been Volta, in the early part of the nineteenth century. Righi had just died – in 1920 – without a clear successor to his position as Italy's leading physicist. At the very time when physics in Germany, France, Great Britain, and the United States was proceeding in a lively manner, and remarkable, even revolutionary ideas were being brought forth elsewhere, Italy was, from that point of view, a "provincial backwater in which practically no one was active in current developments."[80]

Physics in Italy was indeed weak institutionally by the usual measures, including variety and strength of the professional societies in that particular field, numbers of its members, and number of chairs at its universities; except for Rome, where there were two physicists, there was only one chair in each of the other major Italian universities, and the turnover of professors was about one in a decade. Institutional weakness was also evident in support of research, flow of students, quality of research journals, number and distinction of national and international conferences, influence of a substantial number of scientists of that field within government, and attention given them by the press and public. It is not without significance that the only Nobel Prize in physics awarded to an Italian before Fermi in 1938 was the prize – half share – for 1909 to G. Marconi for radiotelegraphy – and that, if anything, was something of an embarrassment to many Italians, since he had not found support in Italy.

Fermi was probably aware of some good and even distinguished work done, at that time or in the recent past, by certain Italian physicists such as Bartoli, Puccianti in Pisa, Garbasso in

Florence, and Corbino, who had been a student of Macaluso in Palermo and who was brought from his native Sicily to Rome by Blaserna. Of these, Corbino was perhaps the most outstanding. But while knowledgeable and interested in many branches of physics, he too was no longer active at the time. Fermi must have seen fairly clearly from the beginning that the state of Italian physics was on the whole not excellent, for on entering the Scuola Normale Superiore in Pisa in 1918 he found that he already knew most of the subjects which were being taught, that he had to learn modern physics on his own, and that there were only two other students in physics – Franco Rasetti and Nello Carrara.

It is, therefore, indicative of his promise as a future statesman of science that immediately after graduation from Pisa, on returning to Rome, he associated himself with and was indeed acceptable within a circle of distinguished mathematicians which centered on Castelnuovo, Levi-Civita, and Enriques.[81]

The state of excellence in mathematics in Italy at the time, as indicated by names such as Cremona, Peano, Severi, Volterra, Enriques, Ricci-Curbastro, Levi-Civita, and Castelnuovo, was indeed high. These men had advanced Italy into the front rank of mathematics and brought it international recognition. Most universities had several chairs in mathematics. The number and quality of students and of journals were good. Therefore, just as Fermi's early students and colleagues were to come from a field outside physics which had promised a better career – Segrè, Rasetti, Amaldi, Persico, and Majorana were all initially headed for engineering – Fermi himself had to make his way into a tradionally weak academic field in Italy, and had to do so through support derived from a strong and vigorous one.

Dealing with academic politics and national ambition

Fermi's ability to use opportunities offered by academic politics and by current nationalistic ambitions was also of some significance. It is commonly held that Fermi was deeply apolitical. But this estimate cannot apply to the politics of university life on which, for better or worse, depended the creation and maintenance of a new and relatively costly establishment such as Fermi's group.

We know from Fermi himself that almost immediately after he returned to Rome following graduation from Pisa, he arranged to present himself to the director of the Physics Laboratory at the University of Rome, the man then thought to be not only the most prominent Italian physicist, but also the foremost Italian statesman of science – Orso Mario Corbino. In Fermi's words: "I was then 20 years old; Corbino was 46. He was a senator of the Kingdom, had been a minister of public instruction, and was universally known as one of the most eminent scholars."[82] Fermi found himself immediately welcome, and they discussed Fermi's studies: "In that period we had almost daily conversations and discussions which not only clarified many of my confused ideas, but aroused in me the deeply felt reverence of the pupil for the master. This reverence steadily increased during the years I was privileged to work in his laboratory."[83]

For the subsequent history of this case, it was a crucial encounter. Corbino[84] understood physics well, having done interesting experimental work himself; he was now effectively involved in industrial consultation, administration, and political work; he had become senator in 1920 and minister of public instruction in 1921; although not a member of the Fascist Party, he was also briefly minister of national economy until 1923. In Rome, where all the power in Italy was centered, Corbino understood and enjoyed the wise exercise of power. "An excellent speaker, he was endowed with sparkling wit, and his intellectual traits were complemented by a warm and generous personality and a propensity to academic maneuvers. He liked to arrange promotions, transfers, and the like, and usually succeeded in these attempts."[85]

But most significantly, by 1922, after nearly a decade of having made no contributions to physics himself despite his exceptional ability and deep, continuing interest, Corbino was passing through a serious personal reassessment. During a discussion, as minister of public instruction in the Senate in 1922, he exclaimed:

> Now, Honorable Senators, I also have passed through a crisis which I want to prevent for my colleagues of

> tomorrow . . . I still yearn for science. I yearn above all, in the bitterness of political action, for the peaceful days passed among experiments and apparatus. And I regret that, after the death of Augusto Righi, Italian physics has been unable to find his successor.[86]

It is not difficult to see that on encountering the evident genius of Fermi within this framework of his own preoccupation, Corbino, the politician, entered into a state of symbiotic collaboration with Fermi, the scientist; this was not less effective for being probably largely unconscious. It was as a result of Corbino's efforts that, despite the essential rigid structure of the Italian university, a chair in theoretical physics was created at the University of Rome, and in such a way that it could be filled only by Fermi. Soon thereafter, again through Corbino's influence, Fermi became the only physicist elected to the Royal Academy of Italy, at the age of twenty-eight.[87] Similarly, Corbino was the moving force behind the administrative actions necessary to obtain appointments and funds for every one of the members of Fermi's circle, including the creation of yet another chair in physics, in 1930, for Rasetti[88] – in spectroscopy, at precisely the time when Rasetti was leaving spectroscopy and going into nuclear physics.

The support provided by Corbino, and, through him, by the administration of the Italian state, was such that Fermi's group became essentially a protected operation. Without this protection it is almost inconceivable that young Fermi could have weathered the opposition from hostile forces such as the other professor of physics at Rome, Lo Surdo, and from other established persons who, for many years, did not look kindly upon Fermi's growing activities. For Fermi, Corbino was a godsend not only as a protector but also as administrator – just as Fermi later had George Pegram as a sympathetic administrator at Columbia, Arthur H. Compton at the Metallurgical Laboratory at Chicago, and Samuel Allison at the University of Chicago. The appointment of Lo Surdo as director of the Physics Laboratory after the death of Corbino in January 1937 signaled danger to the institutionally fragile group, and may have had an effect in turning Fermi's thoughts more strongly to emigration.

Corbino's influence on the growth and activities of the Fermi group is everywhere evident – for example, in the change of direction of the group's work as indicated in his speech of 1929, and on such occasions as his opening speech in 1931 at the Rome Conference on Nuclear Physics or his talk at the royal session of the Accademia dei Lincei in 1934. Often he spoke at the risk of offending other interests; thus Segrè recalls that Corbino "was jumped on terrifically by all the physicists in Italy" – except for Fermi's collaborators – after his speech in 1929.

Corbino even busied himself with drawing students' attention to Fermi at a time when there were almost no students in physics at the University of Rome. As Amaldi recalls:

> He started to make propaganda . . . For instance, in my case, I was Corbino's student in my second year . . . Then he stopped five minutes before the end [of his lecture] and said: "Now some of you should stop your study of engineering and you should go into physics, because we now have a new professor of physics here in Rome . . . I can assure you that this is the man who can bring physics to a high level in Italy. In this moment, the young people should go into physics."[89]

Corbino also was of considerable importance to Fermi's group since he talked frequently and informally with all of them as friends; thereby he "contributed materially to their education, and not only as scientists."[90] Last, but not least, Corbino either outlined a sort of epistemology for Fermi and his group, or at least in part adopted and echoed Fermi's. For example, Corbino in 1929 pointed out some of the weaknesses of the traditional Italian ways of doing physics, including the habit of looking for new and unforeseen phenomena "without any aid from the theories which existed." He predicted that the future development in experimental physics must be coupled more strongly with the state of the theory, and also that modern research in any country must be carried on in full awareness of the deepest tendencies of scientific research elsewhere in the world.

While not a nationalist, Fermi was not insensitive to the opportunities opened up by national ambitions. By temperament and training, he saw himself as a member of the world community of scientists. But his early, somewhat discouraging experiences during his stay in Göttingen with Born, Heisenberg,

Jordan, and Pauli in 1922–23 may have shown him that his expectations for his own career, and the hope that Italy might achieve first rank in physics, could be brought together to mutual benefit.

Fermi considered the possibility of emigrating in 1922[91] when he saw the Fascists becoming well entrenched. Upon his return from his trip to Germany and the Netherlands, he appears to have brought back an appreciation for the need to supplement and replace the old Italian tradition, one dissociated from theoretical advances which were mostly made abroad. He would therefore have agreed fully with Corbino – if, indeed, he was not himself responsible for many of the ideas in that speech – when Corbino said in 1929:

> Considering that collaboration between theoretical and experimental physicists in Italy is only now beginning, and that we are far from having the lavish means in our laboratories which other countries possess, it is not surprising that Italian physics was not able to contribute more to scientific progress in this great period of renewal. If we can correct these two deficiencies, Italy will regain with honour its lost eminence.[92]

This theme was of course a factor operating not only in the genesis of Fermi's group. Indeed, it formed a very integral part of contemporary Italian life and culture. Every child was being exposed to an atmosphere of grandiloquent nationalism which extolled the glories of the past, stressed analogies between the hoped-for future and the Roman Empire, and dwelt on the list of exceptional figures in recent Italian history. But the focus of national hopes on Fermi was nevertheless deliberate and early. It had found expression in the report of the committee which, on November 7, 1926, awarded to Fermi the first place in the competition for the new chair of theoretical physics at the University of Rome. The report was Corbino's work, and he wrote there about the person whom he probably regarded as the new Righi:

> He moves with complete assurance in the most difficult questions of modern theoretical physics, in such a way that he is the best-prepared and most worthy person to

> represent our country in this field of intense scientific activity that ranges the entire world. The Committee thus unanimously finds that Professor Fermi highly deserves to have the chair of theoretical physics, the object of this competition, and feels it can put in him the best hopes for the establishment and development of theoretical physics in Italy.[93]

The degree to which free and rational science conflicts with the ideology of totalitarian states has often been the subject of inquiry. In Italy, at any rate, the C.N.R. and the Royal Academy, in both of which Fermi served and which in turn supported his work, were used by the Fascist government to supplement the existing Italian Physical Society and the Italian Association for the Advancement of Science. One experienced observer has gone so far as to say that "Fermi would have been almost impossible in Italy without Mussolini."[94] But whatever the character of a regime, if it was desirous of supporting science, it would have had to invent some way of dealing with Fermi, given the absence of a clear tradition for his subspeciality.

For his part, Fermi was in principle not interested in ideologies. One component of the scientific tradition in every country, for better or worse, has been for the members of the profession to accommodate themselves by and large to whatever government is in power. Robert Hooke's draft preamble to the statutes of the Royal Society of London declared that the scientists of that day did not intend "meddling with Divinity, Metaphysics, Morals, Politics, Grammar, Rhetoric, or Logics." When Grand Duke Leopold II, himself elected a fellow of the Royal Society of London in 1838, consented to the calling of the influential first Riunione degli Scienziati Italiani at Pisa in 1839, he granted permission only on the condition that "neither philosophy, nor politics, nor history, nor eloquence, nor poetry, nor legislation, nor public economy, nor public administration" be discussed.[95]

Fermi probably would have found this natural and acceptable, as did, for other reasons, those in the state administration who had to support him. Nor did Fermi care to take up causes not directly related to his work. For example, in Italy there was a popular prejudice against science as a possible source of irreligious beliefs and as antithetical to genuine culture. Fermi did not enter into this debate.

On the other hand, there was a long tradition in Italy of having scientists drawn into the councils of government, which resulted in a freer and more natural association of political leaders with Italian scientists and scholars than was the case in many other Western countries.[96] The line leading from Blaserna to Corbino and then to Fermi was only one example of such cases.

Another example of the deliberate effort to connect the work of the Fermi group with the desire to meet national needs may be seen in the passage toward the end of Corbino's speech of 1929 in which he pointed out supplementary benefits which might accrue from greater support of modern physics in Italy. He noted that the theory of relativity – of which Fermi himself was one of the chief exponents in Italy – could be turned to great use to provide abundant energy: "In these nuclear phenomena, the transcendental importance of which does not need to be emphasized, one would transform matter into energy and vice versa to the tune of 25 million kilowatt hours for every gram of transformed matter."[97] If this should not come quickly, Corbino hinted at other advantages: "Thus, even if physics were to move toward a saturation level, the study of its application to other disciplines, such as biology, if conducted by true experts with the resources of modern physics, could bring results of the greatest scientific and practical value."

These have, of course, been the arguments brought out quite regularly by physicists when they have sought financial support from society. What is impressive, however, is the foresight of Corbino, whose remarks precede those of other physicists such as E. O. Lawrence in their search for finding a legitimate basis for the support of research in physics in the solution of practical problems. Fermi rarely had to speak up this way on his own behalf, as long as Corbino was at hand to do it for the group.[98]

Finally, in listing the various ways in which Fermi's career and Italy's national ambitions coincided, we cannot avoid mentioning the hypothesis that, somewhere in the minds of the supporters of Fermi's group, there may have been an unexpressed hope that they would gain the kind of recognition which comes with the award of a Nobel Prize to a country which is regarded as scientifically of the "second rank." The local effects of such an award as that for physics in 1907 to a physicist in the United States and in 1949 to one in Japan have not been "measured," but they were considerable in terms of the great increase in national self-confidence in that field of science. The appropriateness of Fermi for the eventual receipt of such a prize must have become more and more obvious in early and mid-1930s, and may well have been entertained even earlier by alert, politically active persons such as Corbino.[99] To be sure, if this example may seem to provide a sort of recipe for the way a scientifically "developing" nation might go about securing a chance of winning a Nobel Prize in a field of science, it is still obvious that by far the chief requirement is first to have a Fermi appear – and then to hold him.

The internationality of Fermi's group

Charles Weiner has coined the useful term "the travelling seminar" to describe the lively interaction of European and American physicists in the 1920s and 1930s, even before the exodus from German universities as the result of the rise of Nazism.[100] Scientists, and particularly physicists, had developed to a remarkable degree a sense of being engaged in an international undertaking. Weiner writes:

> During the years immediately preceding the rise of the Third Reich, Europe was bubbling with intellectual activity in many fields of scholarship. Physics was exceptionally ebullient . . . European physicists and their students were constantly in motion, travelling back and forth to exchange newly born ideas. As today, travel and communication were essential aspects of the life of the physicists, contrary to the folkloric image of the scientist locked up in his laboratory, uninterested in personal interactions. Major stops on their itineraries included universities and laboratories in Munich, Leipzig, Göttingen, Leyden, Zurich, Cambridge, Copenhagen, and

> Berlin. In the 1920s it was essential for physicists to have first-hand experience of the work being performed in the capitals of research.[101]

At the Cavendish Laboratory, about 1930, for example, there was a continuous stream of visitors. Some came for two or three days to give a talk, and some would come for two or three years to take a degree. John Cockcroft recalled that "roughly half the laboratory group were from overseas."[102]

Otto Frisch, one of the several hundred German scholars, scientists, and advanced students who found themselves abruptly dismissed when the racial laws were put into effect in Germany in the spring of 1933, went to work in the laboratory of Blackett in London, joining there a group of physicists. "We called ourselves the League of Nations, because there were so many different nations represented. There was a German and a Swiss and an Italian [Occhialini], and one chap was a Greek from France, and an Indian, and so on."[103]

Fermi seems to have understood the international character of physics from the beginning. For example, he undertook the study of languages for this reason. Amidei recalls: "I advised him [about 1917] – and he immediately followed my suggestion – to study the German language, because I foresaw that it would be very useful for reading scientific publications in German without having to wait until they were translated into French or Italian."[104] His own trips to Göttingen, Leiden, Copenhagen, and the United States at early stages of his career helped to express and reinforce that sense of belonging to an international community of physicists – a community in which Italy still had to make its way.

In fact, good use was made by Fermi's circle of all four chief modes – embodied more or less in formal institutional arrangements – for expressing and using the internationality of science. One was traveling abroad, if necessary with a fellowship such as one from the Rockefeller Foundation or from the Italian government, in order to learn techniques and to obtain speedy publication abroad – this was part of the reason for the trip of Segrè and Amaldi to Rutherford's laboratory in the summer of 1934. There was also the need to measure oneself against the knowledge and plans of others, and to strengthen the bonds of international colleagueship. Even D'Agostino, the chemist who had been induced to join Fermi's group, had learned his techniques at the Joliot-Curies' laboratory in Paris.

A second institutional mode consisted of the attendance at and the conduct of international conferences, starting from that in 1927 at Como, and including the conference on nuclear physics in Rome in 1931. For the latter the funds had come from Italian industry, and the ubiquitous Corbino had presided over the negotiations for the grant. Many of those who came to Rome had been, or in turn would soon be, hosts to visits from members of the Fermi group. These facts indicate a peculiar aspect of the Roman internationalism. The collaboration between Fermi and Corbino produced a kind of ad hoc internationalism, as homemade – and as effective – as their radioactive sources used for research, and rather different from the formalized internationalism in other parts of Europe, such as at the Solvay Congress, and the conferences of international unions or professional societies.

A third mode was the provision of hospitality in Rome to visitors from other countries. Sergè recalls: "We had had a tremendous number of visitors who had brought in a lot of life. We had had Bethe, Bloch, Peierls, London, Feenberg [and Placzek and Bhabha]. By 'visitor' I mean someone who spent some time really and participated in the life of the place [rather than someone who came just for a day, as Raman and many others did]."[105] Others who belong to that list include S. Goudsmit, C. Møller, G. Uhlenbeck, and E. Teller. There were a number of visitors in Rome on Rockefeller fellowships in physics or mathematics.

It is, however, significant that despite all these visitors, the group in Rome differed fundamentally from most other major physics centers insofar as there was no place in it for long-term participants from other nations. Unlike the situation in London, Cambridge, Paris, Copenhagen, Hamburg, or Berlin, Rome was not a physicist's "league of nations." The work was done by a relatively small, tightly knit group of Italians who, as scientists, had practically grown up together. When in the spring of 1933 a vast number of German scholars and scientists were dismissed and many institutions and individuals outside Germany bestirred themselves to find work and shelter for them[106] – long before the Ethiopian War and the racial agitation overtook Italian politics – the Rome group was not considered, or did not consider itself, ready and able to provide more than a very transient refuge for a few. Nor did it exploit this opportunity to acquire the kinds of talents and expertise which helped to transform many laboratories outside Italy, converting them from ephemeral teams into true institutional centers with an international composition.[107]

The fourth feature of the institutional internationality of science is publication abroad and the sending of preprints to a list of colleagues in many countries. The policy of Fermi's group as to publication was carefully balanced to include publication in foreign, as well as in Italian, languages and journals. Only when great speed of publication became essential in 1934 during the period of the rapid growth of the list of newly identified radioactive isotopes, was publication undertaken in *Ricerca Scientifica*, where there was no "refereeing" to delay publication and where preprints were made available very promptly for distribution abroad.

Apart from these four modes there was pressure owing to the incentive to realize Corbino's – and others' – hope of placing physics in Italy in a prominent position on the world map of science. It may have been partly for this reason that, apart from collaborating with short-term visitors from abroad, the Fermi group chose to remain a purely Italian group.

As we saw, the choice of problem and methodology of research, by which the artificially radioactive isotopes produced by neutrons were discovered, did not spring from some conceptual "crisis" in science. On the contrary, once the Italian group had chosen to enter into international competition for recognition as a major center for nuclear physics – itself a decision prompted not by a "crisis" but by the feeling that the earlier fields of physics and expertise were becoming exhausted and therefore boring – the chief sense of crisis was the threat of missing the main chance, or of being beaten to it. As it turned out, this unspoken pressure, which the Italians had put on themselves, paid off magnificently, not only for science in Italy, but, by the very nature of science, for science everywhere.

Social organization of the group's activity

There was a superb match – and a mutually beneficial influence – between the structure and organization of the group of scientists in Rome on one side, and their scientific ambitions or programs on the national and international level on the other.[108] There appear to have been a half-dozen factors in the operation of the Rome group which, acting together, characterized its social structure and made in a large measure for its effectiveness.

Isolation and protection. The group's relative isolation and protection from outside effects was striking and owed much to Corbino as an administrator and a politician. As Segrè said:

> Perhaps this is an oversimplification, but you should put it so: that Fermi was the scientific and technical leader, and Corbino took care of finding the money and the administration, and so on . . . It implied more politicking than you would expect because Fermi had enemies in Rome, had a lot of people who were his enemies, and since they couldn't hit Fermi, they were rather inclined to hit smaller fry.[109]

For example, Lo Surdo at Rome considered it a personal affront when Fermi was called to Rome. There were others, "and one of the important functions of Corbino was to neutralize all these people."[110]

To a degree, Corbino personally played the role which conventionally, in a better-established scientific discipline, would be taken by institutional arrangements. In any case, Enrico Fermi was able to maintain a personal sense of distance from political combat. All reports agreed that he was content to leave authority unchallenged, as long as a good working atmosphere was available. As the grandson of a farmer who had become county secretary to the Duke of Parma, and the son of an administrator in the railway system who had had a very satisfactory career, Fermi had in some measure the point of view of a civil servant; he had little desire to invest energy in what must have seemed to him the thorny and perhaps insoluble problems of science and society.

There was another benefit the group derived from its isolation and its protected position. This was a period of great achievements and ebullience in physics; but at the same time the social, economic, and political conditions of Europe were in disorder. Not the least problem that anyone interested in or working on physics faced, in Italy as elsewhere, was that there were hardly any career expectations in the field, other than perhaps in secondary school teaching and to a very limited degree in the universities. Most of Fermi's associates had switched from engineering studies, which traditionally led to much more secure employment in Italy.

Under Fermi, they settled for relatively poorly remunerated posts as assistants, and they continued to do that for a considerable time. Amaldi remarked later that they simply did not worry about this as a problem.[111] It probably helped that almost all in the group, excluding Fermi himself, came from professional families which were relatively comfortable financially. They lived modestly – and, in Amaldi's case, by supplementing his income through the income earned by his wife as editor. Amaldi added:

> We were not interested at all in that period to make a quick career. I was not at all interested and this is more or less the general idea. We felt it was so nice what we were doing, so agreeable, so wonderful, that there was no reason to go away and get a better position. I remember this because on this we had very long discussions. One of us said, "Well, after all, what is the reason to get a better position, to go in a place where you don't work! Here it is so nice, this is the thing that we should do, this is where you live and work . . ." We were so convinced that we were doing nice things.[112]

In this particular group, as in some others at the time, research was carried on in the manner which allows it to be regarded as a "charismatic" activity, rather than as a "mere" academic career.[113] Ben-David is, I think, right when he refers to the incidence of high scientific achievement as charismatic in the sense that it possesses the features of an expression of the "deepest and most essential qualities of a specially gifted person."[114]

It is, however, also true that Fermi's pragmatic and skeptical realism modified the "metaphysical pathos" which tended to be associated with the pursuit of science on the other side of the Alps. Nevertheless, the creative and physical energies which Fermi aroused in his associates and in himself shows that, within the protective environment provided for them, in the eye of a gathering hurricane they could carry on science in the most elevated manner.

Small and flexible group. More needs to be said now about the formation of Fermi's group, for it did not follow the lines of a well-accepted model. In 1932, Frisch remarks, "In Europe there were few laboratories in which nuclear physics research was conducted, and I think the word 'team' had not yet been introduced into scientific jargon. Science was still pursued by individual scientists who worked with only one or two students and assistants."[115] By 1932, however, Fermi had already built up a group which, while small (by present standards), was agile, widely trained so as to be ready to pounce on the important problem, and willing to wait for it.

When Fermi had to, he could run a larger group, too, as he showed later at Chicago, although not without grumbling. A small group of high quality seemed to fit his personality better. It was not his style either to build an institute in the manner which has since become fashionable, with a wide range of major facilities which others would come to use for their own research problems, or to build a teaching center which would attract and train large numbers of students from near and far. He liked to invest his energy in a group which provided personal contact on physics problems which were of direct, current relevance to his research interest. The efficiency of such a group tends to decrease, and the organizational problems increase as the group increases in size.

A small group also may have the edge in the ability to exploit a lucky break, or even to recognize it properly when it does come along accidentally. The advantage of Fermi's group in this respect was shown precisely when Amaldi and Pontecorvo were assigned the task of producing reliably reproducible intensities of artificial radioactivity from silver where a neutron source was used to bombard it. It was during this work that they stumbled upon an effect which ultimately, in the mind of Fermi, produced the hypothesis that slow neutrons may have a much bigger capture cross section than anyone had thought. The key observation was that of the "miraculous" effect of a wooden table in one part of the laboratory, on which the activation experiment gave large yields, as against a marble table on which the experiment gave small yields. (The effect was due to the slowing down and scattering back toward the target of neutrons which were heading down into the hydrogenous material, i.e., wood.) It was just this observation which led to the explanation of the effect attributable to slow neutrons, upon which the great advances in the last part of 1934 depended. But it is also just this kind of "lucky break" which, if made in a large and busy laboratory, might well not have come to the attention of the person whose contribution is essential to change the observation from a bothersome irregularity – which indeed could have been easily removed, for example, by carrying on all observations on one table only – to a transforming discovery.

Cohesion by recruitment. From the beginning, Fermi's group was different from any other known to me in the history of physics. It was characterized by a cohesion which stemmed from an extraordinary history of recruitment and apprenticeship. Fermi, of course, had offered himself for, or at least consented to, "recruitment" by Corbino, and after arriving in Rome in 1926, Fermi, with Corbino's support, proceeded to recruit a group of a half-dozen crucial collaborators. Rasetti, a schoolmate and friend of long standing, an ex-engineering student at Pisa, was brought in as Corbino's assistant (*aiuto*). Rasetti has described his earlier intellectual recruitment by Fermi:

> I must say that I discovered Fermi very early . . . perhaps a month or two after we had registered [October 1918] . . . I had seen this student who was sitting next to me. We got started to talking accidentally. It took me just a few days to realize the absolutely exceptional nature of Fermi.[116]

In Florence, Rasetti and Fermi were together for about two

years. "We were practically together from morning till evening."[117]

Segrè was recruited in 1927, and his story tells vividly how acquaintance or friendship preceded the relations built up during active collaboration in physics, rather than the other way round, as is usually the case. Segrè said that, in 1929,

> Several things happened at once . . . Rasetti came to Rome; and Giovanni Enriques, who was the son of the mathematician, decided to bring me and Rasetti [along], because he wanted to explore some mountains in Central Italy – I was the only one with a car . . . so we went, all three, and in this way I got acquainted with Rasetti, who was a young physicist.[118]

Segrè had once heard Fermi speaking about quantum theory and was impressed: "It was absolutely plain that here was a man who knew what he was talking about." Either Giovanni Enriques or Rasetti suggested to Fermi that Segrè was a possible person to bring from engineering into physics; so Fermi and Segrè, in the summer of 1927, met in a very characteristic fashion:

> We went to the seashore together several times, swimming and bathing, and so on. Then he would ask me, "Well, how would you do this?" . . . He had a heavy, dangling rope attached to one end, and I had to study the vibration . . . He wanted to see whether . . . he had found a man to bring in, in physics . . . I was also smelling him, more or less; it was a reciprocal process.[119]

Segrè in turn recruited Ettore Majorana, his schoolmate in engineering, whom he considered a prodigy, "through direct dealings between myself, Fermi and Rasetti."[120]

Persico, who was also an engineering student for the first two years at the University of Rome, had of course been a boyhood friend of Fermi, originally a schoolmate of Giulio, Fermi's brother, who died in January 1915. The loss of his beloved brother affected Fermi deeply, and he turned simultaneously to Persico and to a dedicated study of science, an interest both shared.[121] Pontecorvo was a close family friend of Rasetti, and transferred from Pisa to Rome in order to continue his studies under him.

Amaldi was recruited in part by Corbino's talk about Fermi in Corbino's class, but the ground had been prepared in a characteristic way. Segrè recounts:

> Fermi spent the summer of 1925 in the Dolomites. As usual, several of the mathematicians from Rome and their families were there to escape the heat of the plains, Levi-Civita. Castelnuovo, Bompiani, Ugo Amaldi, and the younger Francesco Tricomi. R de L. Kronig, a young, brilliant physicist, also joined the company, and he and Fermi went on long hikes with Ugo Amaldi's 17-year-old son, Edoardo, who had just finished high school. The boy was fascinated by their conversation, although he understood very little of it. Later, when Kronig left, Fermi and Edoardo Amaldi, the strongest athletes of the company, went off together on a strenuous bicycle tour of the Dolomites.[122]

It is not without significance that the wives of Fermi and Amaldi were both, before their marriages, students at the University of Rome, Laura (Fermi) taking Corbino's course of electricity for engineers, though a student of general science, and Ginestra (Amaldi), an astronomy student, taking a course in the physics laboratory with Amaldi and Segrè. Fermi and Amaldi were both at Laura's parents' house when there was a gathering of young persons. Another meeting place was Professor Castelnuovo's apartment, where there was an open house every Saturday evening to which, in the Italian manner, his friends and colleagues, such as Volterra, Levi-Civita, and Enriques, would come with wives and children for "a few hours of informal chatter among congenial friends."[123] There Laura would meet her classmates and teachers. Fermi, Rasetti, and Segrè also would attend.

It emerges from this description that bonds of cohesion were forged very early, in a setting determined by the particular structure and customs of Italian society at the time as much as by shared interest in the subject of physics. If one compares the recruitment of groups of collaborators in other centers of physics in Europe at the time – where scientists joined predominantly after their professional promise and formative education had already occurred, or where, as at the Cavendish, they came from the United Kingdom, Australia, the Soviet Union, the United States, and the far corners of the earth – one realizes that we are dealing here not so much with a research group as with something closer to a family.[124]

Family-enterprise model. Fermi himself had come from a closely knit family and was very attached to his brother and his sister. His own fate was determined decisively by the decision of a colleague of his father, Adolfo Amidei, intellectually to "adopt" Fermi at the age of thirteen. He had noticed the young boy in a manner which itself gives a glimpse into the structure of society at the time. Amidei recalled that in 1914 he and Alberto Fermi, his colleague in the Ministry of Railways, after leaving the office at the end of the day, "walked together part of the way home, almost always accompanied by the lad Enrico Fermi, my colleague's son, who was in the habit of meeting his father in front of the office. The lad, having learned that I was an avid student of mathematics and physics, took the opportunity of questioning me. He was 13, and I was 37."[125] Amidei's intervention not only provided Fermi with books and someone to talk about science with, but also a mentor to help plan and shape his career, not entirely in agreement with the plans of his parents. One sees here a model not too different from that of the way Segrè and Amaldi were adopted by Fermi.[126]

We may think of Fermi as the center of a set of planetoids, representing the students and collaborators whom he had recruited, Fermi himself having a position like that of Jupiter surrounded by its moons. The Rome group preferred a different model – the light-hearted analogy with the ecclesiastical hierarchy, in which Fermi's central position was indicated by his nickname "il Papa," because of his infallibility in quantum physics. In addition to the chief collaborators, there was also a changing group of students.

This whole system, however, itself rotated about the center of that universe – Corbino, acting as the sun. If Corbino was not the father – the Rome group did refer to him as "Padre Eterno" or "Padreterno"[127] – he was at least the godfather. He was always at hand, dropping in to talk, participating in seminars and important decisions. In fact he was living with his famliy in the same building, one floor above the Physics Institute of which he was the director. Laura Fermi reports the Fermi and his colleagues were called, appropriately, "Corbino's boys." The usual autocratic European academic model of the dogmatic head of an institute assigning work to more or less obedient assistants was entirely out of the question in this familial system.[128] So was the idea of looking far away for recruits. On the contrary, nothing is more remarkable than the fact that Fermi, and to some degree, Corbino, chose from among those who happened to be nearby, and then transformed the raw material.

Laura Fermi's descriptions of the way the whole "family" traveled – with wives and children – on excursions and vacations, only underline the interpenetration of their human and scientific relationships. Yet another piece of evidence is that the members of this group, before others, accepted multiple authorship for their papers – the sharing of the rewards they felt entitled to was made possible by their easy relationships. That this informality and communal feeling could coexist in a context of hard and disciplined work was an outcome of Fermi's genius as the leader of a group.[129]

Frugality and improvisation. Superb use was made of relatively small funds, and of improvisation wherever necessary. Not only were the salaries very low, but so was support for the laboratory's research, considering its enormous output and success. The lack of a good shop and of an adequate machinist was a great handicap. In the early 1930s the Rome group thought about, but – despite Corbino's hint in 1929 that frontier work would require an accelerator – was never able to build a cyclotron.[130] Segrè reports: "You see, it was a different type of physics. It was done on a few tables with string and sealing wax. It was extremely simple. It cost very little." To be sure, it was not very different elsewhere. Rutherford's laboratory had perhaps ten times as large a budget, but that was still little.

There is evidence that the Mediterranean context gave Fermi's associates a certain edge to make up for this; what could not be bought or made could more easily be borrowed, improvised, or arranged on a personal level. G. C. Trabacchi, the chief of the section of physics of the Istituto di Sanità Pubblica – the only person outside his own laboratory group whom Fermi thanked in his Nobel Prize address – could be counted on to be helpful, regardless of detailed official protocol. He had been an assistant to Corbino and his laboratory was next door. The Fermi group called him "the Divine Providence," for through him radioactive sources could be obtained. Trabbachi had one gram or more of radium – a much larger quantity than most other laboratories. The director of his Istituto, Marotta, also preferred a rather informal style of administration and was willing to see that help would be given.[131] Segrè was skillful too in exploiting the fact that Rome was, in a certain sense, a small place in which persons knew each other. He tells how a jeweler whom he happened to know lent him a ten-kilogram ingot of gold for the experiments on radioactivity. "I couldn't have pocketed it and just run away – one knew more or less which family one was from."[132] It was an extension of the way Segrè, and others, had entered into the scientific circle in the first place.

Fermi's personality in many ways coincided with the scientific realities which demanded improvisation and frugality. Thus it is said that he tried to get along on half his salary. Indeed, the frugality of his scientific claims in physics, where he always tended to underclaim – unlike the more effusive Corbino, who sometimes embarrassed the group by premature or excessive announcements – is not unrelated. As with money, so with time. Fermi's sense of concentration and efficient use of time – as in the division of labor – were legendary, and they were imposed on the group as a whole.

Finally, there is a set of other factors which need be touched on only briefly. One must return to the fact that good use was made of what is now called "public relations," through conferences, publication of scientific work, texts, popular articles, and the strategic insertion of reports about the Fermi group. It is clear that Fermi begrudged the time taken by these efforts, and the occasional misunderstandings which they were likely to generate. But he recognized that no science, even in the protected mode in which he was privileged to work, can fail to take the opportunity of achieving public understanding and support. Indeed, he succeeded – some Italians I have spoken with have said he succeeded too well, that his example and the fields he and his associates cultivated are now preempting too large a share of the total funds available. This, too, is part of the inevitable opportunity costs of any success, and similar discussions are taking place in every major scientific country.[133]

Concluding observations

In sum, Fermi's group helped physics in Italy to "come of age" in the 1930s, by all the usual criteria; in terms of support – financial and institutional – recruitment, opportunities for careers, and national and international recognition. This was achieved by a group unlike any other we know of in physics before that time: a group that modeled itself to a large degree on a family; had internal and external aspirations, one of which was to bring honor to Italy through the work of the Italian scientists; tried to operate within a tradition – though this does not mean archaic – using available institutions but on occasion making institutional innovations; was essentially a small-scale operation; and was politically and economically a protected operation.

In her paper "The Uses of the Traditional Sector: Why the Declining Classes Survive,"[134] Professor Suzanne Berger had drawn attention to the hardy survival of a peculiarly Italian institution, "the small-scale, familial, protected economic unit" in the midst of the economic and political life of the advanced industrial state. Her interesting paper points out roots of this phenomenon in the social life and history of the nation. There are suggestive parallels between her findings and those in this paper; perhaps these will intrigue scholars of Italian society and cause them to pursue the further study of this case.

Today, a half-century after scientists of Fermi's generation entered upon the scene, the level and productivity of Italian physicists are exceedingly high and internationally acknowledged. Thus at the ceremonial celebration of the fiftieth anniversary of the International Union for Pure and Applied Physics in September 1972 at the National Academy of Sciences in Washington with the participation of leading physicists from all parts of the world, it was taken as entirely natural that at the opening both the address which set the keynote, and the first major paper, should be given by Italian physicists.

The excellence of current physics in Italy – particularly in a number of subfields which have their roots in the work of Fermi and his students – has itself, however, aspects of anomaly which invite a brief mention, for the advantageous position of this field is by no means shared by most other fields of science in Italy. This fact is amply documented in the report *Review of National Science Policy – Italy*,[135] which provides the most recent available comparative data. Here are some of its sobering findings. Comparing research and development (R & D) expenditures in OECD and other countries, one finds that Italy stood lowest (at $5.80 per capita in 1963 and $6.80 per capita in 1965).[136] In qualified research scientists, engineers, and technicians, Italy is also in the lowest place: 6 per 1,000 population – one-third to one-fourth of that of the others listed.[137] Italy's share of public research expenditure in the total national budget was 2.4 percent in 1967, about half the figure for countries such as France and Germany.[138] Almost half the budget for all the sciences goes to just three overlapping fields: physics, space research, and nuclear research.

Only 1.7 percent of the research expenditures in higher education in 1963 was financed by business firms.[139] The inadequate staffing of modern scientific research in universities in Italy has been often commented upon; one indication[140] is the figure of 4.9 equivalent full-time research workers per assistant, a figure much in excess of what is considered reasonable for adequate research support elsewhere. Among other obstacles, well known to all who are acquainted firsthand with the situation, is the severe lack of space for adequate research and teaching. One is, finally, not surprised by the comment in the report: "To sum up, there is no organized career for researchers in Italian universities."[141]

But at the end of this melancholy account, one suddenly comes upon the one really positive part:

> One cannot fail to be impressed by the magnitude of Italy's effort in favor of fundamental nuclear physics and the development of a national power reactor program . . . Furthermore, in her nuclear physicists, Italy possesses a group of dynamic scientists whose importance, both politically and scientifically, cannot be valued too highly. Nor can it be denied that this group has been a

> psychological and cultural driving force in national life. These are results which more than justify the monies spent.[142]

One need not expect to be able to trace direct lines leading from Fermi's first encounter with Corbino to the favored position of the field of research today which was initiated in Italy by Fermi's group. But despite the unforeseeable adversities of national and international history which interposed themselves, it is clear that today Corbino would have good reason to be satisfied with the continuing success of his optimistic program of 1929.[143]

6

Can science be measured?

"The mind comprehends a thing the more correctly the closer the thing approaches toward pure quantity as its origin."
Johannes Kepler, letter to M. Maestlin, April 19, 1597.

The coming of science indicators

In January 1973 the chairman of the U.S. National Science Board (NSB) sent the report *Science Indicators 1972* [*S.I. 72*][1] to President Richard Nixon for transmittal to Congress, stressing in his covering letter that the document, prepared by the NSB staff in the National Science Foundation (NSF),

> presents the first results from a newly initiated effort to develop indicators of the state of the science enterprise in the United States . . . If such indicators can be developed over the coming years, they should assist in improving the allocation and management of resources for science and technology, and in guiding the Nation's research and development along paths most rewarding for our society.[2]

More than three years later, these hopes had become more concrete. In December 1975 the NSB sent to President Gerald Ford the second such volume, *Science Indicators 1974* [*S.I. 74*].[3] Although it was longer than *S.I. 72*, by about 100 pages, and contained many more sections and analyses, this second report was built on the same model. Yet Ford's letter of transmittal to the Congress (February 23, 1976) made clear that the administration did not view this series of reports as mere academic efforts to develop and perfect indicators for their own sake, but, rather, that the reports would serve to underline that

> The nation's research and development efforts are important to the growth of our economy, the future welfare of our citizens, and the maintenance of a strong defense. The nation must also have a strong effort in basic research to provide the new knowledge which is essential for scientific and technological progress.[4]

It was a unique launching for a young discipline. By gathering in one convenient place widely scattered data from ongoing studies, the *Science Indicators* publications have suddenly provided a tool for high-level scrutiny of the quality and quantity of the research enterprise. Such focused attention was bound to come into being sooner or later, partly because of the national expectations from research, and partly because the scale of support has made science very visible, as shown by some of the "indicators" in the reports: By 1974 the annual cost of basic research in the United States was nearly $4 billion (two-thirds of that supplied by the federal government), and a total of $32 billion was being spent on all national research-and-development (R&D) efforts (including defense and space projects), of which one-half came from federal sources. Clearly, any enterprise that employs over a half-million scientists and engineers and commands 15 percent of the relatively controllable portion of the federal outlay must be subject to calls for accountability of its performance and to justification of the national investment in terms of returns that the taxpayer can appreciate.

The developing science indicators are therefore positioned at the intersection of a variety of pressures and hopes. From the academic's point of view they are merely a new subset in the long-established field of social indicators – a specific form of statistical series that measures changes in significant aspects of society (e.g., measures of population, health, education, income, or crime). But even while they are being shaped, science indicators are not likely to remain unused in policy decisions; for example, *S.I. 72* promised that these indicators can "assist also in setting priorities for the enterprise, in allocating resources for its functions, and in guiding it toward change and new opportunities."[5]

In fact, these two science indicator publications received a considerable amount of public attention in the press; the content of *S.I. 74*, in particular, lent itself to headlines such as "U.S. Science Lead Is Found Eroding."[6] Already in 1974, the chairman of the NSB noted, "The indicators appear to be of increasing usefulness to the executive and legislative branches of the government in dealing with science policy issues, and in providing an objective basis for discussions of public policy affecting R&D activity."[7] Indeed, he indicated that they had played a role in the president's budget decisions concerning basic research funding for fiscal year 1977.[8]

These two publications, therefore, are of extraordinary interest not only to policy-oriented decision makers – who may have been the primary audience the authors had in mind from the beginning – but equally to working scientists and scholars. They will be interested, first because of the wealth of data (usually in the form of time series) and what they reveal, on subjects ranging from the relative share of research support, publications, or students in different research fields, to the measures of public attitudes toward science and technology, but also because the use of science indicators will undoubtedly come to influence their professional lives deeply. No matter how carefully quantitative reports from professional or governmental agencies are hedged in their release, they become normative. Therefore, during this crucial period in the development of science indicators, we should examine critically the concepts and methodologies used in the measurement of the quantity and quality of science and technology.[9] On the one hand, the need for sound indicators cannot be doubted, and their usefulness will depend on the credibility they have within the scientific community. On the other hand, this is the time to discern and correct shortcomings, and to argue for humility in the face of sometimes rather crude quantifications, before the science indicators are so firmly set that they will be difficult to change or dislodge. One recalls here Einstein's remark:

> Concepts which have proved useful for ordering things easily assume so great an authority over us that we forget their terrestrial origin and accept them as unalterable facts. They then become labeled as "conceptual neces-

> sities," "*a priori* situations," etc. The road of scientific progress is frequently blocked for long periods by such errors. It is therefore not just an idle game to exercise our ability to analyze familiar concepts, and to demonstrate the conditions under which their justification and usefulness depend.[10]

Without foregoing entirely a detailed analysis of selected portions of *S.I.72* and *S.I.74*, I shall concentrate in the following remarks chiefly on some fundamental epistemological problems adhering to any attempt to make "indicators," no matter for what purpose, while referring to the two publications, particularly the stage-setting *S.I.72*, for actual examples.

Quantitative versus qualitative indicators

The idea of making quantitative indicators of anything at all fascinates some persons and repels others as dangerous or absurd. This difference is caused largely by thematically incompatible – and therefore often unresolvable – personal views concerning the ability of quantifiables to lead to or attest to the deepest reality. The history of science is full of cases in which some purists claimed that only when you measure something do you really know what you are talking about, whereas others held that quantification (other than for classification or similar purposes) distorts the full, natural sense of things by confining phenomena in a strait jacket. Like all thematic tensions, this one is healthful for the eventual progress of a science because it tends to juxtapose deeply felt, articulated models and methods.[11]

Most scientists today, if forced to choose, would tend to lean toward the first of these two positions. But thoughtful spokesmen would also perceive and verbalize the consensus on a hierarchical order: The ultimate use of quantitative measures within science is that they can guide one eventually to an understanding of the basic features associated with conceptions expressible in numeric terms, but *not* exhaustively motivated, described, or explained by them – basic features described in terms such as simplicity, symmetry, harmony, order, and coherence.

Similar polarizations concerning quantification, and a similar attempt at resolution, may be noted in the new field of technology assessment. In his article, "Technology Assessment from the Stance of a Medieval Historian," Lynn White has put it memorably in these words:

> Some of the most perceptive systems analysts are pondering today how to incorporate into their procedures for decision the so-called fragile or nonquantifiable values to supplement and rectify their traditional quantifications. Unhappy clashes with aroused groups of ecologists have proved that when a dam is being proposed, kingfishers may have as much political clout as kilowatts. How do you apply cost-benefit analysis to kingfishers? Systems analysts are caught in Descartes's dualism between the measurable *res extensa* and the incommensurable *res cogitans*, but they lack his pineal gland to connect what he thought were two sorts of reality. In the long run the entire Cartesian assumption must be abandoned for recognition that quantity is only one of the qualities and that all decisions, including the quantitative, are inherently qualitative. That such a statement to some ears has an ominously Aristotelian ring does not automatically refute it.[12]

Since the most respectable parts of the "hard" sciences, at least on the surface, seem to be preoccupied chiefly with quantifiable measures, nothing is more natural than the development of "quantifiable" indicators about science. This tendency also parallels both the ongoing "scientification" of modern social studies through the introduction of more sophisticated mathematical methods and operationally demonstrable concepts, and the ever-growing desire of administrators to rationalize the allocation and use of resources. In any case, the desire to formulate some indicator measures, on any basis, is almost irresistible. Who in this field would not want to find out how many scientists there are, how much money they spend, and how many papers they publish? Who in the Congress or in science administration would not like to see how the science enterprise in this country compares with that of other nations, or to what extent the expenditures for basic research bear fruit in terms of industrial developments that can claim to be "outputs" of basic science? And, above all, who would not like to have some, *any* measure of how "good" the efforts of our scientists and engineers are?

Fortunately, *S.I.72* took note of the need for developing indicators of *quality*, stipulating on the very first page of its text that intrinsic measures should include those of the "quantity and quality of associated human resources," and that extrinsic indices would center on "the achievement of national goals . . . and the consequent impacts on that elusive entity, the 'quality of life.' " One cannot help but applaud this ambition, the more so as the quantification of qualities, ever since Nicole Oresme's attempt in the fourteenth century, has been a precarious task.

In measuring qualities, three types of opportunities exist that are not available when the usual quantitative indicator is made by folding a large number of individual contributions into one grand total. The first is the identification of *one or a few crucial events*, perhaps in a mass of noisy data. The quality and health of a science, as we know from personal experience and the abundant evidence of historical study, not infrequently depend on nothing so much as the appearance of one paper of high imagination or the founding of a professional society, the accession of a new journal editor or the entrance of a new patron. (For example, nothing was more significant for the quality of science in Italy in the 1920s than the appointment of young Enrico Fermi to the University of Rome.) Insofar as singular events that have the potential of transforming a field can be identified quickly, they should find some place in any publication that claims to indicate the quality or health of science at a given time – even if such events are not quantifiable and the potential not quickly realizable.

Second, even in a consensual activity like science, contemporaneous estimations of quality are difficult. The full significance of an important event does not necessarily appear immediately, and identification of quality often requires historical perspective. Very few physicists in 1905 were aware that it was one of the greatest years in the history of their field. Sadi Carnot's fundamental paper of 1824 on thermodynamics was not recognized as important until 1834, and then only by one scientist, Emile Clapeyron; after that, it took another decade before the scientific world began to appreciate, through William Thomson's work, the merits of the 1824 publication. Moreover, details of the intellectual structure of Carnot's work have been adequately understood by historians of science only in the past fifteen years.

Similarly, the profound implications of Max Planck's work of 1900 did not become clear quickly, even to Planck himself. It began to be understood generally – at different rates in different countries – only after the 1911 Solvay Congress. And suggestions of the existence of what are now called "Black Holes" lay about, essentially neglected, in the literature for decades before current interest focused on them. (Conversely, some sensations of a moment have a habit of disappearing in time.) Instances can be adduced ad infinitum.

Although on the whole we may be more alert to "significance" today than in the nineteenth and early twentieth centuries, it is not realistic to hope that the quality of scientific work and of scientific life will be estimated safely right away, "in real time." Therefore the assessment of the state of science for a given year, while useful and interesting, must be reexamined and updated in the light of changing results of informed historical scholarship.

Third, we are still very far from solving the methodological problem of how to determine the quality of a subject field. Some attempts exist, such as the recent review by a panel of peers of the quality of previously granted research proposals;[13] but the methodology for developing measures of quality requires much further experimentation. One possibility[14] may be to evaluate the quality of publications in a field, or the quality of a journal, by appropriate random sampling and evaluation, analogous to what is often done in scientific research itself. Thus a properly chosen, rather small sample of, say, one hundred articles in a field, or even a few hundred pages of a journal volume, might be evaluated in depth by a high-grade panel of assessors, somewhat the way a set of research-funding proposals is ranked both relatively and absolutely.

We can find very few indices of quality in *S.I. 72* and *S.I. 74*. One is the listing of the number of Nobel Prizes received by U.S. scientists from 1901 to 1974. There are, of course, difficulties with this simple quantification of quality. The assumption that the award is given on the basis of an international search for work of highest quality, uninfluenced by political or other extraneous considerations, is certainly sounder in science than in some other fields; but the number of awards per year is so small that fluctuations over a short period are not likely to be very meaningful. There are obvious other problems – how to count prizes when they are shared, to which country to credit a prize given to a migrating scientist, or how to treat the time delay between the publication and the honoring of it. Should one pay more attention to the fact that the United States has the largest number of prizes since 1901, or that the rate of receiving prizes in the United States has become smaller since 1951–60, or that in relationship to the population as a whole, U.S. scientists rank far below those of the Netherlands, Switzerland, and the United Kingdom? In fact, what precisely is the link between quality of research and the Nobel Prize?[15] All these questions hint at the larger problem: that numbers by themselves may or may not indicate anything meaningful.

The *Science Indicators* volumes also attempt to measure quality – with due acknowledgment of the difficulties and dangers – by means of literature indicators, comparing "the relative standing of a given nation's literature in the bibliographic references of a large sample of the world's scientific literature. The assumption is that the most significant literature will be most frequently cited by subsequent investigators . . . By this measure the United States ranked or tied for first place in each of eight major scientific fields."[16] *S.I.74* does, however, acknowledge that this method of assessing the quality of scientific literature is subject to factors that may severely slant the results – for example, differences in publishing habits in different countries, editorial policies, or the representativeness of journals in the sample being analyzed.[17]

There are additional attempts at quality measures, as in the interesting distinction between degrees of innovations calculated for several hundred specific innovations, from improvements of existing technology to "radical breakthroughs" (and the resulting finding that the "radicalness" of major U.S. innovations has been slipping markedly over the past two decades);[18] but, on the whole, it is clear that quality measures are still in an early stage indeed; hence the opportunity for improving them is that much larger.

Bulk measures versus fine-structure spectroscopy

Enterprises such as the making of economic indicators or science indicators – almost by definition, and surely by virtue of the vastness of the task – tend to gravitate to the use of aggregative, composite indicators, to bulk or gross measures rather than to attempts to look for fine or detailed structure. To use an analogy, the preference is to measure the intensity of a source by cumulating the energy emitted at all wavelengths, rather than passing the beam through gratings or prisms to lay out the spectrum and measuring the detailed energies radiated at different wavelengths. Of course, important questions exist for which it turns out no fine details are needed, in which therefore a "pre-Newtonian" treatment of the beam (i.e., without spectral resolution) is adequate; in any case, that is usually the far simpler measurement to make. But such bulk measurements could not have led to the optical discovery of Fraunhofer lines, Balmer lines, black-body radiation, hyperfine structure, and so on – all important to the advance of physics.

Here again, contrary thematic preferences enter the researcher's design. Some are "lumpers" and will be naturally predisposed to seek bulk measures; others are "splitters" and will seek fine structure. To some, the decay of a society may be indicated by, say, the rate of fall of the gross national product or the net trade balance; to others, the same information is contained in significant details, such as the observation that the mastiff is starving at the master's gate. To some, the rather smooth exponential rise of the curve of the *cumulated* number of science abstracts as a function of time is the raw material for speculation; to others, the essence of the work lies in filtering out the few, sporadically appearing significant articles from the steady stream of the banal, or even in the strange, sometimes large fluctuations of the annual output of articles.[19]

By training and preference, research scientists tend to be more interested in identifying significant detail than in aggregating data. Their interest in a new idea is often aroused by what might seem a specific, minor anomaly, an unexpected mismatch along the edges when expectations and reported findings are superimposed; and they usually test their theories by referring to carefully selected phenomena in a well-defined corner of phase space.

Thus a peculiarity noted in the "fine structure" of phenomena (e.g., the observed differences in the intensity of artificial radioactivity found on two different laboratory tables in Fermi's group in 1934) is more likely to signal the existence of a fruitful new area of work. On the other hand, gross properties are more likely to be useful for exhibiting, confirming, or summarizing the state of an existing, perhaps well-established, theory. They also succeed in a field such as economics because of the possibility of converting very different products to the common denominator of price.[20]

When bulk measures are called upon in scientific research, it is generally assumed that they do not mix together incommensurable populations. In fact this can be known only post hoc, that is, when an adequate theory has determined what the pure and the mixed populations, respectively, consist of. I greatly doubt that we are near enough to such an understanding in the case of science indicators, and I am struck by the tendency throughout most of *S.I.72* and *S.I.74* to opt for bulk measures (such as time-line series for "mathematics" or "engineering" or "social sciences," taken as a whole), even when more detailed spectroscopy of data was available in the literature.[21]

The *Science Indicators* volumes seem to be looking primarily to thermodynamics for their model. That is a dangerous exemplar. Thermodynamic variables such as pressure, volume, temperature, and compressibility are indeed wonderful indicators of the state of a sample – up to a point. They have clear operational meanings; they are connected to a body of knowledge that allows these measures to achieve interest by demonstrating that gross matter behaves according to laws; above all, they are independent of any detailed understanding (or ignorance) of the microstructure of matter. But even after the resolution of those enormous conceptual struggles (spanning more than a century) that were needed to develop notions by which the laws of thermodynamics could be formulated and tested,[22] thermodynamics remains not such a clean and simple model on its own terms, for example, when dealing with mixed rather than pure systems.

To illustrate the tendency toward bulk measures and the predominant influence of the "lumpers," consider the very title of the books, *Science Indicators.* According to a doubtlessly apocryphal story, when Jerome Wiesner was in charge of the Office of Science and Technology, he had his staff run a contest to discover a less clumsy, shorter term for identifying the activities referred to as "sciences and technologies." The largest number of proposals were that the single word "science" be used. To be sure, calling it "science and technology" instead of "science" is still a gross oversimplification, the gathering into an arbitrary two sections a large set of different activities that exist in a more complex space and do not fit along one line. The very fact that there is no good name for the set of activities for which indicators are here sought is an indicator of the complexity of the activities.

If nothing else can be done about this problem quickly, at least a note of caution is needed when bulk measures and global statements are used: for example, qualifying the claim that the indicators can be used for describing "*the* state of *the* science enterprise";[23] or listing all the separate sciences and technologies included in the gross measures. In addition, even at the risk of making science indicator publications more bulky, there should be a balancing effort at disaggregation and at giving detailed reference to and summaries of main publications that do contain fine structure.

Employment statistics

A significant example of insufficient fine structure in *S.I.72* is found in the section "Science and Engineering Personnel" – "Supply and Utilization." There, "science indicators" merge into the more general "social indicators." This section offers a remarkably brief treatment of one of the chief preoccupations of many scientists in the very year for which these indicators were designed (as well as the previous two or three years) – namely, the substantial decrease of financial support for science and technology, with consequent unemployment, underemployment, and holding actions. A report issued by this nation's chief research support agency and addressed to the President for transmittal to the Congress bears an extra measure of obligation to present in the clearest light just those problems that may be most difficult or embarrassing for the life of science. Obviously, few official "indicators" will be scrutinized more skeptically by scientists, scholars, and their professional societies than employment data. To be sure, there is no *proof* that a nation's science effort as a whole is necessarily of lower quality when there are fewer employed scientists and few funds, but the human costs of unemployment require that we act on the basis of this assumption until contradictory evidence exists. A report that underplays the highly visible and personally often tragic problem of unemployment and underemployment risks the label of cover-up, the more so as the claim was made that the indicators would reflect impacts on the "quality of life."[24]

One notes, for example, that space was available in *S.I.72* to state that "the overall science and engineering unemployment rate was still only about half of that for all workers"[25] – a euphemism produced by mixing into one gross indicator two differently prepared populations (not to speak of the fact that the high unemployment rate in the general work force – officially between 5 and 6 percent throughout 1972 – was neither humanly nor politically satisfactory). But in order to allow a full picture of the national situation to emerge, and to permit the identification of specific disaster areas which merit special help – surely an important function of indicators – it would have been more to the point to analyze the details of unemployment and underemployment, including those who have been excluded from the employment statistics or those who are subprofessionally employed. All this is the more urgent as the effects of insufficient employment on the life of the scientist often are unusually severe – in terms of the rapid loss of a scientist's or an engineer's capacity to stay in the profession once the ability to remain updated is impaired. Moreover, if the categories of the unemployed or underemployed are not in line with generally "accepted" definitions of unemployment, this might cast doubt on those definitions.

Alan C. Nixon, recently president of the American Chemical Society (ACS), has pointed out[26] the discrepancies between simplified aggregative figures and figures based on detailed analysis. For example, if the definition of people with employment problems is allowed to include transient or part-time employees, and so on, the unemployment figure was 8.3 percent of the ACS membership, and 9.5 percent for the Ph.D. members. Nixon's analysis is only one example highlighting the artificiality of some of the concepts in general use for official employment statistics. Thus the fine report *Work in America*, prepared by a special task force to the Secretary of Health, Education, and Welfare,[27] warns that

> The statistical artifact of a "labor force" conceals the fluidity of the employment market and shifts attention from those who are not "workers" – the millions of people who are not in the "labor force" because they cannot find work . . . Although this narrower concept of a "labor force" is useful for many economic indices, it is inadequate as a tool for creating employment policy.

For example, it excludes from consideration "people who answer 'no' to the question 'Are you seeking work?' but who would in fact desire a job if one were available and under reasonably satisfactory conditions." Such people, the task force report indicates, range from those in school or training programs because they have been unable to find suitable jobs, to many women and older people who no longer even look for jobs.[28]

Although the dismaying details of the employment situation in science were apparently not considered sufficiently urgent to prompt a warning flag in *S.I. 72*, they *were* significant in shaping the career decisions of many young people to whom the portents were clear. To give an example that cannot be found in the global indicators of *S.I.72*:[29] From 1961 to 1970 the number of doctorates granted in physics rose, roughly parallel to the rise in the total student population, to about 1,550 in 1970; then the curve turned down, and the projection of new physics Ph.D.'s in 1978, based on students in the pipeline, is now 950, that is, 60 percent of the output of 1970.[30] Such trends lead to contrary but equally arresting predictions. Some now warn of a severe shortage of trained physical scientists and engineers in the 1980s; others see in the recent downturn an early response to the general exacerbation of the employment problem in the next decade.

One of the reasons announced for the existence of *S.I.72* was to "provide an early warning of events and trends which might reduce the capacity of science . . . to meet the needs of the

Nation."[31] Lack of careful attention to the employment picture is, therefore, a serious matter if one places reasonably healthy employment opportunities and, hence, efficient utilization of a national resource, high on a list of the needs of the nation. The concept of a policy of reasonably full employment at or near a level of trained competence is neither simple to implement nor generally agreed upon. Yet by 1973 it was no longer merely a matter for politicians of different ideologies to disagree about in the United States. The "full employment" movement was becoming stronger, as measured by the bills introduced in Congress, and conclusions, such as that in the HEW report *Work in America*, that the provision of the opportunity for each family's "central provider to work full-time at a living wage . . . should be the *first goal* of public policy."[32] However, as measured by *S.I.72* and *S.I.74*, the time for the idea had not yet come for policy planners concerned with scientists and engineers.[33]

Another detail of the employment situation that concerned many scientists in the early 1970s was, of course, the employment, and training for employment, of women and minorities. It is remarkable that *S.I.72* included little on this subject, even though other NSF surveys had gathered much useful data.[34] (*S.I.74* does have four pages on the employment of women and minorities in science and engineering.) Even though women scientists constituted only 9 percent of the total number of physical scientists (chemistry, physics, and atmospheric and space science) in 1970, their rate of unemployment was almost double that of men.[35] The rate of inflow of women and minority scientists would have been useful indicators in *S.I.72* and even the modest starts made toward improvement, such as the National Institutes of Health (NIH) Minority Biomedical Support Program begun in 1972, would have merited notice as an example of the impact of policy changes on the fine structure of the life of science.

To summarize, then, the global and optimistically slanted indicators put out by a government agency may often clash with the soberer details found in the personal lives of scientists and in surveys made by professional societies or other organizations. Therefore, the danger exists that "science" and its preoccupying problems, as they emerge from "official" science indicators, may constitute an entity qualitatively different from science and its problems as perceived by individual working scientists or by the governing boards of professional societies.

Political realities

To avoid the charge of naïveté, anyone reviewing a U.S. government publication such as *S.I.72* must keep in mind two political realities. Everything that a government appointee wishes to publish involving the mission of his agency must be cleared by the Office of Management and Budget (OMB). Moreover, in the specific case of *S.I.72*, that report had to undergo scrutiny by OMB and other White House staff in the days when OMB was not loath to impound funds appropriated for agencies such as the NSF and when OMB's deputy director was Frederick V. Malek, an aggressive member of H. R. Haldeman's White House "management team," with wide-ranging mission and power. (Malek had earlier been special assistant to President Nixon and has been identified as the developer of the "Departmental Responsiveness" program that aimed at obtaining political leverage through grants and contract programs of federal agencies.[36]) Many of the weaknesses and omissions in the published version of *S.I.72* may well stem from these two historical facts. Conversely, the disappearance of Nixon, Haldeman, and Malek from the governmental scene before the publication of *S.I.74* may explain in part why *S.I.74*, although cautiously worded, is a substantially more useful and explicit document; changes in the political situation, as well as critiques received after the publication of *S.I.72*, may have helped toward this end.

An inherent generic limitation in such government reports in general and a special ambiguity in the status of the NSB deserve comment here. On the one hand, the scientific community, and to some extent the Congress, see the NSB as an independent board of scientific statesmen, in a position to speak out forthrightly to the President and the Congress on the state of science and the interests of the scientific community, and whose scholarly and scientific training and credentials should certify the independence and quality of any report they issue. On the other hand, the administration sees the NSB as a part of the administration, subject to the political discipline of the administration, and therefore required to conform in any of its public pronouncements to "the President's program," just as is the case for full-time government administrators.

The result is evidently an uneasy accommodation in a constant tug of war between the NSB and the OMB, the agency in practice responsible for enforcing conformity with the president's program on the part of all presidential appointees in the executive branch. The final text of a report is the result of long and complex negotiations between the NSB, the NSF director, and the OMB. To put the matter bluntly, it appears that in preparing the report *S.I. 72* the NSB tried to seek a compromise between what it would have liked to say and what it thought could get passed, feeling that even a necessarily watered-down or negotiated version of the report would have a considerable positive public impact with respect to the support of science.

In addition, another issue may have inhibited any desire by the authors to stress questions of unemployment – that such discussions might open the board to the charge of being "self-serving" in its report, especially with respect to any statement which could be interpreted as favoring the economic interests of the scientific community.

To be sure, there is a dilemma here. My criticisms of the report do imply that in the current state of the world it is absurd, if not contrary to the obligations of a modern society, to have no plans for putting trained scientists to work on socially useful jobs that take advantage of their skills. They also imply that the existence of a body of scientists whose talents are seriously underutilized represents a notable symptom of the lack of health in the scientific enterprise unless proved otherwise. But how far can this be generalized? Is the case of science greatly different from that of, say, scholars and artists?[37] In either case, is it appropriate for a group of scientists to speak up only for scientists, or conversely to arrogate to themselves the task of speaking up also for nonscientists? Questions such as these, and the more serious problems associated with the ambiguous positions of the NSB, suggest that the public record cannot by itself be an adequate basis for assessing the health of science. Once again we see merit in the argument for stimulating the academic sector to provide assessments that are "independent," or at least numerous and varied enough so that their biases become evident in the ensuing discussions.

Numerical data versus indicators

Gathering censuslike data on the state and performance of the professions is obviously useful and interesting, regardless of the various ultimate uses of the data, and this is, in fact, done regularly by NSF, the National Academy of Sciences–National Research Council, and the professional societies. Indeed, it appears that most if not all the data used in *S.I.72* (with the exception of the Delphi Experiment and the commissioned study on public attitudes) were taken from previously available sources. What, then, is the difference between data and indicators?

To put it in terms of an analogy – measurements of a patient's rapidly declining blood pressure remain data until they are seen by someone with enough understanding of physiology to recognize them as indicators of a change in the state of health. I propose that the term "indicator" is properly reserved for a measure that explicitly tests some assumption, hypothesis, or theory; for mere data, these usually remain implicit. Indicators are the more sophisticated result of a complex interaction between theory and measurement.

Even "taking data," as is well known, is not possible without at least an implicit theory. But an implicit theory is often fallacious, as the discovery of inadequacies in "simple" census data has shown: Significantly more people in certain categories were discovered to have been invisible to the takers of the U.S. census than had been allowed for in the hypotheses underlying the method of measurement.

Data may be believed to be necessarily objectively factual, and yet it may also be possible that the existence of an implicit and possibly erroneous theory has been overlooked. For example, I once worked near a superb experimental physicist, one of whose publications typically was on the tensile strength and other properties of fifty alloys. He ground out data at a phenomenal rate. It turned out, however, that what he was doing was testing an implicit theory that the small amount of impurities in these particular samples of the alloys has a negligible effect on the properties of the metal. His measurements were very sound; that is, they were repeatable on these particular samples. But the level and type of impurities turned out later to have a large effect on the properties, severely limiting the applicability of the measurements to the general case.

At an early stage the plausibility of some data for use as indicators may be a good guide. However, eventually there must be an explicit theoretical base for choosing some data, discarding others, and noting the absence of needed data or of needed fine structure. No such *explicit* theoretical base appears in *S.I.* 72 or *S.I.* 74. Instead, one can occasionally glimpse an *implicit* theory of science (as we shall see in the last section of this chapter).

The attempt to put data ahead of an explicit theory reminds me of a discussion reported by Werner Heisenberg. In 1926, just after his first major contribution to quantum physics, Heisenberg met Einstein. Heisenberg explained that he had abandoned such hypothetical conceptions as the unobservable electron orbits; "since a good theory must be based on directly observable magnitudes," he thought it more fitting to base his theory on observable frequencies of the emitted light instead. Moreover, Heisenberg thought he was there following faithfully Einstein's own model: "Isn't that precisely what you have done with relativity?" Einstein's reply took Heisenberg completely aback:

> Possibly I did use this kind of reasoning . . . but it is nonsense all the same. Perhaps I could put it more diplomatically by saying that it may be heuristically useful to keep in mind what one has actually observed. But on principle, it is quite wrong to try founding a theory on observable magnitudes alone. In reality the very opposite happens. It is the theory which decides what we can observe. You must appreciate that observation is a very complicated process. The phenomenon under observation produces certain events in our measuring apparatus. As a result, further processes take place in the apparatus, which eventually and by complicated paths produce sense impressions and help us to fix the effects in our consciousness. Along this whole path – from the phenomenon to

> its fixation in our consciousness – we must be able to tell how nature functions, must know the natural laws at least in practical terms, before we can claim to have observed anything at all.[38]

To be sure, those who make science indicators need not regard themselves bound by the credo formulated by theoretical physicists; yet the story has some application to the *Science Indicators* volumes, which are full of observations but studiously avoid discussing an underlying theory.

It is significant that the term "science indicators" was, to my knowledge, first published in a paper by Harvey Brooks[39] in which he proposed "four models of the research system" – the Polanyi-Price model, the Weinberg model, the social overhead investment model, and the Toulmin model – and linked these models with "indicators for more quantitative planning and for providing a means by which the output of the system might be measured or assessed." In a related paper two years later,[40] Brooks noted a number of "unresolved issues of science policy" – overspecialization, centralized versus pluralistic management, the place of engineering, mathematics, and physical sciences in new national priorities, critical size versus dispersion in research, the integrity of the self-regulatory systems of science, models for the support of graduate education, and the relation between American and world research efforts. Each of these is, of course, susceptible to discussion in terms of indicators, developed for the purpose and shaped by the model of the research system.

For example, even the carefully more "upbeat" *S.I.72* (and certainly the *S.I.74*) gives evidence of a sense of the ending of the "endless frontier" in the United States, the decrease in momentum and advantage with respect to other countries, and the glum future of contracting prospects for political and financial support for science and technology. (Even in the 1930s, the average annual growth of academic science in the United States was 6 percent, and of R & D funds as a whole, 9 percent.) If one believes, therefore, that industry depends on, and has extra responsibility for, basic research, one will notice with dismay that in real dollars the expenditure by American industry in support of its own basic research declined so markedly that in 1974 it was down to 68 percent of the level of 1966, having thereby returned to the level of 1960 or 1961.[41] An obvious next step would be to obtain indicators to test the economic forces, among others, that have progressively discouraged long-term research in industry – for example, how support for research in industry has been related to profitability and tax structure – none of which was discussed in these volumes.

Moreover, with the burden shifting correspondingly to the universities, it is the more ominous to see that expenditures for basic research per scientist and engineer in doctorate-granting universities were almost 30 percent lower in constant dollars in 1974 than in 1968. The total R&D expenditure from all sources, as a percent of GNP in the United States, had dropped correspondingly, from 3 percent in 1964 to an estimated 2.2 percent in 1976. Of that total, the share provided by federal funds for the "Advancement of Science" – fundamental research and science instruction – in the early 1970s was 3 percent of the R&D total in the United States (and thus only 0.06 percent of GNP), versus 16 percent of R&D for the United Kingdom, 25 percent for France, 41 percent for West Germany. On the other side of the coin, the United States allotted 71 percent of its federal R&D funds to "National Defense" and "Space" – versus 46 percent for the United Kingdom, 35 percent for France, and 22 percent for West Germany. Since these NSB volumes frankly put the main emphasis on the normative aspects of indicators, one hopes that in the future there will be detailed work to analyze such trends, for example, by presenting the details of basic research allocations by fields in the budget of the various sponsoring federal agencies. In this manner one might test to what degree the original expectations and mandates of the various agencies are continuing to be fulfilled and to what degree their missions have been changing.

Precisely because science indicators will be used, for better or worse, by science policy makers, they must not in their present early stage of development be directed only or primarily to such an audience. Rather, indicators are to be thought of as more meaningful the more they lend themselves to be tools for confirmation or refutation in handling questions, hypotheses, or theories that interest scholars and practitioners concerned with the state of the sciences or engineering, from their various platforms in science policy studies, history or sociology of science, and so on. Indicators must not be thought of as given from "above" or detached from the theoretical framework, or as unable to undergo changes in actual use. They should preferably be developed *in response* to and as aids in the solution of interesting questions and problems. (To be sure, the possibility cannot be dismissed that the theoretical basis eventually may need to be amended – useful concepts can survive theory shifts, as we know from many cases in the history of science – or that a "natural history" sort of data accumulation, with weak grounding in theory, may occasionally provide measures useful for later, serious work.)

Different models or perceptions (of the development of science, of the science-society feedback loop, and even of the prognosis concerning the good or bad uses to which science indicators might be put) will produce spectra of different indicators, as well as different views of the limitation inherent in some indicators or of the way in which indicators are functionally dependent on the input and output expectations. This prospect of diversity is not dismaying in the least. The major attraction of science indicators is the possibility that eventually they may help test aspects of different theories, and models of the scientific process. If *S.I.72* or *S.I.74* had begun from some of the current theories, for example those of Brooks, Elkana, Merton, Popper, Price, Toulmin, or one of the Marxist theorists, significantly different indicators would have been developed. Thus, starting from Robert K. Merton's categories, indicators would be required to help measure the degree of universalism, organized skepticism, and so on, and the effort would be to determine whether the system in fact attends to these norms and to its own claims or pretensions regarding them – measured by the degree of openness to (including funding of) young talent, the degree of international collaboration in science, the health of the peer-review process, and the like. It is chiefly these comparative tests of different models that will permit development of better models concerning the life of science.

I do not believe the NSB would strongly resist such ideas. In his testimony, the NSB chairman recognized that "Output indicators . . . are closely associated with the objectives of R&D

and are intimately bound to the social systems that convert and incorporate their results."[42] In fact, *S.I.74* reports that in the United States the percentage distribution of scientific literature for selected fields shows that clinical medicine has the largest share and chemistry and physics nearly the smallest share, whereas in the USSR this is exactly reversed.

There are two other reasons for welcoming the admission of a diversity of models and of corresponding indicators. One is that in the absence of conscious pluralism, *one* theory is likely to establish itself, or at least discourage the others. The possibility that the evaluation of science may be captured by one ideological group is by no means an idle fancy. In fact, the philosopher of science Imre Lakatos, the leader of a sizable group of adherents, repeatedly and frankly announced such a goal. In his most recently published discussion,[43] Lakatos identified his own school of thought, and especially his "methodology of research programmes," as a version of "Demarcationism." A chief role of the "demarcationist philosophy of science" is to "reconstruct *universal* criteria" that help to distinguish "progressive" research programs from "degenerating" ones. Thus Lakatos warned, not only "medieval 'science' " but also "contemporary elementary particle physics and environmentalist theories of intelligence might turn out not to meet these criteria. In such cases, the philosophy of science attempts to overrule the apologetic efforts of degenerating programmes."

Lakatos gave the label "Elitism" to the contrary conception, that is, the notion that in distinguishing between worthy and unworthy work ("progress and degeneration, science and pseudoscience"), the "only judges are the scientists themselves." That idea he scorned as stemming from an "undemocratic, authoritarian" school of philosophy, whereas his own demarcationists "share a democratic respect for the layman." The layman must be helped, of course; for this very purpose "a statute book, written by the demarcationist philosopher of science, is there to guide the outsider's judgment." But *guidance* is leaving too much to chance; perhaps *direction* is safer after all: The demarcationists, Lakatos announced, will "lay down *statute law* of rational appraisal which can direct a lay jury in passing judgement."

It is a breathtaking ambition. Far from being dismissed out of hand, the idea was considered for a time by a major, private research funding agency in the United States for use in selecting among applicants' proposals. Undoubtedly, there would not be universal acclaim if this (or any other) particular philosophy of science takes over the indicators by which the NSF measures the quality of science. The absence of any explicit theory to guide the making and use of indicators may not be good; but the adoption of a single one is likely to be worse.

Another argument for plurality is that, precisely as in the sciences themselves, indices may be developed eventually that are *invariant* with respect to theoretical models. These indices will be the most useful ones: among other advantages, they and only they allow rival theories to be put to meaningful quantitative tests.

Even now, quite opposite theories of science can lead to demands for the same indicators. For example, in the early 1970s the OMB's view was evidently to regard the professional person as a statistic in the free market; from that viewpoint, a rapid decrease in the expected number of new physics Ph.D.'s could be seen as a salutary adjustment of an "oversupply" situation.[44] On the opposite side, to someone concerned that the intellectual and industrial bases provided by science are being eroded, the same curve has a different and much more ominous message. But both parties will agree on the *need* for such manpower indicators.

Advocates of different theories of science will also agree in their concern for the institutional strengths of science. *S.I.72* included an "Institutional Capabilities" section[45] – to which one would naturally look to find measures of the diversity of centers of research and of the strength of the professional societies and their scientific journals. However, this first effort at compiling an "official" array of science indicators did not address such questions; for example, the report does not even hint at the fact that a number of professional societies are in severe financial trouble.[46] Development of such indicators on professional health is badly needed, and it may well be possible to phrase such indicators in a form invariant with respect to theoretical frameworks.

As a practical matter, an NSB publication by itself cannot be expected to go very far in using and comparing a large number of conceptual frameworks (although one would hope that their existence would be noticed). Whatever the NSB elected to do in this connection, scholars not on the staffs of Washington agencies should be encouraged to work on such problems on a proper scale. Science indicators will be far healthier when institutional, programmatic steps are taken to support academically centered research.

The model of science implicit in S.I.72

What, then, is the theoretical framework implied in the *Science Indicators* volumes? Some of the answer comes from what is said there, some from what is not said; some comes from the sequence, and the rest from the style, of the contents. Since on this score *S.I.72* and *S.I.74* are quite similar, examples will be drawn from the model-setting first volume.

By way of background it may be useful to recall that the year 1972 was one of heavy, worldwide scientific activity. A few examples from physics and astronomy only will suffice, using a listing of scientific advances prepared in late 1972 by the American Institute of Physics:

A Columbia University team identified the strong cosmic ray source in the Crab Nebula as identical with the pulsar there. Canadians recorded remarkable bursts of radio emission from Cygnus X-3. At Brookhaven, progress was made toward superconducting electric-power transmission. Fusion reactions became a more probable source of power owing to successful experiments in laser-produced implosion and adiabatic compression of the plasma. The world's largest particle accelerator became operational and gave its first results; so did two new colliding-beam storage rings. A beautiful puzzle was found in neutrino astronomy, where the difference in observed and expected neutrino flux from the sun is so large that the model either of the sun or of the neutrino's stability must be revised. Discovery of traces of naturally occurring plutonium-244 further extended the periodic table. By laser-ranging, the distance to the moon was determined to within 6 inches; and the method could now be applied to measuring continental drift. Accurate radio ranging of Venus, Mercury, and Mars gave preliminary data for another test of gravitational effects in Einstein's theory of general relativity.

More Black Hole theories attracted fascinated attention. And three American physicists shared the 1972 Nobel Prize for the development of a theory to explain superconductivity, one of them (John Bardeen) becoming the first person to win two Nobel Prizes in the same field.

S.I.72 mentions none of these findings or events, thus illustrating perhaps the difference between "knowledge indicators" and science indicators.[47] Another "big story" in science in 1972 – the continuing decrease of support in funds, facilities, educational prospects, and so on – does appear there in aggregate indicators, presented in deadpan style. Thus the curve for federal obligations for academic R&D plants is shown to drop precipitously, from about $105 million in 1965–66 to about $17 million (in constant dollars) in 1971.[48] This decline has had enormous impact on the lives of scientists in the United States – but only one paragraph in *S.I.72* acknowledges the existence of this problem. (Nor, of course, does *S.I.72* mention the dismissal in 1972 of the president's science adviser and the dismantling of the science advisory apparatus, set up by President Eisenhower some fifteen years earlier.) As in other pages, it is as though the authors had been on the verge of telling what some of these indicators in fact *indicate*, but then hesitated and drew back.

No such hesitation attaches to two other aspects of the image of science implicit in *S.I.72*. One is the primarily Baconian, rather than Newtonian (and primarily "hard sciences" rather than natural *cum* social sciences), view of "science." From the outset (in fact in its second paragraph), the report speaks of "the capacity of science – and subsequently technology [thereby revealing its trickle-down theory of science-technology relations] – to meet the needs of the Nation." It is interesting to note that the experts polled in the Delphi Experiment (see previous discussion, "Numerical Data Versus Indicators") listed their priorities the opposite way: Their desired "criteria for use in determining total funding levels for basic research" were found to be, first, the "potential for fundamental new insights," and, second, "science needed to generate technological solutions to major societal needs."[49]

In the same Delphi Experiment, the tenth and last of the "criteria" for funding considered important among scientists was found to be "competitive pressure: activity of greater emphasis existing in other nations." But whereas international competition is the least and smallest part of the image of science reflected by the panel of scientists, the text of *S.I.72* indicates that it is the first and most significant part. A long section entitled "International Position of U.S. Science and Technology" is placed at the very beginning of the book. (In *S.I.74* the section is retitled "International Indicators of Science and Technology.") Emphasis is placed on "international comparisons" – for example, in expenditures; manpower; national origins of literature; literature citations; the "Patent Balance" as a measure of "inventive output" by American nationals versus patents awarded by the United States to nationals of certain foreign countries; relative productivity in manufacturing industries; balance of trade in technology-intensive products; and similar indicators for transnational comparison. A motivating theme for this section can be seen in such findings as "the deficit balance in the high-technology area developed with Japan in the mid-1960's and persisted in the following years, with the largest increase (almost 120%) occurring in 1971."[50]

What this view of science and technology stresses, of course, is the competitive, not the cooperative, element. To be sure, scientists have been known to champion this view too, particularly when trying to get their governments to underwrite large expenditures for scientific facilities. What is, however, rather strange is that we find in these reports no counterweight of the kind so familiar to working scientists, namely, the evidence that actual scientific research is not a national but a transnational enterprise.

Let me give a concrete example. *The Physical Review Abstracts*, a semimonthly publication of the American Physical Society, prints abstracts of articles that have been forwarded from the editorial office of *The Physical Review* to the printer for early publication. A quick check of a typical recent issue[51] shows that – leaving aside minor items such as "Comments" or "Addenda," – 140 research papers are listed. Of these, 51, or about 36 percent, are collaborations with persons at foreign addresses or are written entirely by foreign authors. If one also allows for the fact that many an author listed at a U.S. address appears to be a visitor rather than a "U.S. national," one finds an even more gratifying degree of internationalism emerging from the publication of a U.S. professional society.

However, when the country of origin of the *journal* is chosen as a measure of the "National Origins of Literature" (as in the citation data given on pages 7–11 in *S.I. 72*), not only does a rather different view of the U.S. share of the literature result, but one is almost automatically prevented from seeing or asking about measures of transnational *sharing* in actual scientific work. Future issues of the *Science Indicators* volumes could well try to define indices of such cooperation, among both individuals and institutions, and the relation of such cooperation to the viability of science itself. In this regard, *S.I.74* demonstrates much improvement over *S.I. 72* – for example, in the interesting tabulation that in the science journals of the six major R & D performing nations taken together, almost 60 percent of all citations to the science literature, in eight selected scientific fields, are to articles published *outside* the country of the particular journal.[52]

Eventually, one may even hope that analogous efforts in other countries, of which the OECD surveys of science activity of the 1960s were precursors, would produce a body of work that allows the sciences and technologies to be studied as specific examples of a "world system." We know that from the vantage point of many sciences, each location on earth is only a specialized sample of the universe as a whole. There are common physical laws and uniform physical properties of matter, stretching from one end of our globe to the other, as indeed they do from one end of the galactic cluster to the other. This fact imposes some necessary features on the organization of science as done in different parts of our own world, and makes science an interesting candidate for the study of world systems. The sciences and technologies may already be the best operational exemplification of the concept itself. Moreover, in terms of training, research models, communication, prestige sharing, and so forth, the larger scientific community in the twentieth century is organized on transnational lines and has aspects of universality analogous to those characterizing the phenomena it studies. The system of recruitment and colleagueship at the best universities is a case in point. Scientific data collections are also increasingly produced by international effort. One can well imagine science indicators specially designed to measure these and other properties of the world system.

While the "official" view or model of science that emerges from *S.I. 72* understandably does not stress these aspects, it also omits some indicators that one would expect to find in any model. For example, the relatively low place assigned to education is striking. We are told that federal R&D expenditures for education (in constant 1958 dollars) rose from an almost negligible amount in 1963 to about $100 million in 1972 (or 0.8 percent of total federal R&D).[53] But the report nowhere indicates that this amount was only about 0.1 percent of the annual expenditure of the whole education "industry" (now over $100 billion, of which the federal contribution is still somewhat less than 10 percent) and that very little educational research was being done other than that supported by federal funds. Symbolic perhaps of the ever-ambivalent attitude of NSF to supporting science education research, the NSB report even neglected to mention the substantial and useful role the NSF itself has played in fostering such research, for the NSF was not even on the *S.I.72* list of sponsors of education R&D.[54]

On the other hand, education (and dissatisfaction with its state) was an important aspect of the conception of science that emerged in the poll of scientists and other experts in the Delphi Experiment, recounted in Chapter 6 of *S.I.72*. In the list of "National Problems Warranting Greater R&D Effort," the "inappropriateness and expense of education" fell in the eighth category (out of twenty-one, ranked in terms of "areas which could benefit from science and technology"); 62 percent of the panelists – most of them distinguished scientists, educators, or administrators – thought it an area warranting major increases in R&D.[55] Similarly, 92 percent of the Delphi Panel thought that one of the primary "Factors Impeding Technological Innovation" is that the "education of scientists and engineers [is] inappropriate for innovation."[56] Indicators that would respond to such concerns would have to be far more extensive and more detailed than the scanty ones now given.

S.I.72 also includes discussion of a survey of public attitudes toward science and technology, commissioned by the NSB. Among the interesting preliminary results is the early warning indication of public concern with the ethical and human values impacts of science, and the basic ambivalence or multivalency about science. Thus although a large and rising majority agrees "that science and technology have changed life for the better," 39 percent of the sample of the U.S. population expressing an opinion also did not think that "overall, science and technology do *more* good than harm," and 34 percent of the group expressing an opinion felt "the degree of control that society has over science and technology should be increased."[57]

It would seem that the public, the Delphi Panel, and the NSF authors of *S.I.72* are synchronized at least on the need for a better public understanding of science. The Delphi Panel listed "negative public attitudes toward technology" high among the factors impeding technological innovation, and expressed great concern with the "lack of understanding of [the] process of discovery" and with the "lack of public understanding of [the] role of basic research."[58] Among the criteria for allocating basic research funds among scientific fields, the panel of experts gave considerable weight to whether the work "fosters public understanding of basic research."[59]

Considering this concern, it seems reasonable to develop more detailed indicators of public attitudes toward science, as well as some assessment of the actual level of public understanding of specific scientific knowledge and processes. An educational assessment of scientific understanding by various age groups in the United States already exists in the literature.[60] Such findings might well serve as one element of such a survey. Ultimately, any study of science and technology must allow a visible place for the intellectual benefits reaped by the population at large, and not merely because that population pays for these developments.

The Baconian view of science implicit in the *Science Indicators* reports, and the preoccupation with the "harder" sciences and their technological fruits, prevented any examination of whether there is an appropriate balance of funding between the natural and the social sciences. Many other indications tell us that there is not. For example, the "success ratio" (the ratio of total funding applications to the successful ones) is generally far lower in the NSF programs in the social and behavioral sciences than in the "hard" sciences programs. Only 5 percent of the

NSF's total research obligations (of $591 million) in fiscal year 1975 was made available for the NSF for basic research in all of its thirteen social and behavioral sciences programs, from anthropology and psychology to the history of science.[61] Since the NSF provides one-third or more of all available basic research funds from federal sources for these fields, few alternatives exist for applicants, and the health of the sciences as a *balanced whole* greatly depends, therefore, on the willingness and success of NSF to obtain the needed funds for the large and growing group of America's social and behavioral science researchers. This is obviously a highly charged topic, one that soon would raise the question whether the Congress was wise to be persuaded some years ago not to fund a separate social sciences foundation.

One could go further, praising other individual aspects of the *Science Indicators* reports or identifying needed improvements[62] and criticizing questionable treatment of data.[63] But, in sum, what do the first issues of the *Science Indicators* volumes indicate? The volumes already mark the significant beginning of an important enterprise that will have repercussions on science policy, on scholarly work in the history and sociology of science, and above all on the life of science itself. But, in addition to giving a wealth of data, *do they really show that the sciences and technologies, their quantities and qualities, can be "measured" in a meaningful way?*

I am left with the strong feeling that while a favorable answer to the question is not yet guaranteed, it is highly probable that future issues in this series can converge to a positive result. The problem of making and testing indicators is an immensely attractive one. Good researchers are entering the field. And the volumes themselves have identified and can aid the development of the bases and scope of the enterprise – in strengthening the theoretical foundations and making these more explicit; in grounding indicators in stated problems; in perceiving the fine structure and the qualitative changes more clearly; in infusing a greater spirit of courage and independence; and in bringing together independent research scholars in the sociology and history of science and related areas, in order to cross-check, extend, and professionalize further the making of science indicators.

7

On the psychology of scientists, and their social concerns

Scientific optimism and its costs

There is much discussion these days on ethical problems of scientific advance, the social responsibility of scientists, the participation of scientists in public policy discussions, and the need to bring the scientist and "citizen" together to clarify their mutual expectations.[1] These worthy moves seem to depend on harnessing the natural scientific optimism of academic researchers in the service of their societal concerns. At first glance, these qualities appear to be natural allies – the former reigning inside the laboratory and the latter outside – and we might therefore expect them to be allied in some way, perhaps even stemming from a common trait that, if nurtured, would make both flourish equally.

I suspect, however, that on the contrary we may be dealing with characteristics that, at least for the large majority of scientists in academe, are inherently antithetical. It is perhaps not an accident that scientific optimism is as old as science itself, whereas examples of societal concerns on the part of scientists acting *as* scientists are much more recent. Another clue is that basic researchers in the physical and biological sciences have only rarely looked for their puzzles among the predicaments of society, even though it is not difficult to show that the *lack of relevant scientific knowledge* in such "pure" fields as physics, chemistry, or biology is among the central causes of almost any major societal problems – whether in nutrition, population, pollution, mental health, occupational disease, and so on. (For example, a better understanding of the physics, chemistry, and biology of the detailed processes of conception is still fundamental to the formulation of sounder strategies for dealing with overpopulation and family planning.)

On this point, of course, scientists will defend themselves at once with some clarity and much passion. They will say the discoverers of the laws of thermodynamics that allowed the design of efficient machines, taking over burdens previously carried on the backs of men and animals, were not motivated by anguish over the lot of humanity, nor would such motivation have likely led to the discoveries. It is not a lack of compassion but the nature of science itself that requires us to take the circuitous route of research. Sentimentality will lead not to humanitarian advance but to bad science.

A great deal of this argument is undoubtedly just. Yet what Whitehead called the "celibacy of the intellect" also becomes apparent there. Pessimism about the efficacy of societal concerns, rarely tested, has deeper roots in the psychology of scientists – a subject rarely considered when hopes are voiced for "science-and-society" links. It seems to me that the best way to begin to understand the sources of that pessimism is to look at the opposite side of the coin. A comment attributed to Anne Roe comes to mind: On looking back at her long and distinguished studies on the psychology of scientists, she is said to have commented that the one thing all of these very different people had in common was an *unreasonable amount of optimism* concerning the ultimately successful outcome of their research. Whereas the stereotype of the humanists is that of a rear-guard group, gallantly holding up the flag of a civilization that is now being destroyed by barbarians, the scientists tend to feel that the most glorious period in intellectual history is about to dawn, and *they* will be there to make it happen. As C. P. Snow has said, they have the future in their bones. In truth, everyone who has done something basic in science treasures these memorable periods of individual euphoria when one has found a problem – one that may be tormenting in its subtlety, and all too slow to crack open – but at least one that promises to be delimitable, "analyzable," *solvable*, and therefore worthy of throwing one's whole being into it.

This hopeful, most absorbing and rewarding aspect of scientific research – and also another side to which I shall soon turn – is, however, precisely in opposition to the demands of an outer-directed humanism that, one may assume, is at the base of a person's societal concerns. Indeed, the psychodynamic vectors that propel a scientist into the bright world of solvable problems often turn out, on examination, to have components originating in the flight from the dark world of anguished compromises and makeshift improvisations that commonly characterize the human situation. (To a large degree this is undoubtedly true also for humanists and other scholars. But that is suspiciously easy to assert, and one therefore hopes that the kind of research results I shall summarize for the case of scientists will some day soon be done for the other groups.)

The whole dilemma has never been put better than in a passage in Einstein's writings of which I have long been fond.[2] He was considering who these people are who aspire to live in the "Temple of Science," and acknowledged first of all that they are mostly "rather odd, uncommunicative, solitary fellows, who despite these common characteristics resemble one another really less than the host of the banished." And then he asked:

> What led them into the Temple? The answer is not easy to give, and can certainly not apply uniformly. To begin with, I believe with Schopenhauer that one of the strongest motives that lead persons to art and science is flight from the everyday life, with its painful harshness and wretched dreariness, and from the fetters of one's own shifting desires. One who is more finely tempered is driven to escape from personal existence and to the world of objective observing and understanding. This motive can be compared with the longing that irresistibly pulls the town dweller away from his noisy, cramped quarters and toward the silent, high mountains, where the eye ranges freely through the still, pure air and traces the calm contours that seem to be made for eternity.
>
> With this negative motive there goes a positive one. Man seeks to form for himself, in whatever manner is suitable for him, a simplified and lucid image of the world, and so to overcome the world of experience by striving to replace it to some extent by this image. This is what the painter does, and the poet, the speculative philosopher, the natural scientist, each in his own way. Into this image and its formation, he places the center of

> gravity of his emotional life, in order to attain the peace and serenity that he cannot find within the narrow confines of swirling, personal experience.

What a splendid image: science as self-transcendence, as an act of lifting oneself into a purer state of being! Eloquent support for this view is not difficult to find. James Clark Maxwell confessed in his inaugural lecture at Cambridge (1871):

> When the action of the mind passes out of the intellectual stage, in which truth and error are the alternatives, into the more violently emotional states of anger and passion, malice and envy, fury and madness; the student of science, though he is obliged to recognise the powerful influence which these wild forces have exercised on mankind, is perhaps in some measure disqualified from pursuing the study of this part of human nature.
>
> But then how few of us are capable of deriving profit from such studies. We cannot enter into full sympathy with these lower phases of our nature without losing some of that antipathy to them which is our surest safeguard against a reversion to a meaner type, and we gladly return to the company of those illustrious men [of science] who by aspiring to noble ends, whether intellectual or practical, have risen above the region of storms into a clearer atmosphere, where there is no misrepresentation of opinion, nor ambiguity of expression, but where one mind comes into closest contact with another at the point where both approach nearest to the truth.

There we have indeed a clarion call to move to the high, pure empyrean where one can work undisturbed on what counts for science, while at the same time (and, it would seem, as a necessary consequence) withdrawing from the scene of mankind's problems and concerns.[3]

There are, to be sure, also other sources for scientific optimism. One is what Boris Kuznetsov[4] has called "epistemological optimism": Most scientists feel there is now no *ignorabimus* to be feared, that at least in principle knowledge can be gained indefinitely, without limits. Even the occasional mountainous obstacle will turn into a monument to a memorable victory. Remember the "physics of despair" which, Max Planck confessed, drove him to the quantum, or Niels Bohr's allegiance to the aphorism that "truth lies in the abyss." When the idea of Laplace's deterministic supreme mind failed, it led not to disaster and demoralization, but to Heisenberg's even more fruitful indeterminism. *There* is ignorance one can live with – so unlike the "narrow confines of swirling, personal experience," by its very nature ever unrequited, unsolved, unconquerable, unyielding to reductionistic subdivision into "simplified and lucid," manageable reconstruction.

Even as the laboratory is a microcosm for harnessing the puzzle of phenomena captured within its walls, the uses made of the scientific findings by others, unknown and far away, but presumably for the alleviation of the sorry conditions of human life, is the macroscopic equivalent. The one is the scientific imperative, the other the technological imperative. Both are powered by the same optimistic dynamism. William James in his lectures on pragmatism in 1906 recognized the proper scale of that ambition when he noted that the division between *optimists* (that is, the type he also characterized as rationalistic, intellectual, idealistic, and monistic) and *pessimists* to hinge on nothing less than the most far-reaching matters: "There are unhappy men who think the salvation of the world impossible. Theirs is the doctrine known as pessimism. Optimism in turn would be the doctrine that thinks the world's salvation inevitable" – although, one must add in the case of the scientist, inevitable not by his direct intervention but rather as a by-product, through the working out of the consequences of his ideas by others who will follow him.

At least until lately it has been taken for granted that science and optimism are virtually synonymous throughout modern society, that somehow both the doing and the findings of science will be for the good of mankind – for science's own sake, for its antimetaphysical and liberating philosophical message, for its eventual technological fruits. We know that these beliefs are now held far more widely in the "second" and "third" worlds, both of which endorse science as a path to a better society, than they are held in our own world. The public's ambivalence about the beneficence of science and technology is amply documented[5] – now that the impact of science, medicine, and technology is larger and more visible than ever before, it also is more worrisome and confusing to the wider public.

Of course, not all scientists have been hiding from the swirling fortunes of civic involvement and the forums of public discussion. On the time scale of history, social responsibility and other social concerns as a topic of active introspection by even a small percentage of practicing scientists is a recent notion, largely a post-Hiroshima conception; yet there are already some honorable institutional inventions and landmarks – from the founding of the Federation of Atomic Scientists and the appeals of Leo Szilard and James Franck in 1945 to the Asilomar Conference on Recombinant DNA Research (1975). Other achievements for which research scientists can take a large share of the credit and which must at least be mentioned include the Pugwash movement, the Committee on Science and Public Policy (COSPUP) of the National Academy of Sciences, the codes of ethics now adopted or under discussion among scientific societies, the relevant activities of the National Institutes of Health (NIH) including the institutionalization of concern with experimentation on human subjects, the AAAS Committee on Ethics and Science, and the development within NSF and NEH of programs for the study of the ethical and human values impact of science and technology. These institutions, and the few real statesmen of science we *do* have now, are evidence of goodwill and – one fervently hopes – of a reservoir of talent and interest among hitherto uncommitted scientists.

But there's the rub. The vast majority of working scientists in fact are quite happy to leave the discussion of societal concerns to the small minority. Most in the silent majority and even some in the vocal minority, although willing to tolerate such discussions up to a point, would not go as far as to make such concerns a matter of personal, active participation. I would estimate that the fraction of scientists in the U.S. active in such matters (e.g., helping to formulate or administer science policy on matters of explicit societal concerns or even writing or teaching occasionally on the topic) is of the order of 1 percent.

On the assumption that scientists will have to be involved in larger numbers, we thus come to the crucial questions: Do the public's expressions of ambivalence with respect to science, together with the existence of a core of concerned scientists, constitute harbingers of a substantial increase in the participation of scientists on issues where science and societal concerns come together? Or, on the contrary, could it be that we have already gathered up the largest fraction of that small minority of scientists which is susceptible to such considerations – that from now on we shall find it ever harder to add more recruits from the ranks of working physical and biological scientists? To use an analogy from physics, does there exist some inherent barrier so high and wide that the small quantum leakage through it cannot be expected to become much larger?

I believe the more pessimistic of these two possibilities is closer to the truth. We shall have to face the possibility that for the large majority of scientists euphoric personal commitment to and pursuit of science *as currently fostered and understood*, and the hoped-for societal concerns needed on the part of professionals in today's world, are at bottom orthogonal or possibly even largely antithetical traits – antithetical both in terms of the psychodynamics of the majority of individual scientists *and* in terms of the social structure of science as a profession. A chief aim in this chapter will be to scrutinize and juxtapose widely scattered data now in the literature, and to see what causal connection may exist among the personal-psychological, institutional, and sociopolitical factors in the scientific profession that have fostered the current state. We shall find some evidence that the selection, training, and socialization of scientists are biased in just the direction where the requirements of a code of explicit, personal societal concern are least easy to fulfill.

We may not have to go as far as believing seriously the anguished exclamation of Albert Szent-Györgyi, who was recently quoted to have advised: "If any student comes to me and says he wants to be useful to mankind and go into research to alleviate human suffering, I advise him to go into charity instead. Research wants real egotists who seek their own pleasure and satisfaction, but find it in solving the puzzles of nature."[6] However, it may be true that, for many, scientific work, as shaped by present practices and images and from a very early point in the career, is now perceived by the scientists themselves as being located at one end of a seesaw, rising only at the cost of paying correspondingly less attention to practical societal concerns. I shall argue that a better understanding of the psychodynamics of young scientists and of the institutional pressures on them is needed to show us how to bring out more of the socially aware side of scientists than we thought was needed in our golden age of innocence, so recently past.

Stereotypes and prototypes

Making due allowance for the fact that the study of personality and character structure is not one of the "hard" sciences, the way to begin is to look at what data there are concerning the raw material, the scientists themselves. How flexible is their frame of thought and behavior? How willing are they to struggle with opinions and beliefs that, in the nature of the case, are inherently unprovable, or with problems that are inherently resistant to simplification, or even to moderately satisfactory solutions? How able are they to deal with the personal stresses and strains that are part of any participation in serious debates on such unverifiable and unyielding problems as societal concerns?

There are a few useful published studies on the psychology of scientists. Some are retrospective analyses on groups of children and young persons who later become scientists; some are on samples of adult members of the profession; and a very few are individual psychohistorical case studies of scientists such as Newton, Darwin, Einstein, and Fermi. Most of the studies that have been made focus a great deal on what may be called the *styles* of creative scientists, the kind of behavior they exhibit fairly consistently.[7] It is easy enough to list the main features that define scientific style as commonly understood at present; the point to keep in mind is, however, that this common understanding is shared not only among the scientists themselves but also, in about the same terms, among the wider public. Thus, on the basis of a survey of the existing literature in the field, B. T. Eiduson concluded: "Scientists as a group seem to be caught up in the same stereotypes that the public holds about them, *and, in fact, the researchers seem to have been drawn into science by some of the same fantasies and stereotypes.*"[8]

1. The scientist typically insists that in written reports of work – from students' laboratory write-ups to the papers we referee and the textbooks we write – the individual traces of the personal self be attenuated as far as that can be done. Hence the impersonal style of the scientist, for which there are only few counterexamples. The aim is to make one's work seem "objective," repeatable by anyone, or, as Louis Pasteur said, *inevitable*. This is part of the scientist's value of "other-orientation," to use Talcott Parsons's term, and at the same time is another way of transcending the world of personal experience. Through the analyses of Robert K. Merton and others we know that disputes and priority fights exist as intense undercurrents. But their importance is generally disavowed by scientists, as are the deep thematic presuppositions underlying apparently "neutral" presentations.

This eradication of individualistic elements from publication is of course exceedingly functional insofar as it helps to minimize personal disputes of the unresolvable kind and removes interpersonal obstacles to consensus. But it is also seen as responding to the epistemological demands of science itself. With typical succinctness, Einstein summarized this view of reality in a manner that shows how the concept of the individual self sinks into the shadows: "Physics is an attempt conceptually to grasp reality as it is thought independently of its being observed. In this sense one speaks of 'physical reality.' "[9]

2. The second, related commandment of the scientific ethos, as commonly understood, is to be logical, not emotional. Statements that fall into areas with a large component of not easily verifiable or of falsifiable content are frowned upon, and issues dealing with ethical conflict, responsibility, or even long-range prediction of the technological applications of scientific findings are therefore not expected to be raised in scientific meetings. Mere opinions, preferences, emotions, and instincts must be repressed, and even the exhilarating flights of intuitive imagination must be recast in deductive style to be respectable.

3. Errors or unlikely hypotheses are to be avoided at all cost, not least because the scientific community is far less tolerant or forgiving on that score than almost any other group. A batting average of only, say, 0.500 would not look good at all in this

league. One is taught to repeat, verify, and repeat again before going public, not the least to avoid endangering one's dignity and credibility in the experimental field.

4. The desired outcome is the simple, not the complex. One aspires to economy of thought and uses Occam's Razor to help achieve it. Therefore in most cases, and for most scientists, the reductionistic strategy is far safer than a synthesis-seeking one. The main charm of work in a physical or biological science is that it ideally permits one to formulate statements of simple lawfulness that is the very opposite of the complexity characterizing most social questions and interactions.

5. As with the content of science itself, the setting in which one does one's science is ideally as removed from interpersonal disputes as possible. (This is a significant point to which we shall return.) As the psychologist David C. McClelland put it with ample research documentation, "Scientists avoid interpersonal contact," and "scientists avoid and are disturbed by complex human emotions, perhaps particularly interpersonal aggression."[10] Even recent changes in the self-identification of young scientists only reinforce this old theme:

> There is evidence, however, that differences in the way science is being practiced today are accompanied by certain differences in the identifications that scientists have with other scientists. An example of this changing trend is the researcher's shying away from identification with the "great but maladjusted" or "eccentric" scientist. Reverence for forefathers whose outstanding minds were sometimes housed in very peculiar and odd personalities still exists, and yet the newer scientists seem consciously to be dissociating themselves from peculiar and difficult associates or students, knowing full well that they may be thus shutting themselves off from some very creative workers in their own laboratory. These men nowadays prefer to depend for progress on well-organized, smooth-running, large-scale operations, whose stability demands the minimum of interpersonal relationships, especially disturbed ones.[11]

A paradigmatic hero who fits most of these five traits superbly and who remains one of the widely admired scientists and role models is Enrico Fermi – ever "cool" and rational, virtually always right, generous in sharing his intellectual self, always focused on the pioneer problems of physics of the day, finding the unraveling simplicity in the most surprising yet convincing way but also, to very nearly the end of his life, studiously avoiding participation in any of the political or science-society questions that raged through both history and science in his time.[12] The hero among most scientists is clearly *not*, say, a Leo Szilard – as "hot" as Fermi was "cool," also a superb scientist but spreading himself over many fields from physics to biology to cybernetics, fond of speculative and daring proposals, and above all deeply and successfully occupied with the need to make his life and influence count in matters concerning the impact of science and technology on political (national and international), ethical, and social problems. Fermi and Szilard may be seen as prototypes that define two extremes of the range of models of scientific behavior, one widely accepted, admired, and imitated, the other regarded with a certain affection but deeply appreciated by only a relative few.

We shall leave them for a time. For the moment we should note that the *differences* between these or any other scientists are not accounted for in the public stereotypes of working scientists. The same problem adheres to some of the psychoanalytical studies of scientists, for example, that of Lawrence Kubie,[13] who held that their work can be coordinated with symptoms of what Kubie called "masked neurosis." The psychoanalytical perspective invites (perhaps all too readily) the construction of parallels between the prototypical behavior of scientists (as well as other scholars) and that of the obsessive-compulsive personality type.[14] If one believed the parallelism fully, it would raise the insuperable puzzle how and why scientists for the most part do in fact preserve themselves sufficiently to be as effective as they are, as both professionals and human beings. Yet there are perhaps certain analogous characteristics:

(a) There is narrow, intensely focused attention, almost constant concentration on one area within which the more creative persons roam freely. (The powers of concentration in scientists such as Gauss were legendary. At moments they have amounted to voluntary sensory deprivation – for example, Fermi reported that he had failed altogether to hear the blast of the atomic bomb at the Alamogordo test because he was working on a problem of measuring the intensity of the blast by a simple, improvised method.) Attention is diverted and dedicated to the high drama and wonderful entertainment going on in one's own brain.

(b) One is driven to succeed and therefore keeps away from situations that are unlikely to yield success. There is a tense deliberateness to clear up puzzles and ambiguities or to remove the threat of the unknown, and restless intellectual (or physical) activity.

(c) One is preoccupied with control, with the need for order and structure that extends from regularity of schedule to ready acceptance of authority outside one's own field of expertise.

(d) And finally, there is only a limited need or ability to deal with affective experience, hence – as we have noted – the avoidance of open battle and other social risks. (One remembers Newton's remark that he would be rid of that litigious lady, Natural Philosophy, promising in a letter to Oldenburg during the controversy following his early work on optics, "I will resolutely bid adieu to it eternally, except what I do for my own private satisfaction"; and later: "I desire to decline being involved in such troublesome and insignificant disputes.")

Findings of sociological-psychological studies

Emile Durkheim warned, "Every time that a social phenomenon is *directly* explained by a psychological phenomenon, we may be sure that the explanation is false."[15] But one need not demand the exclusion of other than psychological causes, or the acceptance of the framework of orthodox psychoanalytic theory, or the postulation of parallels with obsessive-compulsive types, and yet allow the possibility that the scientific-societal polarization many scientists seem to experience may well have psychological roots. In the absence of one clearly fruitful framework, the safest entry point for the time being seems to be through a study of the results of the more sociological-psychological investigation.

Despite all the changes in the profession, a major classic of this kind is still Anne Roe's *The Making of a Scientist*,[16] published over two decades ago. A clinical psychologist, she studied the careers of sixty-four leading, male, U.S.-born scientists (in the biological, physical, and social sciences), concentrating on the most eminent in each field. Separately, her findings within this group were checked to some degree against a larger sample of more ordinary scientists. Her chief aim was to study the relation between vocation and personality structure. In an earlier study on artists, she had discovered that definite and "very direct relationships [exist] . . . between what and how the man painted and what sort of person he was and what sort of problems he had." As to scientists, she significantly found that "although such relationships pertain among them also, they are usually much more obscure and the technique of reporting scientific results serves rather to hide the man than to display him."

Her findings do not add up to a profile of part-time social reformers, nor does the total pattern seem altogether unfamiliar to the nonspecialist or in drastic conflict with the popular stereotype. Not unexpectedly, curiosity about a special area, to the exclusion of all else, was the chief characteristic, evidently transcending even the intense devotion to work (that was also found among her group of artists) and the willingness to select one variable for study, holding the rest constant where possible. "There is probably no more important factor in the achievement of this group of [physical] scientists than the depth of their absorption in their work." This trait is of course of high utility because, as Roe understands, it is unlikely that one can achieve anything of great value if one does not dedicate oneself deeply, even passionately to the work.[17]

A whole set of her other findings is naturally correlated. The age at marriage was rather late for that group – twenty-seven years on the average. (The divorce rate among physical and biological scientists was markedly below average, 5 percent among the physicists and 15 percent among the biologists in her sample.) She found her scientists often to be working right through a seven-day week and during sleepless nights. Even when they engaged in sports, it was preferably individual rather than team play. Significantly, only four in her study group of sixty-four leading scientists "played any active part in political or civic organizations." Most of the physicists and biologists "disliked social occasions" and avoided them "as much as possible." Only three out of the sixty-four were seriously concerned with organized religion. Far more than the rest of the population, her scientists came from families where "value is placed [on learning] *for its own sake* . . . not just for economic or social rewards." A decisive event in their career choice was early research experience or, even before that, "the discovery of the possibility of finding out things for oneself." One might say that their first and most interesting scientific discovery may have been their own selves.[18]

Many of the scientists had "quite specific and fairly strong feelings of personal isolation" when they were children. A strikingly low number of theoretical physicists, only three out of twelve, reported that they had "good health and normal physical development" as children. Loneliness or the existence of very few friends was commonly reported by both the physicists and biologists. A main hobby during childhood was solitary reading. "Many of them [physicists and biologists] were slow to develop socially and to go out with girls." Whereas this was painful to some, it did not seem "to matter enough to most of them to do anything about it." "There is a characteristic pattern of growing up among biologists and physical scientists . . . The pattern is that of the rather shy boy, sometimes with intense personal interests, usually intellectual or mechanical, who plays with one or two like-minded companions rather than with a gang, and who does not start dating until well into college years." Even then, dating may be a very secondary matter. "This is in great contrast to the social scientists . . ."

None of the scientists, particularly the physicists, liked business. "The extreme competitiveness, the indifference to fact, the difficulty of doing things personally, all were distasteful to them." Coming now closer to the finding reported also earlier, Roe notes that "the biologists and physicists . . . are strongly inclined to keep away from intense emotional situations as much as possible." On the basis of the Thematic Apperception Test, she reports that "none of these groups is particularly aggressive."

Summing up, she finds that with all the variety in her subjects, "there are patterns, patterns in their life histories, patterns of intellectual abilities, patterns of personality structure, which are more characteristic of scientists than they are of people at large . . ." The lack of early and continued attention to societal concerns is part of that general pattern, both of life history, of work, and of personality structure.[19] She contrasts this with the ancient stereotype that the "scientist is a completely altruistic being, devoting himself selflessly to the pursuit of truth, solely in order to contribute to the welfare of humanity." To be sure, from World War II on, "scientists are being forced to consider the social repercussions of their work," and, she agrees, it is "an excellent thing both for them and for society." But the characteristics of early career development and personality formation show the built-in obstacles to the easy fulfillment of that hope.

Roe's study was perhaps the first popularly noted, scientific psychological work that bore out by research results the long-suspected basic dichotomy implied for many in their choice between a scientific career and a career that concerns itself with the world of personal relations. In later studies, Roe accumulated the evidence for a powerful generalization that has proved to be of value for other career-development studies: "Apparently one of the earliest differentiations, if not the first one, in the orientation of attention is between persons and nonpersons."[20]

This fundamental perception of a thematic antithesis, also familiar from other psychological studies, was bolstered and documented by subsequent research, perhaps most strikingly by the study of William W. Cooley and Paul R. Lohnes. Their work, *Predicting Development of Young Adults*, was a five-year follow-up study based on the 1960 Project TALENT Test-Inventory of almost a half-million young people in the United States, grades 9 through 12.[21] Cooley and Lohnes quote Roe's finding on the earliest differentiation and add, "This observed polarization of interest in childhood is built into the [career development] tree structure as the first branching, presumably occurring in males round the fifth or sixth grade."[22] (Indeed, the dichotomy may in part be established much earlier, and a genetically determined component is of course also not out of the question.) They found that the polarization of general interests, which has also been called the "people-versus-thing" (or "people-versus-ideas") polarization, is

> the earliest trait for which there is fairly solid research evidence . . . The distinction is whether the boy's primary orientation is toward people or toward science-technology. This orientation first appears in student talk about vocations somewhere around fifth or sixth grade, when the student passes from the fantasy to the interest stage in his career development. In interest inventories this basic orientation shows up as interest in the various science, technology, mechanical scales *versus* interest in areas which involve more people contact, business, humanistic or cultural concerns. This is the earliest variable on which we find we can classify careers, and thus in our tree structure this variable defines the two branches at the first level of branching.[23]

No matter that counterexamples exist among one's colleagues, and no matter how skeptical one may be about the methodologies in some psychometric studies, the solemn warning implied in the findings of Cooley and Lohnes seems to me sufficiently bolstered by other careful research, such as that of James A. Davis,[24] and must be taken seriously. The "people" versus "thing" potential is undoubtedly connected functionally to differences in styles of effective thinking and feeling – for example, those modes, means, and ends that are fundamentally logical, experimental, invariance- and simplicity-seeking, versus those that are predominantly affective, intuitive, and ambiguity-tolerating. But the danger appears to be that such predominant preferences, which can be made to coexist tolerably, are allowed to degenerate into polar opponents that tend to become mutually exclusive. The "interests" that show up early in these interest inventories are, at this time in history, still cast and perhaps nurtured in ways that permit a young person all too easily to make the "people" versus "thing" choice – and a whole set of institutions, in education, guidance, and peer group support, is designed to help make a choice, and to make it comfortable enough to live with it. Moreover, as those of us can attest who have worked hard to modify the present institutional context – for example, by inventing and injecting a combination of scientific, humanistic, and cultural concerns into the educational system for young people, or by trying to bring to the attention of scientists and humanists alike the ethical and human value implications of science – such efforts must all overcome an immense amount of disbelief, resistance, or hostility. Indeed, the chief obstacles are those professionals in decision-making positions who have "made it" precisely by sticking to their early choice between things and people on reaching the bifurcation of interests and who are now successfully barricaded behind that self-chosen barrier. Here lies the crux of the problem of interesting more people in "people-*and*-thing" subjects such as research on the public policy or the ethical impacts of science.[25]

In a study of seventy-nine eminent scientists, Warren O. Hagstrom stressed that a key element of internal social control in science is peer recognition.[26] It is essential to keep in mind that the risk of ostracism or isolation is difficult to bear for scientists, who depend greatly on a somewhat distant and abstract form of peer approval. As the scientific community is presently constituted, preoccupation with social action invites just such risks, at least before one is amply fortified with professional honors. Receiving adequate recognition is less likely when traditional role behavior is violated. One must therefore have a clear mandate before one introduces major changes into such a carefully structured social entity, for anything that isolates a scientist tends to lower his reputation and productivity greatly.

In this roundup of survey studies, let us finally look at statistics gathered specifically on the scientist as a young person, particularly William W. Cooley's work, published in 1963.[27] This was an investigation – again only on males – over a five year period (1958–63) "toward the goal of a better understanding of how, when, and why boys enter careers in science and technology." The longitudinal study involved 700 persons, in groups from grade 5 through Ph.D., in overlapping sets, to find the attributes of boys at different ages who will remain in the "potential scientists' pool" (PSP), at least in commitment if not in performance. Cooley finds the PSPs to be high on many good qualities but markedly low on social interest, AVL (Allport-Vernon-Lindzey Scale), welfare interests, and political interests or interest in "power over others."

This finding invites a comment on the obvious fact that more than a few good scientists have nevertheless chosen to go into positions of political prominence and power and that a few have even done a very good job of it by any standard. This fact however does not cancel the pervasive declarations of disapproval or even contempt among scientists (though not without ambivalence) when the suggestion comes up of participating in affairs having to do with public policy or political power.[28] Evidence of the dismay many scientists have experienced in participating in or merely watching the essential business of rendering scientific advice on politically sensitive matters is not difficult to find. Case studies in the weekly "News and Comments" section of the magazine *Science* or in J. Primack and F. von Hippel, *Advice and Dissent*,[29] and Philip Boffey, *The Brain Bank of America*,[30] contain enough "horrors" to discourage precisely some of the most sensitive and creative scientists from participation. Of course this is not the intent of the authors of these important publications, and they would be the first to urge that it would be entirely counterproductive to turn the battle over to the hardened politicos and bishops – to be found among scientists as in any other academic group – who are ever willing to shoulder the necessary tasks all by themselves.

In looking for a theoretical basis for distinguishing between the careers of scientists and nonscientists, Cooley comes on a significant point: "Outside of the emphasis upon the superior abilities of scientists, introversion[31] is perhaps the most frequently cited 'personal characteristic of scientists.'" Cooley hastens to reassure us that "although scientists are far less interested in dealing with people in their day-to-day work, their concern for humanity and human rights appears to be no different from that of the other groups studied."[32] Cooley's remark on the scientist's "concern" is not presented as a research finding, and I would like to think of it as a hopeful perception of a trait to be nurtured, even if it starts out as something of an abstraction. In the extreme case it reminds one again of the scientist who remains one of the best role models, not the least for his ability of frank self-assessment. Einstein wrote:

> My passionate sense of social justice and social responsibility has always contrasted oddly with my pronounced lack of need for direct contact with human beings and human communities. I am truly a "lone traveler" and

> have never belonged to my country, my home, my friends, or even my immediate family, with my whole heart; in the face of all these ties, I have never lost a sense of distance and need for solitude – feelings which increase with the years. One becomes sharply aware, but without regret, of the limits of mutual understanding and consonance with other people. No doubt, such a person loses some of his innocence and unconcern; on the other hand, he is largely independent of the opinions, habits, and judgments of his fellows and avoids the temptation to build his inner equilibrium upon such insecure foundations.[33]

Conclusions

Still other studies could be discussed here if space allowed, and we would find again and again confirmation of the general picture that we have obtained here concerning the interaction between the research scientists' cognitive characteristics, developmental and background characteristics, and personality patterns. By way of summary, some conclusions and caveats:

1. Optimism and human goodwill are well-known personality characteristics of scientists. But the optimism about the role and eventual success of science is quite possibly correlated with a pessimism or skepticism about the chance of doing much personally to guide human affairs. As a rule, a scientist's goodwill seems to be held abstractly (and is expected by others so to be held in their stereotypical image of scientists), rather than being acted out operationally in the sphere of social interactions – except, it should always be emphasized, in the area of collaboration among scientists themselves (in laboratories, teams, student-teacher exchanges, and so on), which is generally as humane as one is apt to find anywhere else.

2. The coherence of a scientist's identity, as presently constituted and reinforced by social role models and by educational and other institutions, may well depend on this antithetical casting of scientific optimism and wider social concerns.

3. Whether it is psychologically or socially imposed, an early choice seems to be made between people and things as antitheti-

cal entities when young people reach certain branching points in the career tree. The stereotypes associated with these choices are shared by scientists and nonscientists alike and are thereby reinforced. The system is "in resonance," with personality traits and institutional settings supporting one another. If it errs, it is on the conservative side. Thus Cooley[34] shows that the norms now used for guidance in high school are out of date even with respect to recent changes in the content and institutions of science itself.

4. We must remember Thoreau's remark: "Winter, with its ice and snow, is not an evil to be cured." My intention has not been to deplore or castigate or moralize. Nor do I think lecturing at scientists about social responsibility, ethics, and morals will do much to shift the mean value of the response curve of scientists toward more societal concerns. A head-on attack of trying to increase the average score on some "social interest" scales is surely doomed to failure.[35] Rather, I want to fasten on an essential fact that is often not stressed sufficiently: that there is here, as in all natural populations, a *spread,* a normal or quasi-Gaussian-type distribution.

Imagine a bell-shaped curve in a plane where the horizontal axis indicates the present degree of responsiveness of physical and biological scientists to societal concerns. The highest vertical point on the curve indicates the most probable value for this population. (From all we have heard so far, this peak is likely to occur to the left, at fairly low values along the horizontal scale, compared with the curve for others – nonscientists, for example.) Educational systems, professional societies, and other social institutions tend to be designed so as to "resonate" with the mean or most probable value of a characteristic on which the institution depends. Normally, it is feedback from the peak value that stabilizes a system.

Different scientists are of course located on different parts of the bell-shaped curve; a Fermi would presumably appear near the low, left end, and a Leo Szilard or a Linus Pauling on the other wing, near the high end. This perception suggests some strategies for change. One is to increase the amount of feedback coupling of the system at the "Szilard end" – for example, most simply by beginning to make teachers, guidance counselors, and scientists at their more accessible moments (e.g., during attendance at professional scientific meetings) more aware of the mere *existence* of the right wing of the curve, the historical existence of successful scientists who, in their very lives, exhibited the possibility of combining *rather than choosing between* people and things. The public "image" of scientists often needs correction; thus Madame Curie's reported remark "Science deals with things, not people" has to be supplemented by her own dedicated participation in groups such as the League of Nations Committee on Intellectual Cooperation. Another strategy is structuring opportunities so that participation is possible or choices are confronted within the setting of a professional society. The formation of the Forum on Physics and Society within the American Physical Society is a relevant example of such an institutional development. The availability of mid-career opportunities (such as fellowships or seminars) also needs to be enlarged greatly for scientists who are ready for and interested in the interactions between science and the world of human affairs.

5. This is not the place for a detailed discussion of strategies to bring about major changes by changes in the classroom, or even to summarize all the likely "first-aid" tactics, but some points are so clear that they must at least be mentioned. Particularly at the precollege level we need more large-scale curriculum innovations opposing the pervasive dichotomy that Cooley and Lohnes (among many others) quite rightly identified as interest in the various sciences versus interest in areas which involve humanistic and cultural concerns. One way is to put at least an effective modicum of history of science, epistemology, and discussion of the social impact of science and technology right into the educational material used in the science classes. This was a chief motivation for those of us who have worked on one such program.[36] Other educational experiments along these lines, difficult though they may be to execute in detail, include more positive and more aggressive college-level programs designed to encourage actively what has been called "double literacy."[37] I have no fear that such arrangements will scare off from a science career those who, for good reasons of their own, must do science as their flight to the pure empyrean. I do expect that the mere visibility of the option, properly presented, can attract to

science a latent population, including socially concerned persons who, given the current self-image of the scientist, would normally not think of science as their proper career field. Scientists (and nonscientists) now surely need also more exposure to substantial discussions of the ethical and human value impacts of science and technology. The flight of most members of a profession to the high empyrean, where they can work peacefully on purely scientific problems, isolated from the turmoil of real life, was perhaps quite appropriate at an earlier stage of science; but in today's world it is a luxury we cannot afford.

It may, of course, turn out that we must settle for the "seesaw" for many if not most of the productive research scientists. High scientific (or any scholarly) performance is often obtained at the cost of low social involvement, and the very identity formation of the scientist may depend on his carving out a manageable portion of the world for himself and reducing, therefore, his ties with the rest. This is the classic model, and, so far, educators not only have not resisted it but have built it into their assumptions. We clearly can no longer do this, and we must try to do something about it. As an educator and optimist, I have a professional affirmation that education will find in the "Szilard" tail of the curve, not Szilards – because that would be too much to hope for – but at least people who have enough of the Szilard component to meliorate the stern demands of science. I place much hope in finding incentives that will influence students to do excellent science and yet to be interested actively in societal concerns, or at the very least to be able to work together and sympathize with those who do.

6. A further strategy is this: Since it is not likely that the mean value of the distribution can be shifted by converting those now on the low, left wing of the curve, we must try to add to the population of the right, high wing of the curve by increased visibility and effectiveness, in addition to increased recruitment. By a kind of affirmative-action process, the members at that end of the spectrum should be given more than their usual low share in matters such as filling leadership positions in professional societies (as the American Chemical Society did a few years ago in electing Alan C. Nixon as its president, the American

Physical Society in electing V. Weisskopf and W. Panofsky, and Margaret Mead's election to the presidency of the AAAS).

7. I believe it is reasonable to hope that the scientists' fear of the intractibility and seamless quality of societal problems – which has made the flight to simpler models of the world functional and reasonable – can be shown, in sufficiently many interesting instances, not to be borne out by the facts. Except in biomedical areas, we have had few pioneers willing to map out plausible routes to the kernel of scientific ignorance at the center of complex societal difficulties.[38] Nor has there been adequate financial support for work that is unlikely to give early results. Hence the attitude of the profession may be a self-fulfilling fear. But once even a few scientists have dared to identify and successfully worked on such essential questions, the bridge will become progressively more firm, more visible, more traveled, and more plausible. Eventually, the fear of unavoidable failure should be deprived of its rational component. Much is at stake here, and much depends on the generation of a few prototypical successes.

8. Finally, we need more research on this whole topic. In this period of rapid change we need to know far more up-to-date facts, and not only because outdated stereotypes are still being encouraged. Most data available are pre-1968, and there is evidence that the imagination of young people today interested in becoming scientists is more flexible and more ready to risk people-*and*-things thinking than used to be the case.[39] The methodology of research has also been much improved since the days when many of the data still used today were obtained.

There have been a host of other significant changes as well. Consider the situation with respect to women in science. As was evident from my summaries, most of the available data on scientists are on males in the United States; there has been an outstanding lack of research and useful results on women scientists.[40] Cooley and Lohnes had to confide:[41] "If we ever hoped to develop a theory of careers that would apply to both sexes, the experience of wrestling with the follow-up data has disabused us of that notion . . . We have done very little with the problem of female career development, either empirically or theoretically. The Project TALENT girls, in their development toward womanhood, are a challenge we haven't really met." And again: "We subscribe to the suggestion of Roe and of Super that the career process for women needs to be conceptualized differently and researched differently than for men. The process is probably more complex and difficult to study for women . . ."

In the meantime, there are straws in the wind that the career goals and self-identification of young women with respect to science may be shifting rapidly in the United States. For example, Poll No. 101 of the Purdue Opinion Panel[42] shows in a national poll that 5 percent of the girls in high school in spring 1975 indicated a preference for the job family "engineering, physical science, mathematics, and architecture," *up* by a factor of 2½ from the results for the poll of 1972. Moreover the old stereotypical discrimination against women with respect to scientific ability and mathematical interest may be fading fast: "The results of this study indicated that high school students hold positive attitudes toward science career roles for females, with nearly unanimous consensus on this issue. High school students as a whole believe that females are equivalent to males in scientific ability and interest in mathematics."[43] It is a testable hypothesis that the new influx of young women into science who are coming from outside the older, stereotypical population may yield more representatives from the right, high wing of the bell-shaped curve referred to earlier – at least if they find role models and other support to encourage the trend.

It is evident that a whole spectrum of interesting research problems suggests itself in the still very obscure area called the psychology of scientists. How is it really possible that such a human enterprise as science, in the hands of such a diversity of people, does in fact yield relatively invariant information? How do the various and perhaps contradictory elements in the psyche of the scientist somehow function together to produce a body of work that transcends the individual's limitations? More understanding of matters of this kind will help us in the task of fusing native scientific optimism and mature societal concerns among scientists.

Part III: Public understanding of science

8

Lewis Mumford on science, technology, and life

For nearly a half-century, Lewis Mumford has been writing on one of the chief questions of our time: how technology can be directed to meet the needs of society. Starting with a book on utopias in 1922, he has developed his message in nearly two dozen books, including the influential *Technics and Civilization* (1934), *The Condition of Man* (1944), and *The City in History* (1961). This message has reached a kind of climax in Mumford's *The Myth of the Machine II: The Pentagon of Power* (1970).[1]

Its chief theme, announced on the opening page, is the "wholesale miscarriage" of technology in its present form – "megatechnics." Mumford wishes to show us how and why this happened, how it undermines "our capacity to live full and spiritually satisfying lives," and what can be done about it. Though the volume is somewhat unwieldly and repetitious, Mumford's style is as fresh and passionate, his prose as persuasive as ever. Moreover, the current of the time has caught up with him. He is now reaching a far greater circle of readers than ever before. Perhaps to his surprise, Mumford now finds himself in the front line of a popular parade that may adopt his message as a battle flag. This success, together with what it implies for the relation between science, technology, and the citizen, seems to me in many ways the most significant – and disturbing – aspect of the book.

First, a summary of its contents – as far as possible using Mumford's own words. In broad strokes, with rarely more than anecdotal backing for his points, he sketches the history of the "myth of the machine," which he finds has obsessed Western society and misdirected our energies. Technology before the eighteenth century, far from being negligible, had a beneficent form that Mumford calls "polytechnics" – variegated, based on handicraft, small in scale and slow in pace of production and consumption, but just for that reason held in check from being rampant and dominant. We are told the pace allowed for chatting and singing, for leisure within work, for time to converse, to ruminate, to contemplate the meaning of life.

But increasingly this stability was undermined by new principles and incentives of capitalist finance and the imperious demands of militarism for weapons. The resulting standardization and mass production subverted the earlier conceptions of diversification, self-control, and voluntary censorship of inventions, as practiced for example by Leonardo da Vinci. Somehow the rise, first in science in the sixteenth century, of what Mumford identifies as the mechanical world picture, gave all these disparate efforts the subjective unity needed to ensure their eventual dominance. It turned out to be an ideology that gave absolute cosmic authority to the machine itself, as against human concerns. That, in short, is the myth of the machine.

Mumford sees this sequence of events as a replay of a similar disaster, discussed in some detail in the previous volume in this series (*The Myth of the Machine I: Technics and Human Development*), which overtook mankind in the Pyramid Age of antiquity. For while there may have been other reasons for the mechanization of the world picture – for instance, the ritualistic routines and careful accountancy of time and money in the monastic orders, or even a profound fault in Christian theology by which God was thought to have given man, as the only creature with a soul, the earth and all nature for his unlimited use and enjoyment – the most ominous influence on man's organization-seeking mind was Copernicus's paradigmatic model of the sun-centered universe. Its eventual acceptance somehow, unexpectedly, brought to the fore the other brutal habits of mind of the Sun God worshipers of the Pyramid Age: a belief in the strictest sort of determination, and hence in an absolute order, political, social, and industrial. Analogous to the large-scale use of slave labor to build monuments in Egypt, there ensued a similar regimentation of factory or labor systems which provided the major impetus for the industrial revolution that followed.

There is now a dubious alliance, Mumford continues, between scientific determinism and authoritarian control which menaces human existence. Unlike Aristotle, who is praised for having taught that living organisms are endowed with autonomy and purpose in mind, Kepler, Galileo, Descartes, and Newton are particularly singled out for condemnation of their methodology. The Sun God dazzled and blinded them, we are told, and made them forget the artist, the subjective side of life, and the very presence of man, even the phenomenon of life itself.

We may thank these men and their followers to this day for the exchange of a Christian's universe, focused on man's existence and his ultimate salvation, for a purely impersonal universe, without a god except the blazing sun itself. Only rarely, as in the America of Emerson, Thoreau, Audubon, and Melville, could the romantic and utilitarian personality live side by side, not merely coexisting but prospering together. Today, Mumford explains, only a handful of heretics, mostly poets and artists, dares to hold out against the machine-conditioned utopia; he also mourns the vanished pleasures of physical work. Human evolution is said to have been arrested in favor of a closed and completely unified system, with an unchecked tempo of mechanical and scientific development in a few selected directions. The all-pervasive technological imperative has brought with it a brutal conquest of land and nature, and "the effecting of all things possible," in the phrase of Francis Bacon. As the great cities grow, the vices, perversions, corruptions, parasitisms and lapses of function increase disproportionately, and finally must terminate in Necropolis, the City of the Dead. (Here, and in many other places throughout the book, we hear Mumford in the role of Isaiah, who thundered: "Woe unto them that go down to Egypt," and "how is the faithful city become a harlot! She that was full of justice? Righteousness lodged in her, but now murderers.")

The most ominous development was the fairly recent consolidation of the various forces, interests, and motives into a system Mumford variously identifies as the megamachine, the power complex, or the Pentagon of Power (the subtitle of the book).

The main components of the system are these five: *power* in the sense of the use of energy, but particularly atomic energy that has given the state a weapon of pharaonic dimensions; *political power* backed by weapons; *productivity* for the sake of *profit;* and *publicity* or propaganda by which the elite obtains authority and credibility. The chief purpose of this system is the aggrandizement of the power complex for its own sake, and it is this purpose that masquerades under the name "progress." The denial of aesthetic impulses or the unchecked exhaust of pollutants are merely by-products of the workings of the power complex: Mumford brings up to date the old saying "haste makes waste" to read now "haste and waste make money."

Serving this coalition of political absolutism, military regimentation, and mechanical invention, Mumford holds, are not only the government office, the bank, and the laboratory, but also the school, the church, the factory, the art museum; they are playing the same power theme, marching to the same beat, saluting the same flag, the new leaders of a parade first marshaled by the kings, the conquistadors, and the financiers of the Renaissance.

It was from about 1940 on that the megamachine really came into its own in modern form, assembled by the military-industrial-scientific establishment that is cut off from inspectional controls by the rest of the community. It was the perfection of nuclear absolute weapons that gave science and technology the chance to back infantile ambitions and psychotic hallucinations, with industry and the academies as willing accomplices in the whole totalitarian process. (I am still summarizing Mumford's book largely in his own language.) In this respect the United States and the Soviet Union are symmetrical mirror images of each other, and the cold war has served both to pursue the chief goal of the megamachine: the development of means of total extermination. War is the body and soul of the megamachine, whose worship demands wholesale human sacrifice. Mumford ends this portion of the apocalyptic vision by closing the circle: ". . . the cult of the Sun God turns out, in its final scientific celebration, to be no less savage and irrational than that of the Aztecs, though infinitely more deadly." We are captives in an empire ruled over by desiccated and sterilized minds to which, thanks to the scientists, are dedicated the immense energies that modern technics has made available. We are denied a better life that, Mumford writes, cherishes love and sex and art and what he calls the pullulating dream world. This, he concludes, is where we have been brought today by the likes of Copernicus, Kepler, Galileo, Descartes, Newton, Fermi, B. F. Skinner, Herman Kahn, Buckminster Fuller, McLuhan, the Rand Corporation, Arthur Clarke, the bureaucrats, the technicians, and, above all, and again and again, "the scientists."

What, then, does he think can be done about it? First of all, Mumford urges that we hush the cry of "forward" until we know where technology is taking us – he wryly recalls that the official slogan of the Century of Progress Exposition in 1933 in Chicago was "Science Finds – Industry Applies – Man Conforms!" – and that somehow we dismantle the military megamachine. To reinstate the list of ultimate realities of life (birth and death, sex and love, family devotion and mutual aid, sacrifice and transcendence, human pride and cosmic awe) there now is needed some religious conversion, a transformation on the large scale that cannot occur by rational thinking and educational indoctrination. The God that will save us will rise in the human soul.

The most palpable evidence that there are already cracks in the megamachine is, in Mumford's opinion, the youth movement. Indeed – and here Mumford is at his most eloquent – some of the young people prefer to live as if the nuclear catastrophe had already occurred: among ruins without permanent shelter, without regular supply of food, without customs or habits except those improvised from day to day, without books or academic credentials, vocations or careers, or any source of knowledge except the inexperience of their own peers. They mass together and touch each other, and only in this way have any sense of security and continuity. The arts, too, mirror the derangement which the megamachine has produced in the human spirit, and thereby may help to awaken man sufficiently to his actual plight.

This great religious transformation, Mumford urges, is within our grasp if we adopt a new world view, one based on the biological study of living organisms in the nineteenth century. The name for this new vision is ecology, and it will lift the burden of the myth of the machine from us. Its hero is Charles Darwin, who, Mumford believes, lacked mechanical interest, who refused to buy a compound microscope, and who in moving outside the mechanical world picture is supposed to have been aided gently by an ineptitude for mathematics.

With an earth-centered organic and human model before us, mere quantitative production will give way to increased variety and "plenitude," not to speak of balance, wholeness, and completeness. This will be a time of reclaiming the planet for life through mutual aid, loving association, and biotechnic cultivation. The power system will be transformed into a beneficent organic complex. Lifetime concentration upon a single occupation or task will give way to a varied existence, for example, a few hours of desk work followed by some gardening or carpentry.

If the chances for this program seem unlikely, Mumford asks that we think of how small the chances were for Christianity at the time of the Romans. Indeed, this is an auspicious moment; the human support of the power system was never more frail. It will not be long before the center surrenders or blows up. Once the myth of the machine is thrown off by individual souls, the gates of the technocratic prison will open automatically, despite their rusty ancient hinges, as soon as we choose to walk out. And with these phrases, the closest the book comes to giving a prescription of just how to end our enslavement, the volume ends.

Turning now to an assessment of the book and its messages, there are many things to praise. The chief concern of Mumford is itself eminently worthy. At the very least, technology is clearly in need of rescue from its chief exploiters – primarily large-scale industry with a vision of its social purpose far too narrow to match its enormous power, and the military with its seemingly uncontrollable appetite and its success in avoiding democratic accountability.

Mumford is undoubtedly right in warning of the shortening of the time interval between scientific achievement and technological exploitation, with the difficulties this creates in obtaining the necessary assessment, and the consideration of those who will have to live with the innovation. Mumford is right, too, in drawing more attention to the emergence of new styles based on individualization, diversification, intimate relationships, and the development of personal choices and initiatives – in styles of education no less than in styles of dress.

I was also delighted with Mumford's all-too-brief acknowledgment that there is a subjective and qualitative side to the doing of science which scientists hardly ever talk about, the "intellectual playfulness and aesthetic delight" in scientific work which can be an enormously important component of scientific choice and motivation. Historians of science and other scholars are only now beginning to dig at the edges of this field. The book also contains marvelously perceptive glimpses and imaginative passages, scintillating wit and sarcasm and, in some chapters, a playful, albeit Dionysian atmosphere.

But the most important good to come from the work, it seems to me, may not be one that Mumford would expect: It is the awakening of those in the scientific establishment who supposedly have been concerned with science education. The preponderance of their time, thought and money has gone into the training of more professional scientists and technologists, and precious little has been left over for serving the vastly larger audience: the Mumfords and their readers. This book, and its success, are exhibits of the failure of educational institutions in the period of rapid increase of scientific knowledge. We have no one to blame but ourselves for the fact that the image Mumford paints of science and technology is so monstrously distorted, and may be so widely believed.

If one looks beyond the seductive style and passionate exhortation one finds that the arguments are built on soft ground. For example, the account of the dramatic fight between the two great world systems – the entrenched mechanical one that has the stage all to itself for so long, and the newly emerging, organic one – is to a large degree a stage device, not backed up by proof or reference. In truth, there has existed in science in almost every period since Thales and Pythagoras a set of parallel, antithetical systems or attitudes, one reductionist and the other holistic, one mechanistic and the other vitalistic, one positivistic and the other teleological. In addition, there has always existed another set of antitheses or polarities, namely, between the Galilean (or more properly Archimedean) attempt at precision and measurement which purged public, "objective" science of those qualitative elements that interfere with the attainment of reasonable agreement among fellow investigators, and, on the other hand, the intuitions, guesses, daydreams and thematic commitments that enter into half the world of science in the form of a personal, private, "subjective" activity.

Science in its full sense has always been propelled by both these forces – using, as it were, both steam and sail. It is, therefore, simply a mistake to dismiss Kepler's notion that the brain is a specialized organ peculiarly adapted to handling mathematical information as "baseless." It was most excellently based in Kepler's – and Galileo's – neo-Platonism, which deeply influenced his science. Similarly, to admit that Galileo was after all "an open-minded naturalistic humanist" does not make up for or explain the many pages of abuse heaped on him earlier. And the world system of Newton was no cold machine, but rather an exquisite structure that swam in nothing less than God's own sensorium; as Koyré has shown, it was just this conception of God's presence and continued action which formed the intellectual basis of the eighteenth-century world, with its characteristic linking of science and the humanities. As to Mumford's imaginative attempt to turn the work of Copernicus, canon of Frauenberg Cathedral and doctor of medicine, into a virus which infected the modern age with an Egyptian Sun-God-cum-machine obsession – for that there is also no solid support.

More curiously still, the book fails to distinguish between healthy and sick science or between healthy and sick technology. All four are mixed together. It also fails to show clearly just where the dangers of misguided technology now lie. Mumford can be forgiven for saying little or nothing about the fact that technology has after all both raised crushing burdens from the backs of mankind and provided tools that opened our eyes to whole new worlds, both animate and inanimate. But his rhetoric carries him far in the other direction. Thus he holds that if one could get into the center of the power system today and "tear(s) aside the curtain," one will find there "the latest model IBM computer, zealously programmed by Dr. Strangelove and his associates"; for the Sun God has now taken on the guise of a central omnicomputer, ready to handle only what is "quantitatively measured and objectively observed."

Here the Sun God fantasy can be badly misleading. The danger the computer poses for democratic decisions and civil liberties does not arise from its capacity for quantitative storage or its objective output, but from the storage and fast distribution of false or private information, furnished and used by the kinds of people who have furnished and used it even in the good old days. The computer must not be used as a lightning rod for this ancient problem, amply described by humanists from Euripides to Dostoevsky to Beckett. The stereotype of the mad science-fiction "scientist" must not divert us from the ordinary human potential for evil.

Among the misunderstandings of the operations of science, I must mention one more because it impinges crucially upon Mumford's discussion of the promised rise of the new world view. He repeatedly expresses the hope that science as a whole will reorganize its basic conceptions, regard the living organism as endowed with autonomy and purpose, in one embrace deal with all the phenomena of nature, including man himself, and abandon the classic scientific style of thinking, whereby the parts are "deliberately isolated, carefully prepared, precisely measured, in order to intepret the whole." We are told that in such a new science there would also be a home for the "theologian, the mystic, the lover, and the poet"; moreover, it is said to be close to Darwin's "complementary ecological approach," where the *whole* reveals the nature, function, and purpose of the parts.

But history speaks against the realistic likelihood of success of this program as stated and speaks strongly for the likelihood of its failure. The Aristotelian method of doing science did come to a halt; and Darwin, far from avoiding deliberately isolated parts and precise measurements, was brought in large measure to the formulation of his doctrine by his detailed observation of small portions of the ecology, indeed of the precise shapes of the beaks of finches. As to theologians and poets, it is not so long ago that they vied with each other, at times, to find in science imaginative support. The much-copied model of the Royal Society was designed to bring together all men interested in knowledge of a sharable kind; thus the diarist Samuel Pepys appears as the president of the Royal Society on the *imprimatur* on the title page of Newton's *Principia.*

To say, as Mumford does, that the everyday world of "blue and yellow, bitter and sweet, beauty, delight, and sorrow" cannot be handled by the sciences based on seventeenth-century methods is of course not altogether untrue, although psychophysics and psychology have something interesting to say about it. What is so wrong is to harp on it without listing the rest of the "everyday world" on which the sciences have shed a light and some compassion: the world of animals and plants, matter and radiation, sickness and health, overpopulation and undernourishment, individual and group life in society. Before anyone dismisses science as the bad habits of inhuman robots rather than seeing it as a successful method – within its limits – for understanding the world we live in, including our minds, bodies and societies, he had better realize that at the center of every major problem upsetting us these days there sits, among other ignorances, also ignorances of scientific facts. Thus, to cite only one example, while more knowledge than we now have about the way palatable food plants may be grown with little fertilizer will by itself not stop worldwide malnutrition, all measures to combat it in the *absence* of such much-needed knowledge will be badly handicapped and probably unsuccessful.

This brings us to the deepest flaw of the book and of its message. From Mumford's utopia, rationality and clear thinking, even if balanced by passionate commitment, seem largely banned – and, with it, one-half of man's potential capacities and faculties. Understanding, even that of the phenomenon of life itself, is nowhere given an important place. That place appears to be reserved for "man's subjective impulses and fantasies," "floating imagery," "bodily impulses," and "spontaneous intangibles." For this reason, too, Mumford does not hope for much from "continued studies in the schools and the university."

Now one does not have to be blind to the many deficiencies of our academic institutions or to the impotence of much that passes for intellectual work. But there is simply no way of avoiding the enormous task of trying to bring about a better coalescence of man's complementary capacities, by an intellectual understanding of the processes of mind, life, society, and cosmos, joined with "capacities for growth, exuberant expression, and transcendence."

If this task is too large for any one person, something may be hoped for from collaboration among thinkers from all fields who respect one another despite their differences. It is precisely here that Mumford's effect may be most destructive. The majority of his readers, whether he likes it or not, will read this book as a diatribe against science and scientists. The words "mechanical," "scientific," and "dehumanized" appear together in tiresome repetition. "Scientific" is followed by "deadly," "scientific progress" by "human regression," with meaningful asides regularly (such as that Hitler operated "in a scientifically advanced country"). Ironically, Mumford does not touch, let alone name and expose, those who are responsible for many of the excesses of uncontrolled technology. Instead – and this is a harbinger of consequences to come from the book – Mumford falls into his own trap and turns on those with whom he should forge an alliance against common enemies.

There is a revealing example of it when Mumford discusses the American geneticist Hermann Muller, a scientist who for all his human faults could never be accused of having been unconcerned with the social responsibility of science. Mumford first quotes Muller's serious manifesto and heartfelt warning in Muller's own words:

> Man as a whole must rise to become worthy of his best achievement. Unless the average man can understand the world that the scientists have discovered, unless he can learn to comprehend the techniques he now uses, and their remote and larger effects, unless he can enter into the thrill of being a conscious participant in the great human enterprise and find genuine fulfillment in playing a constructive part in it, he will fall into the position of an ever less important cog in a vast machine. In this situation, his own powers of determining his fate and his very will to do so will dwindle, and the minority who rule over him will eventually find ways of doing without him.

Then Mumford proceeds to miss the point disastrously. Not seeing that Muller pleads Mumford's own cause, and having already made up his mind that Muller and his fellow scientists belong to that "minority who rule over" the average man, Mumford pounces upon Muller:

> "Find ways of doing without him" . . . Would it not have been more honest to say "do away with him"? Already these faithful servants of the megamachine have taken for granted that there is only one acceptable view of the world, that which they stand for: only one kind of knowledge, only one type of human enterprise has value – their own, or that which derived directly from their own. Ultimately they mean that only one kind of personality can be considered desirable – that established as such by the military-industrial-scientific elite which will operate the megamachine . . . [and will] if necessary "do without" those who may challenge their methods or deny the validity of their ends.

Muller is further berated, and linked with Eichmann and his colleagues who ordered the Jews into the gas chambers because they too were thought to be unfit to participate in Hitler's enterprise and failed to "find genuine fulfillment in playing a constructive part in it." Now one might dismiss such self-hypnosis if it were not sad to see how such prejudices interfere with simple understanding and, indeed colleagueship. The danger is that the same false rejection of a Muller by Mumford, despite his occasional disclaimers to the contrary, will cause among his readers large numbers of potential allies to turn away from one another.[2]

The book will not and does not claim to be able to influence quickly for the better any of the urgent problems of the day: It will not change one vote when the next step in the weapons race is taken: it will not keep one acre of wilderness from being turned into a Disneyland or a stripmine; it will not halt the decay of a single city block, or of a single family hurt by racial prejudice or left hungry or ill-housed in the midst of plenty; it will not lift the spirit of a single talented student in a crowded, neglected school. What the book can do immediately is to aggravate a split which Mumford himself deplores, that is, the conflict between the "abstract-mathematical-technical sphere" versus the "concrete, the organic, and the human" – at precisely the time when thinkers of all styles should close ranks against the common anti-intellectual wave, when they should join the groups – in which scientists have taken the lead – which are now in the forefront of the struggle for better-ordered priorities and against the excesses of technology.

Indeed, the time may have come for a new, naturalistic humanism, as Mumford believes. There also may be hope for the achievement of Mumford's laudable goals: the better integration of rational and emotional capacities; the confidence of the quantitative and qualitative views of man in nature rather than man against nature; and the loosening of the hold of the military machine in most major countries. But if it further splits groups of intellectuals who are potential allies by making them quarrel over false issues, the final irony of the book would be to have hindered rather than helped achieve these goals.

9

Frank E. Manuel's Isaac Newton

When Isaac Newton died in 1727 at the age of eighty-five, his scientific writings were considered the highest achievements of the Age of Reason, the basic text upon which the Age of Enlightenment would build. From the platform of his successes in physics and astronomy in the *Principia* and the *Opticks*, he would rule the imagination for over a century. It was symbolic that the young Voltaire arrived for a visit to England just in time to attend Newton's impressive state funeral. To the skeptic rationalist Voltaire, whose subsequent popularization of Newton's works spread the power and fame of the new science beyond the shores of England, Newton was the happy exemplar he needed: Newton, he said, had the "peculiar felicity, not only to be born in a country of liberty, but in an age when all scholastic impertinences were banished from the world. Reason alone was cultivated, and mankind could only be his pupil . . ."

Until a few decades ago, this is chiefly how Newton continued to be seen. Even his explicit references to the deity in the General Scholium in the second edition of the *Principia* (published in 1713, when Newton was seventy-one) and in the Queries added to the later editions of the *Opticks* could be accommodated; in any case, such references were relatively rare, and although puzzling, their import had somehow been blunted by familiarity over the two centuries or more. Occasionally, to be sure, there was discussion of the "other Newton" who had in fact prepared a *Chronology of Ancient Kingdoms Amended* for publication, and who privately and secretly was caught up much more in religious writings, chronologies, alchemical study, commentaries on biblical prophecies, and the like, than befits the rational scientist. But these intimations passed, and often were put down to bouts of piety, madness, or senility. Newton had kept his guard up, motivated by his fear of controversy, particularly on religious matters. He did let out some hints that he was not spending all his creative energy on physics – for example, in a letter of 1679 to Robert Hooke, in which Newton casually remarked he was doing scientific work only "perhaps at idle hours, sometimes for a diversion" – were easy to neglect. The pile of manuscripts on religion from Newton's pen is now known to amount to more than a million words. But only some thirty pages on theological material had been published by Newton himself in his lifetime.

The turning point to the more modern and complex view of Newton came with three events, all falling within about a decade. The first was the sale in 1936 at Sotheby's of the many bundles of Newton's private papers and theological and alchemical manuscripts that had been kept until then in the Portsmouth family. The papers began to reveal a Newton diametrically opposite from the one so well known from the textbooks. For example, one of the purchasers, the economist John Maynard Keynes, wrote a rather sensational essay on "Newton the Man" for the Newton Tercentenary. He painted Newton as the last of the Magi.

While the popular eye was seeking a new focus, scholarly attention in intellectual history was also being subjected to a reevaluation of the place of the "scientific revolution" that culminated in Newton. Thus, Herbert Butterfield introduced his influential *The Origins of Modern Science* (1949) with the famous remark that the scientific revolution "outshines everything since the rise of Christianity and reduces the Renaissance and Reformation to the rank of mere episodes . . ."

The third ingredient was the rise of sound scholarship in the history of science over the past three decades. A set of well-trained scholars has grown up who concern themselves with the innumerable facets of Newton's life and work. One can speak of an international scholarly industry, and its fruits are in such major contributions as the recent and ongoing publications of Newton's correspondence, his mathematical papers, the introduction to and new edition of the *Principia*, and many more. The greater the number and variety of people joining in, and the further they dig, the more fascinating and limitless the emerging picture of Newton becomes.

Frank E. Manuel's latest book, *The Religion of Isaac Newton* (1974) is the most recent find – and it is a jewel that deserves careful study.[1] Manuel, professor of history at New York University, came to Newton precisely from the "other side," not primarily through the study of the scientific papers. He first wrote on Newton in a chapter of his book *The Eighteenth Century Confronts the Gods* (1959). In 1963 he published *Isaac Newton, Historian.* By that time, chiefly through the writings of the philosopher-historian Alexandre Koyré, there had culminated a trend of modern analyses of Newton's metaphysics that had its beginning with Richard Bentley, Samuel Clarke, and Voltaire. To this, Manuel juxtaposed a Newton who saw himself as a *historical* scholar whose interests comprised "an interpretation of mythology, a theory of Egyptian hieroglyphs, a radical revision of ancient chronology founded on astronomical proof, an independent reading of the sense of the Bible, and circumstantial demonstrations of the fulfillment of prophecy in the historical world."

The next step was Manuel's *A Portrait of Isaac Newton* (1968) – like all Manuel's work, both lively and profound, calculated to arouse debate in the more stuffy seminars. "The pot of Newtonian studies is bubbling vigorously," he noted with satisfaction, even as he was stirring it up more by adding a new ingredient: psychohistorical analysis. Calling himself a "fellow traveller" of Erik Erikson, whose discussions and work had "deeply moved" Manuel, he focused on long-neglected materials such as the private notebooks Newton had kept as a boy and young man. It is a record of psychic terror not explained only by the repressive upbringing of the time. These notes and self-imposed exercises are a record of Newton's anxiety, confessions, loneliness, repression of instinctual desires, shame, fantasies of bringing death to his stepfather, mother, and himself – and the dread of punishment for all these.

In Manuel's view, Newton had only two refuges: "One was the bible literally interpreted as historical fact, de-allegorized, de-mythicized of everything vague and poetic, reduced to the concrete; the other was mathematical proof . . . The discovery of his mathematical genius was his salvation; that the world obeyed mathematical law was his security." Escape from confusion and sin was in subordinating oneself to the task of building systems of order, adhering to ascetic behavior enforced by power of one's will, and submitting to the will of the Lord – the severe, austere Old Testament God, not the metaphysical abstraction but the personal, dominating reality. Thus, Newton held, it is our duty to study the scripture as objective historic record.

On the other hand, the study of natural philosophy also had much to say about God, knowable through the objective historic existence of observable phenomena. Although the study of the causes of natural phenomena, Newton admitted, does not bring one directly to the First Cause and Creator, it does get one as close as one can hope to come in this mortal life. Hence it is our duty to pursue this study also. Science is not some idle pursuit to satisfy one's curiosity. Rather, it has missionary force. Newton said that when Natural Philosophy at length shall be perfected by his method, "the Bounds of Moral Philosophy will also be enlarged" (*Opticks*, Q.31).

Manuel confessed in *A Portrait*: "I steered my way between the Scylla of historians of science and the Charybdis of psychoanalysts." It was a *tour de force*, and as the reviewer in *Isis* observed accurately, "he has covered his flanks so admirably that it will drive some readers to distraction. Historians of science may never forgive him." Tracing Newton's life, Manuel had emerged with a coherent picture. Prematurely born on Christmas Day 1642 as the son of an unlettered yeoman who had died some two or three months earlier, Newton appeared to have come to regard the very circumstances of his birth and survival as a sign that he had been chosen by God for some special task. Manuel showed the effect on Newton's development as a person, on his world outlook, and on his style of life. From his early orphaned state (his mother, Hannah, a strong woman whom Manuel regards as the central figure in Newton's life, left him in the care of the maternal grandmother from the third to the eleventh year of his life, having gone to follow her second husband), he became a Cambridge recluse, a psychologically disturbed man in middle age, the authoritarian Master of the Mint, and finally the dictatorial President of the Royal Society – the "autocrat of science."

In Manuel's new book, some of these earlier findings are assumed, and others are expanded upon. What Manuel calls "my third,. and I hope final, attempt to grapple with the personality and non-scientific thought of Isaac Newton" was triggered by Manuel's analysis of Newton's manuscripts now on deposit in the Jerusalem National and University Library. The manuscripts, part of the Sotheby's sale, had been rescued for scholarship by the well-beloved orientalist of Yale, A. S. Yahuda. In 1940 Yahuda consulted his friend Albert Einstein about them, and Einstein replied with his usual perception that they gave fine insight into Newton's "geistige Werkstatt." The relevant manuscripts are reproduced in the appendix of the four chapters of Manuel's slim book (Appendix A, fragments from a treatise on The Revelation according to Daniel – typical of the fusion of Newton's religious and natural-philosophical concerns – and Appendix B, "Of the Day of Judgment and World to Come").

What, Newton asks, is the design behind the obscure prophecies in the Apocalypse if not "for the use of the Church to guide and direct her in the right way? . . . If it cannot be understood, then why did God give it? Does he trifle?" Not only must one try to understand, and so obtain the blessings promised for it, but conversely one risks the wrath of God for not trying to understand if one was meant to do so. Therefore Newton sets out to construct rules for interpreting these writings – rules for ordering and simplifying the meaning of images such as "the three froggs, the head or horn of any Beast, the whore of Babylon, the woman Jezabel," to see, for example, whether these images may stand for identifiable kingdoms or sects or peoples. Thus Newton analyzes Daniel's "vision of the four Beasts": the fourth being the fiercest of them, it cannot stand, Newton explained, for the kingdom of Antiochus Epiphanes, as some witless earlier scholar seems to have asserted, for that regime was less fierce than the three before it.

A typical rule that Newton announces for dealing with obscure passages is

> to choose those constructions which without straining reduce things to greatest simplicity . . . Truth is ever to be found in simplicity, and not in the multiplicity

> and confusion of things. As the world, which to the naked eye exhibits the greatest variety of objects, appears very simple in its internal constitution when surveyed by a philosophical understanding, and so much the simpler by how much the better it is understood; so it is in these Visions. It is the perfection of God's work that they are done with the greatest simplicity. He is the God of order and not of confusion.

Applying this methodology of maximum simplicity, which worked so well for him in physics, to theology, Newton believed he could reduce mysterious and chaotic passages to confident clarity. Thus his survey of Old Testament prophecies and the Relevation of John led Newton to find proof that the earth will continue to be inhabited by mortals after the Day of Judgment, for ever and ever. On the day of his Coming, Christ will judge not only the dead but also the quick; hence the dimensions and cubic footage of the New Jerusalem will have to be big enough to accommodate the crowd of those who are chosen. In fact its length in each of the three dimensions, Newton finds from his analysis, will be "the cube root of 12,000 furlongs." (Even for the short run, some predictions can be derived from such study: The reign of the Popes, Newton announced, could last at most another sixty years.) Manuel concludes, "In whatever direction he turned, he was searching for a unifying structure. He tried to force everything in the heavens and on earth into a grandiose but tight frame from which the most miniscule detail could not escape."

It must be noted that these manuscripts do not come from the period of Newton's dotage. On the contrary, they may be dated in the 1670s and 1680s, when he was in his prime. It is that much more amazing that Newton's God appears explicitly only once in the first edition of the *Principia* (1687) (the General Scholium was not added until the second edition, twenty-six years later). This single reference occurred almost casually, in Book III, where Newton argues that God must have placed the planets at different distances from the sun for some purpose. Even that, Newton regretted, and he removed the passage from his own interleaved and annotated copy of the first edition, as I. B. Cohen has recently pointed out.

But if God did not appear in the pages of the *Principia* more clearly until the later editions, Newton himself had no illusion that the phenomena themselves could be purely mechanical. For one thing, far too much remained unexplained (and therefore not discussed in the *Principia* at all). To give one example: To David Gregory, astronomer and mathematician, who was close to Newton at the time, Newton is recorded to have said that "a continual miracle is needed to prevent the sun and the fixed stars from rushing together through gravity." (Physicists, incidentally, are still struggling with this problem; Julian Schwinger, in a paper on "Sources and Gravitation" a few years ago, speculated on the possibility of a "dynamical equilibrium," and he ended his paper with the proposal that "the gravitational attraction of two atoms across the universe is balanced by the quantum kinetic energy demanded by localization within the universe. Does the quantum stabilize the cosmos?")

But beyond all the puzzles that seemed to be explainable only by invoking the deity and hence proving the necessity of his existence (as in Newton's letters of 1692 to Richard Bentley), the properties of and necessity for God are built into Newton's very physics, into his conceptions of absolute space, time, the ether, gravitation, and sense perception. To Newton, his work was only the beginning of a grander synthesis that eluded him, one by which he hoped to understand both nature and its creator.

It is a miracle how Newton's psyche, caught in this high service, held together as well as it did. More than once, Newton used the phrase *Jeova sanctus unus* as an anagram for Isaacus Neuutonus. Behind the melancholy countenance, we now can guess, there must have been a great flux of exaltation and terror. For where others saw Newton's unmatched skills, he must have been most aware of how inadequate all these were for the awesome task that had somehow fallen on the shoulders of the lad from Lincolnshire.

10

Ronald Clark and Albert Einstein

Today, more than two decades after his death at the age of seventy-six, Albert Einstein is remarkably alive in contemporary science. I am not speaking about his "practical" influence, those parts of his theories that came to be incorporated into modern gadgets, from TV sets to nuclear-power reactors. Rather, as a quick look as current research journals or at the Science Citation Index can show, his work is still the acknowledged base of new research results in an astonishing variety of problems in physics, cosmology, and to some extent also in chemistry.

Because science grows cumulatively and moves fast, scientists usually do not cite work done more than a few years ago and are even less apt to look back for reasons of sentimentality or hero worship. So, when they keep referring to and using Einstein's results, it is for good reasons. Indeed, during the past decade it has turned out that some of the most exciting frontiers lie exactly in that branch where Einstein, with few followers but with the obstinacy of a prophet, did his work in his last decades, namely, in general relativity theory.

A rapidly growing number of historians of science, too, now work on Einstein and his influence. Today's theory of knowledge also carries his fingerprints, partly because his early publications on relativity and quantum physics helped shape the modern style of doing science – moving ahead with daring, free imagination, but keeping one's rope anchored in a few places to the granite of basic principle.

In short, from this point of view Einstein's chief importance today lies in the fact that his legacy still provides much of the power and direction for modern science and epistemology. But a very different image arises out of the recent Einstein biography by Ronald W. Clark,[1] a correspondent during World War II who since then has written several biographies of British scientists (J. B. S. Haldane, the Huxleys, Sir Henry Tizard). At first glance he seems to be dealing with an entirely different person, "a man," he tells us, "who can, without exaggeration, be called one of the great tragic figures of our time," not least for having become a scientific "museum piece" (quoting one of Einstein's typically self-derogatory remarks).

The paradox is not due to any lack of diligence. Clark has evidently read and digested scores of other biographical works and looked at documents in many archives. On the way to his conclusion, he passes in considerable detail through the chapters of the now rather familiar story: Einstein as the rebellious boy in southern Germany; the student in Switzerland; the patent-office employee writing magnificent scientific papers; the pacifist and academician, who, after moving back to Germany, becomes charismatic overnight on November 7, 1919, when a prediction within his general relativity theory is borne out by measurements of British scientists. Then, the rise of the Nazis; Einstein as refugee, settling in the United States and reaching the status of scientist-philosopher and World Conscience who helped launch the Pugwash movement of scientists working toward arms control.

Throughout, Clark also intersperses his text with anecdotes illustrating Einstein's lively wit and independence. There is a cast of thousands, kings and commoners who happened to interact with Einstein and who shared with him the elations and horrors of the first half of this century.

There is also a special chapter on Einstein's role in alerting Roosevelt in August 1939 to the possibility – realistically perceived, as it turned out – that the Germans would combine their scientific headstart on nuclear-fission work with their access to uranium ore and their ambitions for world conquest and so would be tempted to make a nuclear weapon that, on scientific principles, was generally known to be a possibility. Clark correctly points out that the effect of Einstein's letter is by no means clear even now. Referring to research in the United States and Britain, Vannevar Bush noted previous progress: "This show had been going on long before Einstein's letter," and A. H. Compton held that the intervention may have slowed the work down.

At any rate, as Clark says, the world was going into a nuclear age whether or not Einstein had signed the letter to Roosevelt. Ironically, after December 1941, Einstein was carefully kept insulated from research on the A-bomb – "in view of the attitude of people here in Washington who have studied his whole history," as Bush reported with regret. Einstein got wind of what went on, enough to worry greatly about a postwar weapons race. His plan was to inform and rally scientists in the major countries to press for the internationalization of military power. Niels Bohr himself hurried to Princeton to swear Einstein to silence in order not to "complicate the delicate task of the statesmen."

There is a good deal more of this throughout the book, illustrating defeats of the public and political Einstein. But there is a curious quality of these defeats which earned him the description "naïve" or "tragic." Whereas the "statesmen" of our time have largely failed because they so often implemented fundamentally bad ideas, Einstein failed in this area because there seemed to be no realistic way of implementing his fundamentally good ones.

One should add that Clark's account is on the whole vivid and readable. His encyclopedic approach and spotty use of scholarly advisers have caused him to include some dubious material, to give some poor translations for the German originals, to get interviews mixed up and to err in factual matters such as birth or death dates for Einstein's wife, sister, and son. Flaws at this level could easily be remedied, and they do not explain the paradox between Clark's tragic Einstein as seen from the outside – which is the chief novelty of this biography – and the towering, creative figure of Einstein seen from the inside of science and history of science.

Understanding the deeper reasons for the paradox will tell a great deal about both the book and Einstein. One might start with Einstein's own profound and declared disinterest in the very thing that animates any biography: the whirlpool of transient detail. On a very few occasions, autobiographical fragments were squeezed out of him.

One occasion came on receiving the Nobel Prize in 1922, when he had to furnish an autobiographical essay for official publication. His was embarrassingly short – only fourteen lines. In 1946 he was persuaded to write his famous "Autobiographical Notes" – and devoted virtually all his space to his conceptual development in science and epistemology. Therefore, it is an autobiography containing the names of intellectual ancestors such as Hume, Kant, and Mach, Newton, Faraday, and Maxwell – but not of any of his family relatives.

Nor could he bring himself to read through the many biographies published about him, unless they were so patently outrageous and scurrilous that he was persuaded, in one or two cases, to try to prevent their publication. There is good evidence that he read little, if anything, even of the work of three biographers he respected – Anton Reiser (the pseudonym for his son-in-law, Rudolf Kayser), Carl Seelig, and his successor to his chair in Prague, Philipp Frank, whose book was the most sensitive and reliable of the many biographies available.

Einstein's distaste for the "merely personal" was not just a peculiarity. Like other extraordinary men, he felt that the personal, moment-to-moment existence, dominated by ever-changing wishes, hopes, and primitive feelings, is a chain that one should try to cast off in order to free one for the contemplation of the world "that stands before us like a great, eternal riddle." In the simplified but lucid image of the world that can thereby be gained, he once said, a person could hope to "place the center of gravity of his emotional life, in order to attain the peace and security that one cannot find within the narrow confines of swirling personal experience."

Even when he was eventually heaped with praise, he felt "the only way to escape the personal corruption of praise is to go on working . . . Work. There is nothing else." That is where his integrity lies and hence is the focus of his autobiography. And when asked what he had to say about his favorite composer, Bach, he replied in the same vein: "Listen to his work, play it, love it, honor it . . . and otherwise shut up about it."

Now it is only fair that a biographer must not be limited by his subject's view on biography. And in principle, an "outsider's" view of a scientist can only be welcomed. But once one has chosen him as one's subject, Einstein obliges his biographer to put his science not too far from the center of the work. Philipp Frank's book shows that such a task can be adequately handled for nonscientific readers without becoming a textbook. Even though Einstein's science is proverbially difficult, the essence of Einstein's discoveries is in fact accessible without much mathematics. As Clark reports him to have said about himself, "My power, my particular ability, lies in visualizing the effects, consequences, and possibilities . . . I grasp things in a broad way easily. I cannot do mathematical calculations easily. I do them not willingly and not readily."

Clark's book does not give a particularly skillful or accurate presentation of Einstein's chief scientific contributions, even qualitatively. Nor are matters much helped by evaluations of this kind: "Einstein's new idea appeared to have slipped a disc in the backbone of the universe." Under the circumstances it comes close to writing a *Hamlet* in which the Prince of Denmark is not only absent but is replaced by King Lear.

There is a second reason for the paradox. To any external observer, Einstein's character was full of ambiguities, tensions, and polarities which sometimes produced results that now make amusing reading. But if one digs deeper, one will find that these polarities are essentially connected to his scientific genius. Einstein's disinterest in making sure he would not turn up incorrectly dressed for some formal occasion was not unrelated to his ability to adopt an unconventional point of view when it was needed to expose the key fault in some hoary old problem of science.

The polarities in Einstein's style of life and thought, of which one does catch glimpses in Clark's book, are quite extraordinary, as even a brief list will show. There is of course the folkloric image itself – that of the wisest of old men, who even looked as if he had witnessed Creation itself, but who at the same time also was the almost childlike person. As Einstein himself once said, he succeeded in good part because he kept asking himself questions concerning space and time which only children wonder about. Then there is the legendary, iron ability to concentrate, often for years, on a single basic problem, regardless of contemporary schools or fashion; and, opposite to that, is his ever-ready openness to deal, after all, with the barrage of requests for help and personal involvements that appealed to his fundamental humanity and vulnerability to pity.

Einstein is the apostle of rationality, characterized by clarity of logical construction; and, on the other hand, there is his uncompromising belief in his own aesthetic sense in science, his advocacy of not looking in vain for "logical bridges" from experience to theory, but of making the great "leap" to basic principles, guided only by an intuition that rests on a sympathetic understanding of experience. His personal philosophy of liberal agnosticism and his withering contempt for established religious authority of any sort are well known – and, at the same time, he has also a clear personal religiosity. As he says in one of his letters: "I am a deeply religious unbeliever."

And what is most significant, such polarities also go straight through his scientific work. There is the well-known contradiction between his devotion to the fundamental thema of continuum, as expressed in the field concept, and on the other hand, his brilliant contribution to physics, based on the thema precisely opposite to the continuum, namely, the discrete quantum. In his very first paper on relativity theory in 1905, one can find just such a polarity; the positivism of the operational variety, which Einstein used for defining the concepts, coexists with a rational realism inherent in putting the two basic principles of relativity a priori at the base of the paper.

To be able to see and use such polar opposites lies close to the very meaning of genius. The seemingly ambivalent style of thinking, acting, and living is therefore not merely good "copy," but needs to be considered as one aspect of his unusual ability to deal with the ambiguities inherent in the chief unresolved problems of science. The key to his genius may well lie in the mutual correspondence between his style in thought and act on one side and the chief unresolved puzzles of contemporary science on the other.

Another such correspondence exists between his life style and his search for the great simplifications in science by which he sought, and usually found, connections between previously separate concepts such as matter and energy, space and time, mechanics and electrodynamics, and gravitation and electromagnetic fields. This instinctive desire to remove any unnecessary asymmetry or excess pervaded his behavior no less than his science. At stake was nothing less than finding the most economical, simple, formal principles, the barest bones of nature's frame, cleansed of everything that is ad hoc and redundant. To one of his assistants, Einstein said: "What really interests me is whether God had any choice in the creation of the world."

In fact, sensitivity to previously unperceived formal asymmetries or to incongruities of a predominantly aesthetic kind (rather than, for example, a puzzle posed by unexplained experimental facts) – that is the way each of Einstein's three basic papers of 1905 begin. In all these cases the asymmetries are removed by showing them to be unnecessary, the result of too specialized a point of view. Complexities which do not appear to be inherent in the phenomena should be cast out. Nature does not need them.

And Einstein does not need them. In his own personal life, the legendary simplicity of the man was an integral part of this reaching for the barest minimum on which the world rests. Even people who knew nothing else about Einstein knew that he preferred the simplest possible clothing and that he hated nothing more than artificial restraints of all kinds.

Other connections of this kind may be found, but only the raw material is in the biography. For example, there is constant reference to Einstein's religious feelings. The author indicates they developed in prewar Berlin when, as Clark puts it, "the fact of his Jewishness was brought home" to him. We are told that he is "the most famous Jew in modern history," that he had "the deep respect for learning which the Jew shares with the Celt," and so forth. When Einstein writes to Hans Mühsam, we are told – for no particular reason – that he is writing to "an old Jewish friend." Einstein's interest in Judaism and his strong support of Zionism are amply discussed. But Clark misses the relationship between Einstein's religiosity and his physics.

The biblical God of Law, whom Einstein constantly invoked in his letters, if only in a self-mocking manner, was not fundamentally different from the rational God of causal laws, who does not play dice with the world. Quantum mechanics in its later form, in which chance reigned supreme, filled him with a visceral repugnance that is understood best when one remembers he referred to it as a (false) "religion." To see at first hand the coherence with which Einstein lived and thought, both as a scientist fighting for a causal physics and as a person preoccupied with questions of politics and morality, one must turn to the account in the excellent book, *The Born–Einstein Letters*, with commentaries by Max Born.

Einstein's pacifism also comes in for a great deal of attention in Clark's biography. The author's position is announced in the first paragraph: In his view, Einstein "passionately indulged in pacifism, and as passionately indulged out when Hitler began to show that he really meant what he said about the Jews and the master-race." It was not so simple as that; Einstein's position has been carefully examined and documented in *Einstein on Peace*, by Otto Nathan and Heinz Norden. More curious still is Clark's repeated accusation that a "near paranoia" affected Einstein concerning the Germans – something the author, with considerable inaccuracy and insensitivity, identifies as "an echo of the cry that the only good German is a dead one."

Einstein himself would of course not have been greatly bothered. As he said to Born when informed about a misevaluation of his work, "After all, I do not need to read the thing." But now, nearly a century after Einstein's birth, the time has come for balanced, sound biography. Such an enterprise will be much easier when the extensive archives of Einstein's correspondence and manuscripts at Princeton are released for publication, as now seems likely to happen at long last. Then we shall see more clearly Einstein's tragedies and triumphs – for example, his failure to convert Born and Bohr from their heathen gods of quantum mechanics, and the triumph of a glorious two weeks in Holland in 1916 in the company of H. A. Lorentz and Paul Ehrenfest, fellow physicists whom he truly loved.

Martin Klein, in his recent book *Paul Ehrenfest*, gives an account of that visit. Much of the time there seems to have been devoted to playing the piano and violin duets, and Einstein was most satisfied when he succeeded in weaning Ehrenfest away from a preference for Beethoven. (Ehrenfest's notebooks show that for several months afterward he gave more time to Bach than to physics.) The rest of the visit was full of joy, too. Ehrenfest described an evening when Lorentz, in an after-dinner conversation, presented a finely polished question concerning Einstein's new theory of general relativity:

> When Lorentz had finished, Einstein sat bent over the

slip of paper on which Lorentz had written mathematical formulas to accompany his words as he spoke. The cigar was out, and Einstein pensively twisted his finger in a lock of hair over his right ear. Lorentz, however, sat smiling at an Einstein completely lost in meditation, exactly the way a father looks at a particularly beloved son – full of secure confidence that the youngster will crack the nut he has given him, but eager to see how. It took quite a while, but suddenly Einstein's head shot up joyfully; he "had it." Still a bit of give and take, interrupting one another, a partial disagreement, very quick clarification and a complete mutual understanding, and then both men with beaming eyes skimming over the shining riches of the new theory.[2]

11

On the educational philosophy of the Project Physics Course

In late October 1963 the U.S. National Science Foundation (NSF) held a meeting in Washington of about two dozen scientists in order to persuade some of them to take new approaches to introductory physics teaching.[1] The Physical Science Study Committee's (PSSC) course had been available for some time, but the scientists and officials were concerned that the proportion of students in the United States taking any introductory course in physics, alone among all the sciences, was continuing to decrease.

To me the more important argument was that the scientific community had a duty to provide not merely *one* updated, national course that might come to be regarded generally as the only "modern" way of teaching physical science. Particularly in the context of education in the United States, the proper strategy is that of pluralism. On this point, Alexis de Tocqueville's great work, *Democracy in America*, seems to contain wisdom that still applies.

For this purpose one must not fasten too much on the well-known dispirited conclusion at the end of Tocqueville's book. There he expressed the fear that democracy leads inevitably to a deterioration of the great social examples imposed from above, the models of style and the elevated thought constructs that he thought to be characteristic of aristocratic society in all its endeavors, including the sciences. In America, he said, "everyone is actively in motion," intent on some self-aggrandizement, as is guaranteed by the openness of opportunity to all; and thus "there prevails amongst those populations a . . . sort of incessant jostling of men, which annoys and disturbs the mind, without exciting or elevating it." He found the emphasis in science and the arts to be on the applied, not the theoretical – admittedly sometimes with great success: "These very Americans, who have not discovered one of the general laws of mechanics, have introduced into navigation an engine which changes the aspect of the world." But aside from giving credit to such practical triumphs, Tocqueville nostalgically mourned the passing of the aristocratic age as he saw the new mediocrity sweep over everything: "When I survey this countless multitude of beings, shaped in each other's likeness, amidst whom nothing rises and nothing falls, the sight of such universal uniformity saddens and chills me, and I am tempted to regret the state of society which has ceased to be."

This attitude is still the ideological heart of many plans and programs that attempt to impose a monolithic solution, found by a specific elite group, upon a large fraction of the nation. Occasionally, a large crash program seems the only sensible solution (although in retrospect many of these, from the Manhattan Project on, do not stand as unalloyed blessings). But we must remember the chief reason why Tocqueville's lugubrious prediction turned out to be wrong, why this nation, far from having sunk into some all-pervading mediocrity, has in most of its efforts in our time been setting standards of excellence in science and scholarship which no monolithic or aristocratically oriented society could have hoped to achieve.

The reason follows from a fact that Tocqueville himself identified: that in place of aristocratic excellence – models imposed by small groups upon the large mass – the peculiarly American genius is for pluralistic enterprises. In the pursuit of science, art, or commerce, we have reached our place not by following the hierarchical model, a system where everything radiates from the one Ecole Normale and Ecole Polytechnique or the Big Company, but by encouraging excellence wherever and whenever it appears. By warding off monopolistic and monolithic practice, we have usually managed to keep the social units of action *small* enough to give the exemplars of excellence in each unit a chance to assert themselves against the background noise of indifferent mediocrity. The prototype of action is still that of town meeting government rather than of a one-way flow of directions from a single central Rome or Paris, from one National Ministry of Science or of Education, from one single source that grants funding or prestige. If our commercial and political activities have too often abandoned that peculiar American strength in the recent past, at least the educational enterprises function in this way still and are characterized by a large measure of local control, the exercise of individual authority over manageably large groups, and hence the exercise of *choice*. Choice among alternatives must be preserved, in educational curricula as in all other aspects of life in a democracy.

Tocqueville was in fact fascinated by the peculiarly American form of small-scale government and authority as embodied in the town meeting. His discussion of the American system of townships stands near the beginning of the whole work. The small unit of government seemed to him both to embody and to explain the effectiveness of American life, in contrast to the typically European, large units of government which he saw to be the model for life abroad. This "argument from scale" for the governance of education seems to have certain evident implications.

To others at the 1963 meeting the more urgent and more practical problem was, however, the relative decrease in science enrollments in schools, even though the causes and hence the remedies were (and still are) by no means clear. The proportion of students opting to take physics in the last years of high school in the United States had been dropping ever since 1900, as the base of students going to school was expanding. By 1960, fewer than 20 percent of last-year students in high schools were choosing any physics course. In 1963 it seemed that this trend would continue and, indeed, by 1971 the fraction was down to 16 percent.[2] Moreover, between 1960–70 only about 4 percent (less than 100,000 out of 2.5 million) of our high school seniors per year were enrolling in the only modernized high school course in physics then available. As Figure 11.1 shows, during that decade there was a marked drop in the share of the students taking any physics course – from about nine-tenths of the share we had in 1948–49, down to less than seven-tenths in 1970.

The reasons for this pattern are still not understood, the more so as similar trends have been noted in other countries, including

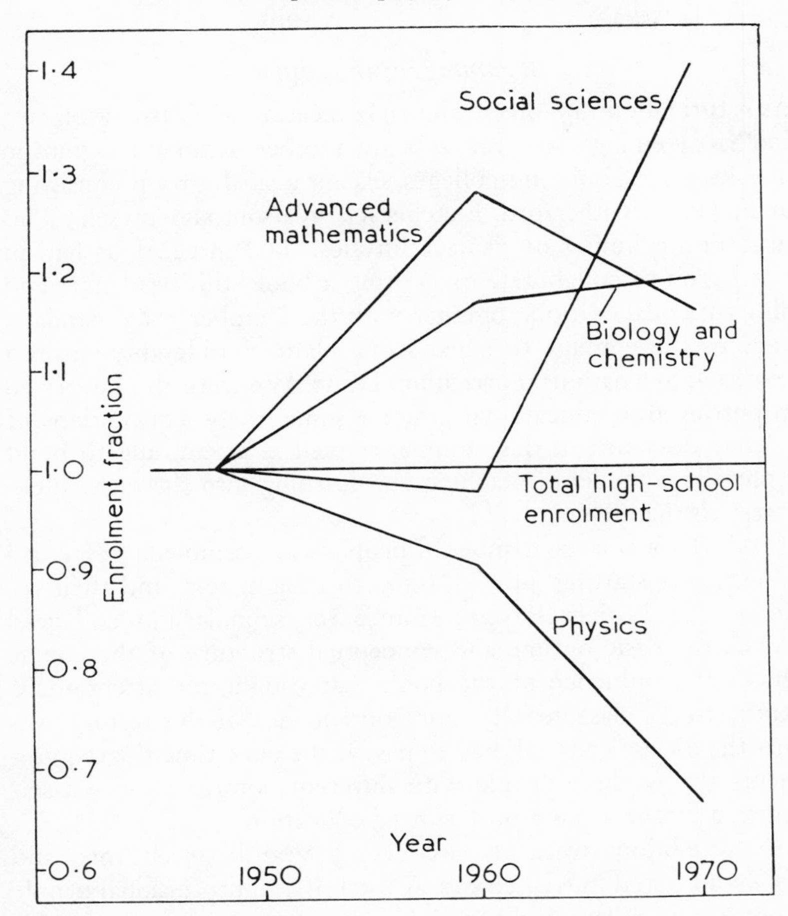

Figure 11.1. "Fractional enrollment in certain high school science courses relative to the fractional enrollment in those courses in 1949. Note that the drop in physics and the lower slope in biology and chemistry took place during the decade following the introduction of the new high-powered curricula. Source: The National Science Foundation." Illustration and caption from the National Academy of Sciences, *Physics in Perspective* ("Bromley Report") (Washington, D.C., 1973), vol. 2, p. 1164.

the USSR. In the United States the increasing difficulty in finding adequate scientific careers and the discontinuation of federal funds for teacher training in new curricula are undoubtedly factors that add to those present in the early 1960s. Other, cultural trends may also play a role, including the rise of antirationalist or "Dionysian" thinking (as discussed in Chapter 3).

Beginnings of the course

As it turned out, no one in the 1963 meeting in Washington was foolhardy enough to agree to begin another national program in physics for schools and colleges, except a small group consisting of F. James Rutherford, Fletcher G. Watson, and myself. This was the beginning of Project Physics. The three of us had in fact begun to collaborate on writing a book; this head start and pleasant collaboration, together with the October 1963 mandate, gave us the courage to expand our plans considerably, from a textbook to a national curriculum effort. We were also interested in putting into educational practice some of the conceptions of science discussed in this volume, as well as attempting to build a pluralistic model of teaching and learning into the curriculum materials themselves.

As a result, a large number of people was assembled at Harvard University, starting July 1, 1964, to design, test, and then redesign the Project Physics Course for schools and colleges.[3] While the basic outline and conceptual structure of the course shows the influence of the book that caught the attention of Rutherford's class initially, one consequence of this history was that the course materials had to pass at the same time the requirements set by three people with different constituencies – a scientist, a teacher, and a professor of education.

New editions were prepared every year between 1964 and 1968 and tested in trial classes. A total of 180 professional people – physicists, college and high school teachers, historians and philosophers of science, psychologists, reading specialists, designers, filmmakers, and so on – collaborated to produce the successive versions of text, anthologies (*Readers*), films, laboratory equipment, transparencies, the test program, and the rest of the course materials, distributed free to the participating trial schools.[4] In addition, the project involved the teachers and students in trial schools, and about a dozen Ph.D. students who based their doctorate theses in education on the evaluation program.[5]

From 1968 to 1970 the three original directors of the project reworked the whole set of materials in the light of the results of the final tests made on 20,000 students. It was no simple undertaking. At last, shortly before Christmas 1970, the final version of all course components became available through the publisher. What had begun in a rather casual way eight years earlier left us, as I recall, too exhausted to celebrate properly on that day. I am not confident we would have responded politely if someone had reminded us then that another revision effort, for the second edition of 1975, had to begin not long afterward.

Current status

It is fair to say the course now flourishes. The publisher estimates that a large fraction of students taking physics in schools and colleges in the United States are using the texts and at least substantial parts of the rest of the course materials. Although it would be difficult to say whether the course was responsible for at least a leveling off of the precipitous drop of students noted up to 1971, it is a pleasant fact that thousands of teachers have undertaken to be trained in modern methods using these materials, and have been able to exercise a choice between alternative courses and choices between materials for different students in the same course, over a large range of types of classrooms. The independent test results obtained by the Educational Testing Service (ETS) of Princeton, New Jersey, on how much physics these students actually learn, show that on the average the Project Physics students do just as well on the ETS tests as do all students nationally in any of the new or old physics courses. Thus it is not to be feared that students are going to be helpless even when given the unmodified national tests. On the contrary, they may well have benefited more from the alternative materials put into the course, not to speak of having a more integrated view of what is important about physics.

One of the significant aspects of the project is the growth of many *adaptations* of the course for schools and universities around the world. From the beginning, we have insisted that we do not wish merely to "export" slavish translations of the U.S. materials. We hoped to provide a model both of *a style of going about making a curriculum development* (e.g., involving scientists, teachers, and historians of science from the beginning, doing careful evaluation of pilot editions, and so on) and of *an approach to the subject matter*. The latter – a humanistic conception of science – is really the heart of the program, rather than any particular piece of equipment, text chapter, topic sequence, use of films or other media, and the like.

Adaptations exist or are in preparation in Australia and New Zealand, India, Israel, Japan, Jordan, and other countries. For Spanish-language and Portuguese-language countries, there are also adaptation teams at work. In Canada, too, we insisted on a thorough adaptation to the local culture and educational context: Hence Canadian groups made two separate adaptations, one in French (published in Quebec) and the other in English (published in Toronto). A great deal of pure and applied science is of course entirely international; yet a student should not be deprived of also seeing the historical connections and present applications of physical science in his or her own country.

Such considerations stem from a concern that has been as important as any in designing the Project Physics Course: the influence of the materials on the total attitude of the student to science itself. Whether or not they will become scientists, it is essential that students have a chance to see the full vision of science and thereby be protected from narrow blinkers or naïve euphoria just as much as from the false and hostile ideas about science and scientists which have been spreading in the past three decades, particularly in industrialized countries. The symptoms are well known, but it has been shown that changes are possible. Thus, on the basis of extensive educational research, Ahlgren and Walberg[6] have published a comparison of the way different physics courses bring to the foreground of the student's consciousness the mathematical and "factual" base, and also the historical and philosophical perspectives, the social context, the humanitarian values, and the artistic aspects. This is a point where a chief aim of introductory physics programs in the United States and in most other countries meet and joint. Wherever knowledge and industry are hoped to be twin pillars of social strength, the base for science is dangerously weak if the vision concerning the place and scope of science is narrow.

We must continue to try to reach a larger proportion of students than would otherwise be taking the initiative to enroll in physics courses as part of their total education. We have found that a humanistic approach to science can enlarge the pool of prospective students. Thus the proportion of young women enrolled in the Project Physics Course is nationally about twice as large as in the traditional physics course. This is an example of a group of students who, for one reason or another, traditionally have tended to avoid any physics course where they did not have this option available to them. (We have always insisted the Project Physics Course should, whenever possible, be *one option*, not the only one. Teachers were not taken into the trial program unless they agreed to continue any PSSC sections they might already be teaching.)

Reaching a more varied audience

A problem common to most countries is how to deal effectively with those who do not necessarily have the motivation or preparation to do very well in the classical, narrowly conceived physics course. A simple but useful way of looking at this problem is given in Figure 11.2. It represents a plane, one axis indicating the students' increasing academic ability, the other their increasing scientific interest. The plane is not populated with equal density; but we know that the student who will become a professional physicist is likely to be in the top right corner. In the United States only about 1,000 students a year become Ph.D.'s in physics, out of an age cohort of 3 million young people. That is a very small yield – about 0.03 percent. But our ideas on how to educate in physics come too often from serving that small group – and from having belonged to it ourselves. In fact, the fraction of the population that took any kind of physics course in U.S. schools was, by and large, concentrated in that top right corner. We do not have to struggle too hard with that audience; the best students will probably survive almost any method used in designing the course, although good teaching will not be wasted and can be very rewarding indeed.

By contrast, one tends easily to dismiss the group in the opposite quadrant. Roughly 20 to 30 percent of each age group is in that group, some perhaps for temporary or spurious reasons. It is a difficult, expensive, and very imporant area – a research subject for entirely different projects.

The rest of the population – the 50 percent or so in the middle,

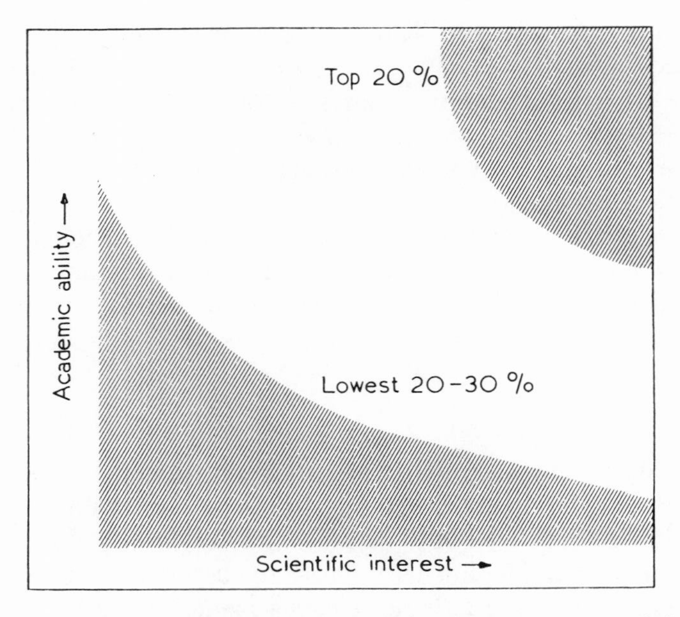

Figure 11.2. A "plane" indicating the two-dimensional gradation of students by academic ability and scientific interest. Adapted from F. Watson, "Why Do We Need More Physics Courses?" *The Physics Teacher* 5 (May 1967): 214. The entire issue of that journal is largely devoted to a report on the Project Physics Course at that stage of its development.

together with the 20 percent at the top right corner – is the group of students from which we have been trying to draw our audience. This plan results in difficulties of two kinds. First, we must make a pedagogical decision, one based on a philosophy of education different from the philosophy applicable to the smaller percent in the top right corner, where there is a rather homogeneous group of people, intellectually and cognitively. The large majority of our intended audience present us with a mixture of very different kinds.

Some are interested in social science, in humanities and the arts, in technology, in "nothing yet in particular," in verbal rather than mathematical learning, and so forth. Some may enjoy working in the laboratory but are poor in verbalizing and writing things down clearly. In aggregate, they are like a gas made up of molecules from the whole periodic table, whereas up there, in the top right corner, we have mainly the Rare Earths. Perhaps the chief trait most students will have in common across the range will be their interest in having the course make *explicit* what the committed prephysicist usually assumes tacitly: that science has an impact on life and thought outside the walls of the laboratory, that science is a cultural force with vast, transforming potential. (I should hasten to add that some of the most thoughtful physicists do believe that this humanistic approach to science education is really just as necessary for future scientists themselves – that those in fact need the humanistic and societal elements in a science course more than anyone else – for reasons discussed in Chapter 7 – because the narrowing spirit of graduate school will descend on them all too quickly.)

If one wishes to engage a great variety of individuals, with all these different "chemical" properties, one must have a course that will be meaningful in a variety of ways, each of which is actually rewarded. Some students will excel in the mathematical or laboratory part, others in the more verbal reports, perhaps connected with their interest in social science or history. Hence the assigned work, and of course the *tests*, must allow some choices or options, to permit different kinds of excellence to show up.

Updating teaching

The second, related consequence of pedagogic importance is that the instructor should not be afraid to experiment with different styles of teaching. After all, the culture of the young and the body of doctrine in education have both moved very fast in the last decade, and we must be ready to update pedagogic ideas as we do scientific ones.

Let me cite one stimulating example. Some schools have inhomogeneous groups of forty or more students in a single classroom. A suitable style for these was developed, called a "modified contract method."[7] The first part of the innovation was to make the whole set of course materials, including the apparatus and instructor's resource book or teacher guide (except of course the model tests and their answers), accessible to the students, who are working in groups of three or four, and to make with them a "contract" that they will take a test on the contents of one four-chapter unit at the end of a fixed period (three to six weeks). The instructor was available as "consultant" to each of these groups or individuals in class on demand, for example, for short lectures. The students frequently decided to split or share the work according to individual skills and interests, for instance, one student being in charge of much of the mathematical work, another of the laboratory work, another taking leadership in working with the film loops or historical essays in the *Readers*.

But the reason this kind of group work was successful was the addition of one more rule: The grade each student receives after taking the unit test is the *average* of all the scores obtained by the members of his own group. Thereby one breaks with the usual classroom behavior, in which good students look out only for themselves and let the others fall by the wayside. In this case the good students must be teachers of the less ingenious members of their group; otherwise, in the final examination, the latter will drag down the average. In short, what happens is that by this method we introduce into the teaching process the *sanctions and rewards that are operating for research teams*, where everybody in the team shares the credit and the blame equally. In this way, the well-tested and effective ethos of the research laboratory is imported into the classroom.

I propose this not as *the* way Project Physics or any course should be taught; rather it is an example of changes in pedagogy that become appropriate when we teach classes different from those we attended in our own student days.

Content and structure

These pedagogic considerations lead us to the question of content and structure.[8] All too often, the selection principle for dealing with the unmanageable total content of physics in an introductory course is that we concentrate on fragments thought to be relatively easy to teach. A great deal more is included simply by habit. Such attitudes are inadequate for a course intended to provide a vision of science at its best. It was decided, therefore, to filter out whatever did not fit into a developing story line that aims to show how the basic parts of physics grew and came together. We hoped, thereby, to develop a sequence of organically related ideas whose pursuit takes a student to an

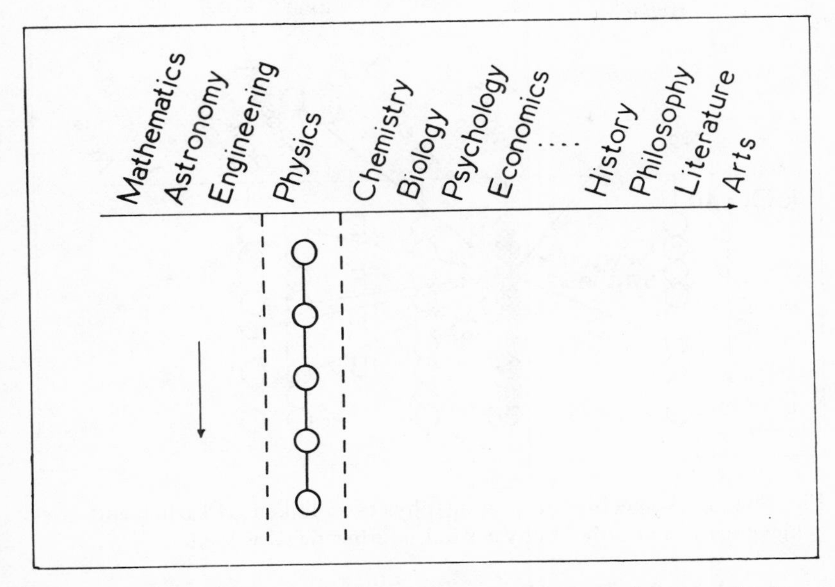

Figure 11.3. Traditional presentation of topics in introductory physics.

ever higher vantage point, a more encompassing view of the working nature, of the style of life of the scientist, and of the power of the human mind.

The traditional way we teach and have been taught is indicated in Figure 11.3. One rationally reconstructed subject (kinematics) is followed by the next (dynamics, waves, etc.); so one pearl after another is put together to set forth physics as it is now known, in a logical way, and rarely with more than a nod in the direction of other fields. Other instructors do the same thing in chemistry, biology, or mathematics. The method has its uses and rewards, though chiefly for the committed specialist.

Our method had to be rather different, since we wanted to illustrate how physical science actually developed as well as the humanistic and societal impact of science – those aspects that are particularly meaningful to students in the large middle group of our audience. We therefore adopted what I prefer to call a *connective approach*. Traditionally, one sees the separate academic subjects arrayed next to one another – mathematics, astronomy, physics, chemistry, engineering, biology, and on to the less and less mathematical fields such as economics, political science,

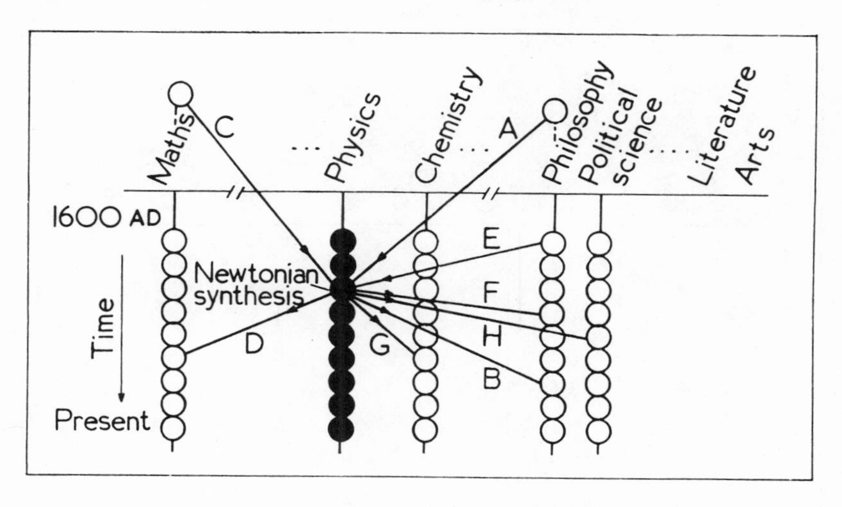

Figure 11.4. A specific advance in physics is linked to earlier and later achievements, not only in physics but in other fields as well.

philosophy, theology, literature, and the arts. Instructors of physics traditionally expect to attend only to a narrow vertical column of items. Yet, historically, almost any of the basic findings or laws in science did develop both vertically *and* horizontally – not linearly, but as part of a constellation, an interdisciplinary network (as indicated in Figure 11.4). This recognition allows us to present a much more meaningful story for our wider audience.

Establishing links

In the first unit of the Project Physics Course we deal with kinematics and dynamics, also noting that the ideas of Galileo and his contemporaries were much influenced by debates (A in Figure 11.4) that go way back in time and over into philosophy. The conceptions of the Greek philosophers certainly played a great role in the fight over the very nature of physical knowledge, a fight that shaped our present ideas of science. Conversely, the success of seventeenth-century physics had a very striking impact on later philosophy (B in Figure 11.4). For example, the conceptions of the separation into primary and secondary qualities and the mathematization of reality, which haunt philosophy to this day, began there, and are links that reach over from physics. I should stress that all such indications should be treated in a serious (not anecdotal) way, but need not be carried to enormous lengths; on the order of 10 percent of class time is enough to legitimize the approach, enough to interest students and to lead them to further research with existing materials.

The next unit is on Newton's synthesis of the mechanics of the earth and of the solar system. There we have a fine opportunity to show that the mathematics Newton used is to a large extent the mathematics of the Greeks (C in Figure 11.4) and that Newton repaid this debt to mathematics by enriching the field with the development of his calculus (D). There are also links with philosophy and theology, for Newton did not take his ideas about space and time out of thin air. Conversely, Voltaire was deeply influenced by Newton's physics, and his antimetaphysical interpretation of it was a strong current in philosophy and theology (F). John Dalton confessed that he had found support for his ideas on atomistic chemistry (1808) in Newton's *Principia* (G), although it appears he based his ideas on a mistaken analogy (offering the teacher a good occasion to demythify and correct popular notions concerning "the scientific method").

Turning to political science, we can find explicit acknowledgment of the debt to Newtonian science and to the Newtonian approach to natural philosophy (H), for example, in the balance-of-power imagery used in Revolutionary America. Other connections (e.g., to literature and the arts) can also be shown easily, and naturally throughout the course there are occasions for mentioning the historic links between the topics and stages of physics itself as it developed.

There are many other such examples.[9] In the unit on energy and thermodynamics, we can and should speak of the industrial revolution and the effects of scientific advance on society. Similarly, in the unit on the nucleus we can talk about the discussions among some scientists concerning responsibility for the ethical and human values impacts of the technological aspects of their work. This discussion connects to an area of strong preoccupation concerning science among some of our students and, in any case, is an obligation for a course that wishes to set science in its fuller cultural context.

Education, not training

With such an approach, one ends up not with a string of separate pearls, all within one field, but with a tapestry of cross connections among many fields. And that seems to me the essential task of education, in contrast to that of mere training. Training is achieved by imparting the most efficient skill for a scientific purpose. Education is achieved by imparting a point of view that allows generalization and application in a wide variety of circumstances in one's later life. This difference explains why the older, linear kind of science course, though perhaps easier to teach, is not appropriate for classes that contain students interested in the power and meaning of science, but who do not all necessarily think themselves ready to be trained as future physicists.

Teachers and scientists, being members of a group that plays a key role in the total cultural life of a nation, should be proud of the existence of this tapestry of interlinking ideas, the more so as their field, physics, has a central place in this total organic structure of intellectual history. It is altogether appropriate that they share this vision of science with their students. In the process of teaching good science, they can also convey a proper sense of the dignity of scientific work as well as of the serious civic responsibilities that are the consequences of its benefits and power.

NOTES

Preface

1 Cambridge, Mass.: Harvard University Press, 1973.

2 Among the more extensive are: Stephen Toulmin, "Science and Scientists: The Problem of Objectivity," *Minerva* 12, no. 4, pp. 522–29; and Robert K. Merton, "Thematic Analysis in Science: Notes on Holton's Concept," *Science* 188 (1975): 335–38.

3 Thus Ernst Mayr, in "The Nature of the Darwinian Revolution," *Science* 176 (1972): 981–89, makes the argument that in the early period of the discussions on the theory of evolution, it was the nonscientist who developed consistent and logical arguments for evolution, whereas important "professional" scientists were prevented from it by the fact that the "cogency of [their] argument relied entirely on the [presumed] validity of silent assumptions" (ibid., p. 982).

4 See, for example, Joseph S. Fruton, "The Emergence of Biochemistry," *Science* 192 (1976): 327–32; Dr. Fruton recalls the debate between adherents of biological reductionism and holism.

5 Compare Robert Nesbit, *Social Science as an Art Form* (Oxford University Press: 1976).

6 The current debate between proponents of "statistical" (or "actuarial") as against "clinical" prediction is an example.

7 Harry A. Wolfson, *Philo, Foundations of Religious Philosophy in Judaism, Christianity, and Islam* (Cambridge, Mass.: Harvard University Press, 1947), vol. 1, pp. 106–7.

8 There is of course much room for personal disagreement in what "simplicity" or "restrictiveness" consists. Thus Einstein and Planck debated strongly in 1914 whether the simplest physics is one that regards as basic *accelerated* motion (as Einstein had come to believe) or *unaccelerated* motion (as Planck insisted). Planck remarked in a comment to Einstein, July 7, 1914: "The laws of nature which we seek always make clear certain delimitations, a special selection from the infinitely manifold class of imaginable, logically noncontradictory relationships. Similarly, we can perhaps bring

about a close correlation between a preference for uniform motion and the particular pre-eminence which in fact characterizes the straight line among all spatial lines." Einstein, on the other hand, found neither simplicity nor necessity in this postulation of a primacy of uniform motion, since, as he countered, "from the point of physics, an absolute significance cannot be ascribed to uniform motion."

9 Steven Weinberg, "Where We Are Now," *Science* 180 (1973): 276–78.

10 Henry A. Murray, in E. G. Boring and G. Lindzey, eds., *A History of Psychology in Autobiography*, vol. 5 (New York: Appleton-Century-Crofts, 1973), p. 288.

11 LeCorbusier, *The Modulor* (Cambridge, Mass.: Harvard University Press, 1954), p. 58.

Chapter 1. Themata in scientific thought

1 This chapter is based on an address given at the twenty-fifth anniversary meeting of the History of Science Society, October 16, 1974. A somewhat condensed version was published in *Science* 188 (April 25, 1974): 328–34. At the History of Science Society meeting, the invited commentator was Robert K. Merton. His discussion of the paper was also published ("Thematic Analysis in *Science:* Notes on Holton's Concept," *Science* 188 [April 25, 1974]: 335–38), and that article should be read in conjunction with this chapter. Merton discusses, among other topics, thematic analysis as both a perspective and a tool for the historiography of science, and the parallels between thematic analysis in history and sociology of science.

2 Cf. G. Holton, *Thematic Origins of Scientific Thought: Kepler to Einstein* (Cambridge, Mass.: Harvard University Press, 1973), 276–78.

3 The concepts of "private science" and "public science" have been discussed at length in G. Holton, ibid., pp. 17–24 387–95. In his published commentary on the earlier form of this essay, R. K. Merton correctly draws attention to other, unrelated usages of similar-sounding terms. What is here called "private science" centers on those aspects "of the 'nascent moment' of discovery which, by convention, ordinarily remain unreported in the 'public science' recorded in scientific journals and monographs" (Merton, op. cit. [n. 1], p. 337). As Merton notes, sociologists investigating types of scientific identities use the similar term "private scientists" to refer to persons who are working principally in industrial research laboratories, who "set little store by publication," and who do not seek recognition and confirmation from the scientific community.

4 See Chapter 3 of this book.
5 M. von Laue and M. Born, *Physikalische Zeitschrift* 24 (1923): 52.
6 For example, S. Weinberg, "Recent Progress in Unified Gauge Theory of the Weak, Electromagnetic, and Strong Interactions," *Reviews of Modern Physics* 46 (1974): 255.
7 The hadron family includes the mesons (e.g., π^+, π^-, π°) and baryons (e.g., proton, neutron, and on to the Omega hyperons). It excludes photons and the lepton family of neutrinos, electrons, and muons, all of which are less massive than the hadrons.
8 One type of IVB is the charged intermediate vector boson, called the W particle, and the other the neutral intermediate vector bosons, called Z particles.
9 S. Weinberg, "Unified Theories of Elementary-Particle Interaction," *Scientific American* 231, no. 1 (1974): 56.
10 F. J. Hasert et al., *Physics Letters* 46B, no. 121 (1973): 138.
11 That is to say, the mu-neutrino hitting a proton in the bubble chamber gives rise to an outgoing mu-neutrino (not seen in the chamber) plus a neutron (also not seen), plus a π^+ particle (whose track is seen). Here there is no net transfer of charges of the kind that happens in the charged-current processes – for example, when a neutrino hitting a proton gives rise to a μ^-, a proton, and a π^+ particle, a process in which a unit of electric charge is exchanged. Rather, the photographs indicate a new kind of weak interaction such as could be mediated by a neutral IVB.
12 Not all themata appear in so many words; therefore, we would properly need a second scanning, proceeding in larger units. Also, the fact that this paper by Weinberg appeared in a more popular journal rather than a professional archival publication helps our purpose; when addressing general audiences, scientists are somehow more likely to reveal their otherwise usually unverbalized thematic assumptions.
13 The atom as thema does not have to refer to a natural physical object such as the discrete elementary entities, the gamma particle, the neutron, and the proton. It can be an element from which much more formalistic entities are constructed. For example, later (Weinberg, op. cit. [n. 9], p. 58) Weinberg notes that the weak interactions, if they really have an intrinsic strength comparable to that of the electromagnetic interactions, "can provide additional corrections to isotopic spin symmetry." Theoretical entities, no less than the now more palpable nuclei or atoms or, for that matter, crystals, can be thought of as a sum or aggregate composed of various terms, for example, a core term and a number of correction terms.
14 As for the graviton, Weinberg remarks parenthetically that it "interacts too weakly with matter for it to have been observed yet, but there is no serious reason to doubt its existence." It is a splendid

and daring dismissal of the weary Simplicio in Galileo's dialogues, who, since 1632, has been exclaiming at just such a point: "What? So you have not made a hundred tests, or even one? And yet you so freely declare it to be certain?" (Galileo Galilei, *Dialogue Concerning the Two Chief World Systems*, S. Drake, trans. [Berkeley: University of California Press, 1953], p. 145).

15 "Families of elementary particles are believed to be a consequence of a symmetry principle known as isotopic spin symmetry, analogous to the rotational symmetry that produced the family of quantum states within the hydrogen atom. The grouping of these families of elementary particles into superfamilies (octets, decimets, and so on) was proposed independently in the early 1960s by Murray Gell-Mann and Yuval Ne'eman" (Weinberg, op. cit. [n. 9], p. 55).

16 A more detailed discussion of the role of anthropomorphic projection and retrojection is given in Holton, op. cit. (n. 2), pp. 100–109.

17 Symmetry principles provide "information about the laws of nature on the deepest possible level"; symmetries that are "broken"; and chiral symmetries (Weinberg, op. cit. [n. 9], pp. 55–56).

18 W. Heisenberg, "Development of Concepts in the History of Quantum Theory," lecture at Harvard University, May 1973, reprinted in the *American Journal of Physics* 43, no. 5 (1975): 392.

19 W. Heisenberg, *Physics and Beyond* (New York: Harper & Row, 1971), p. 241; see also the final chapter, "Elementary Particles and Platonic Philosophy." Heisenberg elaborated these ideas in his essay "Tradition in Science" and its discussion, both in O. Gingerich, ed., *The Nature of Scientific Discovery* (Washington, D.C.: Smithsonian Institution Press, 1975), pp. 219–36, 556–73; and again in "The Nature of Elementary Particles," *Physics Today* 29 (March 1976): pp. 32–39.

20 I have traced the beginnings of Einstein's thematic commitments in his reports of his earliest memories; see Holton, op. cit. (n. 2), Essay 10. For a study of thematic choices implicit in early decision on the road to a scientific career, see Chapter 7 of this book.

21 A. Einstein, in P. Schilpp, ed., *Albert Einstein, Philosopher-Scientist* (New York: Harper & Row, 1959), 2: 686.

Chapter 2. Subelectrons, presuppositions, and the Millikan-Ehrenhaft dispute

Acknowledgments. It is a pleasure to acknowledge the help in the search for documents received from Judith R. Goodstein and Daniel J. Kevles of the California Institute of Technology; Charles Weiner of M.I.T. and Joan Warnow of the Center for History of Physics at the American Institute of Physics in New York; Erwin

Hiebert of Harvard University; Michael J. Higatsberger of the University of Vienna; and the Ernst Mach Archives in Freiburg. I am grateful to John Ehrenhaft for making available to me documents from the *Nachlass* of Felix Ehrenhaft. I thank P. A. M. Dirac for permission to publish excerpts from his letters. Material from the Millikan Collection is being published by courtesy of the California Institute of Technology Archives.

I have discussed this case in my History of Science Seminar, and thank my students, including Bruce Collier and Patri J. Pugliese, for useful comments and assistance in bibliographical searches. I acknowledge gratefully research support during the early stages of this paper from the National Science Foundation Program for History and Philosophy of Science.

Earlier versions of Chapter 2 have been presented at the History of Science Society meeting, Washington, D.C., at the International School of Physics, Varenna, and at Departmental Colloquia at Johns Hopkins and Stanford Universities.

1 P. B. Medawar, *The Art of the Soluble* (London: Methuen & Co., Ltd., 1967), p. 7.

2 Ibid., pp. 151, 155.

3 Hans Reichenbach, "The Philosophical Significance of the Theory of Relativity," in Paul Arthur Schilpp, ed., *Albert Einstein: Philosopher-Scientist* (Evanston, Ill.: Library of Living Philosophers, 1949), p. 292.

4 K. R. Popper, *The Logic of Scientific Discovery*, new ed. (New York: Basic Books, Inc., 1959), p. 31.

5 The best available bibliographies are: For Millikan see *Biographical Memoirs of the National Academy of Sciences* 33: 270–82 (1959). For Ehrenhaft see Lotte Bittner, "Geschichte des Studienfaches Physik an der Wiener Universität in den letzten hundert Jahren" (thesis, University of Vienna, 1950?). Though useful, both listings have similar gaps. For example, a key paper by each man is not listed: Millikan's "A New Modification of the Cloud Method of Measuring the Elementary Electrical Charge, and the Most Probable Value of That Charge," *Physical Review* 29 (1909): 560–61; and Ehrenhaft's "Über eine neue Methode zur Messung von Elektrizitätsmengen, die kleiner zu sein scheinen als die Ladung des einwertigen Wasserstoffions oder Elektrons und von dessen Vielfachen abweichen," *Physikalische Zeitschrift* 11 (1910): 940–52.

6 R. A. Millikan, "A New Modification of the Cloud Method of Determining the Elementary Electrical Charge and the Most Probable Value of That Charge," *Philosophical Magazine*, 19 (1910): 209–28. (Submitted October 9, 1909.)

7 R. A. Millikan, *The Electron: Its Isolation and Measurement and the Determination of Some of Its Properties* (Chicago: University

of Chicago Press, 1917), p. 158 (hereafter cited as *The Electron*). In the revised edition of 1924, p. 161, the statement is repeated unmodified.

8 For example, by R. Pohl, 1911, by R. Bär, 1922, and by O. D. Chwolson, 1927, not to speak of the protagonists themselves.

9 O. D. Chwolson, *Die Physik, 1914–1926*, George Kluge, trans. (Braunschweig: F. Vieweg & Sohn, 1927), pp. 17–18.

10 H. Kruglak, "Another Look at the Pasco-Millikan Oil Drop Apparatus," *American Journal of Physics* 40 (1972): 768–69. See also M. A. Heald, "Millikan Oil-Drop Experiment in the Introductory Laboratory," *American Journal of Physics* 42 (1974): 244–46, and J. L. Kapusta, "Best Measuring Time for a Millikan Oil Drop Experiment," *American Journal of Physics* 43 (1975): 799–800.

11 Millikan, op. cit. (n. 6), p. 220.

12 Ibid., p. 223: "The single observation mentioned above was probably on such a drop [a singly charged and very small drop of water or alcohol], but it was evaporating so rapidly that I obtained a poor value of *e*." This explanation coincides with the opinion recently expressed by P. A. M. Dirac, who among contemporary physicists is probably the least unsympathetic with respect to Ehrenhaft's hope (although not to his technique). In a letter of October 11, 1972, Dirac has written to the author that: "It seems that Millikan's anomalous drop was singly charged while all the others were doubly or triply charged. This puts Millikan in the same position as Ehrenhaft. So far as one can find from the published information, both Millikan and Ehrenhaft (in his more recent work) find an anomalous charge for all their smaller particles, and no anomalous charge for any of their larger particles. The conclusions are 1. There are no quarks. 2. There is some experimental error which makes all the smaller particles appear to have an anomalous charge. 3. By some unexplained coincidence, the anomalous charge is always about ⅔ *e*. I think this is the correct assessment of the historical information."

In a letter to me of December 4, 1972, Professor Dirac added: ". . . I just wanted to make the point that there is a similarity between Millikan and Ehrenhaft. They both found anomalous charges for their smallest particles, and in both cases their anomalous charge was about ⅔ *e*. One cannot suppose that quarks would just attach themselves to the smallest particles, so one must suppose there was a common error affecting their experiments and the factor ⅔ was a strange coincidence." Professor Dirac has elaborated on these points in C. Weiner (ed.), *History of Twentieth Century Physics* (New York: Academic Press, 1977), pp. 290–3.

Recently, Professor W. M. Fairbank, with A. F. Hebard and

G. S. LaRue, has designed experiments attempting to detect the presence of free quarks, and hence of long-lived fractional charges (½ and ⅔ e) on small bodies such as superconducting niobium spheres of mass 7×10^{-5} grams. Although in layout the experimental arrangement resembles Millikan's in many ways, the physical effects being exploited are quite different. At this writing, the data are not yet conclusive; but it will be wise to have done the experiment. See Arthur F. Hebard, "Search for Fractional Charges Using Low Temperature Techniques," Ph.D. Thesis, Stanford University Physics Department, December 1970, and Gloria B. Lubkin, "Stanford Group Shows Apparent Evidence for Quarks," *Physics Today* 30 (1977): 17–20.

13 A review on colloid research in that period is being prepared by R. M. McCormmach. Felix Exner of the University of Vienna, apparently one of Ehrenhaft's early mentors, published observations on the size and motion of colloidal particles in 1900. After the introduction of the ultramicroscope and the theories of Brownian movement, the "colloidal state" became a frontier for pure and applied science. In 1908 Wilhelm Ostwald added a chapter on colloid chemistry to the new edition of his influential textbook, *Allgemeine Chemie*. His son Wolfgang, editor of the new *Zeitschrift für Chemie und Industrie der Colloide* from 1907, published two texts on colloid chemistry. And Einstein thought it worth trying to reach readers outside "pure" physics, by rendering his work on the Brownian movement of small particles in two articles (1907, 1908) in the *Zeitschrift für Elektrochemie*.

14 Biographies of Millikan include the *Biographical Memoirs of the National Academy of Sciences* by L. A. DuBridge and Paul S. Epstein (op. cit., n. 5); the article on Millikan in *The Dictionary of Scientific Biography* by D. J. Kevles (note the bibliography); D. J. Kevles, "Millikan: Spokesman for Science in the Twenties," *Engineering and Science* (California Institute of Technology, April 1969), pp. 17–22; H. V. Neher, "Millikan – Teacher and Friend," *Amer. J. Phys.* 32 (1964): 868–77; and Millikan's own *Autobiography* (New York: Prentice-Hall, 1950) (hereafter cited as *Autobiography*).

15 There are ninety-nine file boxes, well arranged and indexed by Alfred F. Gunns and Judith R. Goodstein, with the assistance of Daniel J. Kevles and with partial funding from the Center for the History of Physics of the American Institute of Physics. For a description and listings see A. F. Gunns and J. R. Goodstein, *Guide to the R. A. Millikan Collection at the California Institute of Technology* (New York: American Institute of Physics, 1975), Publication No. R-269.

16 *Autobiography*, pp. 58–60.

17 In the *Autobiography*, p. 67, Millikan noted that he was "just beginning to get some good leads in 1908" on the "evaluation of the charge of the electron." A remark in *The Electron*, pp. 54–55, puts the beginning date two years earlier: "In 1906, being dissatisfied with the variability of these results [published by H. A. Wilson in 1903 on *e* as determined by observations on falling clouds of charged drops], the author repeated Wilson's experiment without obtaining any greater consistency . . . the results were not considered worth publishing."

18 This information comes from an unpublished autobiographical account of his early years. Partly because of the forced flight from his homeland and World War II, many of Ehrenhaft's documents are apparently lost. Some of F. Ehrenhaft's correspondence is in Archives of the American Institute of Physics, the California Institute of Technology, and the Burndy Library.

19 P. Frank, *Einstein: His Life and Times* (New York: Alfred A. Knopf, 1947), p. 175. This book was translated by George Rosen and edited and revised by Shuichi Kusaka. It is always best to go back to Frank's original, German version, published as *Einstein, sein Leben und seine Zeit* (Munich: Paul List Verlag, 1949). This is the case here also (q.v. p. 289).

20 A report on Ehrenhaft's lectures in Austria in 1947 has been prepared by Paul K. Feyerabend, who was present as a student at the time ("Ehrenhaft in Post-War Vienna," mimeo, 1967).

21 This topic has been explored by Brush, Buchdahl, Hiebert, Laudan, Scott, Brock and Knight, and others. See particularly Mary Jo Nye, *Molecular Reality* (New York: American Elsevier Publishing Co., 1972).

22 *Autobiography*, p. 99. Millikan reports "most of us had viewed [them] for the first time with amazement and thrill at the Dundee meeting [1912]."

23 A. Lampa, *Ernst Mach* (Prague: Verlag Deutscher Arbeit, 1918), pp. 40–41, see also p. 28. (As in all cases where foreign-language originals are cited, the translation is my own.) Another source on the same point, closely overlapping, is a later essay by the positivistic philosopher of science Moritz Schlick, presented in June 1926 on the occasion of the tenth anniversary of Mach's death: "Ernst Mach, der Philosoph," *Neue Freie Presse* (Supplement), Vienna, 12 June 1926. I have translated a portion of Schlick's essay on p. 222 of *Thematic Origins of Scientific Thought: Kepler to Einstein* (Cambridge, Mass.: Harvard University Press, 1973) (hereafter cited as *Thematic Origins*), as part of an essay on the relation between Mach and Einstein.

24 A good indication can be found in J. T. Blackmore, *Ernst Mach, His Life, Work, and Influence* (Berkeley: University of California Press, 1972), chs. 13 and 19. Mach has recently been the subject of

a number of interesting works. See especially Stephen G. Brusch, *The Kind of Motion We Call Heat, Vol. I: Physics and the Atomists* (New York: North Holland Publishing Company, 1976), ch. 8 ("Mach"), pp. 274–99, reprinted with additional notes and minor changes from S. G. Brusch in *Synthese* 18 (1968).

25 E. Mach, "Die Leitgedanken meiner naturwissenschaftlichen Erkenntnislehre und ihre Aufnahme durch die Zeitgenossen," *Phys. Zs.* 11 (1910): 599–606; M. Planck, "Zur Machschen Theorie der physikalischen Erkenntnis," *Phys. Zs.* 11 (1910): 1186–90. The Mach–Planck controversy has been analyzed by Blackmore, op. cit. (n. 24), ch. 14. For an English-language translation of the Planck and Mach essays, together with an analysis, see S. Toulmin, ed., *Physical Reality* (New York: Harper and Row, 1970).

26 Frank, *Einstein, sein Leben*, op. cit. (n. 19), p. 135. When Lampa died in January 1938, Ehrenhaft stressed in an obituary that Lampa had just been about to deliver an address commemorating the 100th anniversary of Mach's birth: "Anton Lampa," *Neue Freie Presse*, Vienna, 29 January 1938, p. 6. On Lampa, see also Andreas Kleinert, "Anton Lampa und Albert Einstein," *Gesnerus* 32 (1975): 284–92.

27 Other aspects of the effect of Mach on the selection of the candidate via Lampa and Pick are discussed in Blackmore, op. cit. (n. 24), ch. 17. The Lampa–Mach correspondence from which I quote is located in the Ernst Mach Archives, Freiburg, Germany.

28 Lampa wrote to Mach on May 1, 1910: "I believe that Relativity Theory is the opening of a phenomenological epoch of physics" (Mach Archives, Freiburg, Germany). See also "Mach, Einstein and the Search for Reality," in Holton, *Thematic Origins*, pp. 219–59.

29 Ibid., p. 227. The letter is dated August 17, 1909.

30 When Einstein left Prague a year later for Zürich, his successor was Philipp Frank, whose candidacy was again supported by Mach, Lampa, and Pick, as well as by Einstein.

31 In his formal writings, Millikan does not always make it clear how strongly he opposes the earlier convention that the term "electron" should be reserved for the quantity of charge, regardless of the mass and other properties of the particle carrying the charge. He does state in his essay, "New Proofs of the Kinetic Theory of Matter and the Atomic Theory of Electricity," *The Popular Scientific Monthly* 80 (1912): 417–40, that his electron has charge, mass, and a discrete, small volume ("probably the smallest thing in existence," p. 434). An early draft of a manuscript in the Millikan Archives, California Institute of Technology, File Folder 4.11, carrying the notation "Probably 1921 or prior. H. H. 1/27/54," is almost certainly a draft of at least part of that 1912 paper; it was very probably written in late 1911 or early 1912.

32 *Autobiography*, pp. 80 and 82. Emphases in original.
33 Evidence is provided by the marginal notes on the various drafts of the *Autobiography;* see File Boxes 65–67.
34 R. A. Millikan, "The Electron and the Light-quant from the Experimental Point of View" (May 25, 1924), *Nobel Lectures–Physics, 1922–1941* (Amsterdam: Elsevier Publishing Co., 1965), pp. 58–59. There is a slight difference between the two versions: None of the words is printed in capital letters in the Nobel lecture.
35 *Autobiography*, p. 83.
36 Ibid.
37 In File Box 67, particularly 67.3 and 67.4; but see also boxes 65, 66, and 68. Folder 67.3 begins with "Scientific Recollections of R. A. Millikan. Personal Recollection of R. A. M. on Rise of American Science." On the file folder is a note that this is a first draft of the *Autobiography*, written on board ship while en route to India in 1939 (Millikan Archives, California Institute of Technology).
38 Still earlier writings support the same point of view. For example, "Saw it here . . . pick up two negatives" (Millikan, "The Isolation of an Ion, a Precision Measurement of Its Charge, and the Correction of Stokes' Law," *Science* 32, n.s. [1910]: 439); also see his article in *Popular Scientific Monthly*, op. cit., n. 31.
39 The ideas of Rom Harré on the role of iconic models and the relation between visual and conceptual thinking seem useful here if one wishes to carry the analysis further. See the discussion in E. MacKinnon, "A Reinterpretation of Harré's Copernican Revolution," *Phil. Sci.* 42 (1975): 67–97.
40 "Lord Rutherford of Nelson," *The American Philosophical Society Year Book, 1938* (Philadelphia, 1939), p. 387.
41 Nye, op. cit. (n. 21), p. 65. I have analyzed the role of visualization in the *Gedanken* experiments of Einstein, in Holton, *Thematic Origins*, pp. 353–80.
42 *Autobiography*, pp. 58–59.
43 J. S. Townsend, *Electricity and Gases* (Oxford: Clarendon Press, 1915), pp. 52–53, calls Millikan's results "interesting" and "the most reliable." On the other hand, when Townsend discusses Millikan's well-developed oil drop method, he treats it merely as one of the improvements of the art, as part of his Chapter 7, "The Formation of Clouds and the Determination of the Atomic Charge."
44 *Autobiography*, p. 63.
45 Ibid., p. 69.
46 Ibid.
47 Ibid.
48 Ibid., p. 72.
49 *The Electron*, p. 15. Millikan was always interested in the history

of science, and a useful review of "Early Views of Electricity" forms the first chapter of his book.

50 Apparently first quoted by Millikan in the draft of his 1912 article, op. cit., n. 31, and again often, for example, in *The Electron*, p. 15, in the Nobel Prize talk, op. cit. (n. 34), p. 54, and in his *Autobiography*, pp. 69–70. Franklin's actual wording is slightly different: "The electrical matter consists of particles extremely subtile, since it can permeate common matter, even the densest metals, with such ease and freedom as not to receive any perceptible resistance." The sentence is at the beginning of an essay, "Opinions and Conjectures, concerning the Properties and Effects of the Electrical Matter, Arising from Experiments and Observations, Made at Philadelphia, 1749." See I. B. Cohen, *Benjamin Franklin's Experiments* (Cambridge, Mass.: Harvard University Press, 1941), pp. 212–13. After the publication of Cohen's book, Millikan referred to it and quoted Franklin's sentence correctly – except in his *Autobiography*. For sources of Franklin's atomism, see I. B. Cohen, *Franklin and Newton* (Philadelphia: The American Philosophical Society, 1956), chs. 6–10.

51 Millikan, op. cit., n. 31, and with slight changes often later, for example, *The Electron*, p. 11.

52 *The Electron*, p. 24.

53 Millikan, "Franklin's Discovery of the Electron," *Amer. J. Phys.* 16 (1948): 319.

54 R. A. Millikan and John Mills, *A Short University Course in Electricity, Sound, and Light* (Boston: Ginn and Company, 1908), pp. 6–8. Having written this textbook, Millikan noted in his *Autobiography*, p. 69, that he then turned intensively to his work on *e*. In an earlier textbook, written with H. G. Gale, he had declared there was "much direct experimental evidence for the existence" of electrons (Millikan and Gale, *A First Course in Physics* [Boston: Ginn and Company, 1906], p. 244).

55 E. Rutherford, "The Modern Theories of Electricity and Their Relation to the Franklinian Theory," in *The Record of the Celebration of the Two Hundredth Anniversary of the Birth of Benjamin Franklin* (Philadelphia: The American Philosophical Society, 1906), pp. 123–57, particularly p. 156. Millikan seems to have discovered this address rather late; his first reference appears to be in his obituary for Rutherford in 1938, op. cit., n. 40.

56 Lord Kelvin, "Aepinus Atomized," *Phil. Mag.* 3 (1902): 257–83. Rutherford, op. cit., n. 55, cracked that its title should be changed to "Franklin and Aepinus Kelvinized." Kelvin wrote: "My suggestion is that the Aepinus' fluid [of electricity] consists of exceedingly minute equal and similar atoms, which I call electrions [*sic*], much smaller than the atoms of ponderable matter . . ." (p. 257).

57 Maxwell, *Electricity and Magnetism* (1873), 1: 375 ff.

58 A. Schuster, *The Progress of Physics during 33 Years (1875–1908)* (Cambridge University Press, 1911; New York: Arno Press, 1975), p. 59.

59 Kelvin, "Contact Electricity and Electrolysis According to Father Boscovich," *Nature* 56 (1897): 84–85.

60 There are numerous resources for the history of the theories of electricity. A good annotated guide is David L. Anderson, "Resource Letter (ECAN-1) on the Electronic Charge and Avogadro's Number," *Amer. J. Phys.* 34 (1966): 1–7. The same author has published a useful introductory book, *The Discovery of the Electron* (Princeton, N.J.: D. Van Nostrand Co., 1964); Chapter 4 has a good explanation of Millikan's method of measurement. Among the more recent articles, see A. Pais, "The Early History of the Theory of the Electron, 1897–1947" in *Aspects of Quantum Theory*, A. Salam and E. P. Wigner, eds. (Cambridge University Press, 1972), pp. 79–93, and A. I. Miller, "A Study of Henri Poincaré's 'Sur la Dynamique de l'Electron," *Archive for History of Exact Sciences* 10 (1975): 207–328, secs. 1–4.

61 *Autobiography*, p. 75.

62 See Townsend, op. cit. (n. 43), ch. 7. Research on clouds condensed out on ions was still a young science; it was largely a by-product of the intense interest in ionization of gases, discovered by J. Elster and H. Geitel in 1894, and was vastly propelled by the discovery of the ionizing radiations, x-rays, and radioactivity.

63 Millikan's determination of e from the observation of charged clouds (which we call his Method I) was based on the following: J. J. Thomson, "On the Charge of Electricity Carried by a Gaseous Ion," *Phil. Mag.* 5 (1903): 346–55; J. J. Thomson, "On the Masses of the Ions in Gases at Low Pressures," *Phil. Mag.* 48 (1899): 547–67; and H. A. Wilson, "Determination of the Charge on the Ions Produced in Air by Röntgen Rays," *Phil. Mag.* 5 (1903): 429–40.

First one observes a layer of cloud droplets, each presumed to be of mass m, radius a, and density δ, falling at speed v_1 under their weight mg. Then one observes a similarly formed layer of cloud droplets, falling at v_2 in a superposed electric field E, that now acts on the charge q of each droplet.
Hence

$$\frac{mg}{mg + Eq} = \frac{v_1}{v_2} \tag{1}$$

Assuming Stokes's law holds without modification for the droplets falling in air (viscosity μ),

$$F_{\text{frict}} = 6\,\pi\mu a v_1 = mg \text{ (in equilibrium)} \equiv \left(\frac{4}{3}\,\pi a^3\delta\right) g \tag{2}$$

Therefore,

$$v_1 = \frac{2}{9} \cdot \frac{g\, a^2\, \delta}{\mu} \tag{3}$$

or

$$a^3 = \left(\frac{9\, v_1\, \mu}{2g\delta}\right)^{3/2} \tag{4}$$

Solving for q:

$$q = \frac{mg}{E} \cdot \frac{v_2 - v_1}{v_1} = \frac{4}{3}\,\pi \left(\frac{9\mu}{2g}\right)^{3/2} \cdot \frac{g}{E\delta^{1/2}} \cdot \left(v_2 - v_1\right) \cdot v_1{}^{1/2} \tag{5}$$

Assuming q is an integral multiple of the unitary charge e,

$$e = \frac{q}{n} \text{ (where } n = 1, 2, 3 \ldots \text{)}$$

64 R. A. Millikan and L. Begeman, "On the Charge Carried by the Negative Ion of an Ionized Gas," *Phys. Rev.* 26 (1908): 197–98. Despite the title, "the charge" was an average of the charges on the droplet in the top layer of the falling cloud. Since this paper was delivered around the beginning of January 1908 and published in February 1908, the work cannot have been performed during the summer of 1908, as Millikan states in *The Electron*, p. 55.

65 E. Rutherford and H. Geiger, "The Charge and the Nature of the α-Particle," *Proceedings of the Royal Society* (London) 81 (1908): 168–71. Rutherford slightly misquoted Millikan's mean value as 4.06×10^{-10} instead of 4.03×10^{-10} – and everyone then followed Rutherford's example, including Millikan himself (*The Electron*, p. 55).

Millikan's and Begeman's paper was also referred to in what turned out to be the last flower of the Thomson-Townsend-Wilson cloud method, an article by R. F. Lattey, "The Ionization of Electrolytic Oxygen," *Phil. Mag.* 18 (1909): 26–31. Lattey reported that "Millikan's" value was "not so accurate" when compared with Rutherford's and Lattey's own.

66 John L. Heilbron and Thomas S. Kuhn, "The Genesis of the Bohr Atom," *Hist. Stud. Phys. Sci.* 1 (1969): 251.

67 See *The Electron*, p. 55, and *Autobiography*, p. 72.

68 *The Electron*, p. 56: "The first determination which was made upon the charges carried by individual droplets was carried out in the Spring of 1909." *Autobiography*, p. 75: "I finished the foregoing measurements just prior to . . . September 1909." See also Millikan, "The Existence of a Subelectron?" *Phys. Rev.* 8 (1916): 596.

69 *Autobiography*, p. 72; italics in original.

70 Ibid., p. 73; italics in original.

71 The method for calculating e was now as follows: In the "balanced droplet" method, which I call Millikan's Method II, equation (5) of note 63 applies, except v_2 is now zero.

72 *The Electron*, p. 63.

73 As he described it later, one starts with a drop that is a bit too heavy to be held by the electric field, and which therefore falls. But before it is out of the view field, evaporation causes it to become light enough to stop falling, and finally it rises as it becomes too light. In the middle there is thus a period of some ten to fifteen seconds during which the droplet appears to be essentially stationary. See Millikan, op. cit. (n. 6), pp. 217–18. Millikan also discussed continuing sources of error in this method such as the short time of fall (five to six seconds at most) for determining v_1.

74 *Autobiography*, p. 74; emphasis in original. Similar confidence is expressed in *The Electron*, p. 70: "In no case have I ever found one [an ion] the charge of which, when tested as above, did not have either exactly the value of the smallest charge ever captured or else a very small multiple of that value." See also Millikan, in *Science* op. cit. (n. 38), p. 440. Strictly speaking, his experiments showed not that the elementary charge of electricity itself had to be atomic but only (as he was aware) that the transfer of charges to and from small material bodies occurred in integral multiples of e.

75 The discovery of an experiment has apparently not been treated in the philosophy of science. There are evident differences between the discovery and the design of experiments, but there is also at least this similarity: that both design and discovery generally occur within the framework of a more or less explicit, prior problem. Thus we can accept Ernest Nagel's description of this experiment if we change the word "devised" to "discovered": "It is unlikely that Millikan (or anyone else) would have devised the oil-drop experiment if some atomistic theory of electricity had not first suggested a question that seemed important in the light of the theory and that the experiment was intended to settle." E. Nagel, *The Structure of Science* (New York: Harcourt Brace, 1961), p. 90.

76 Ehrenhaft's early progress can be followed through these selected articles:

(a) "Über Colloidale Metalle," *Anzeiger der k. Akademie der Wissenschaften (Vienna)*, 18 (1902): 241–243.

(b) "Das optische Verhalten der Metallkolloide und deren Teilchengrösse," *Sitzungsberichte der Mathematisch-Naturwissenschaftl. Klasse der k. Akademie der Wissenschaften* (Vienna), IIa, 112 (1903): 181–209; also *Annalen der Physik* 4 (1903), Band 11: 489–514.

(c) "Die diffuse Zerstreuung des Lichtes an kleinen Kugeln," *Anz. Akad. Wiss.* (Vienna) (1905): 213–14; also *Sitzungsber. Akad. Wiss., Math.-Naturw. Kl.* (Vienna) 114 (1905): 1115–1141.

(d) "Die Brownsche Molekularbewegung in Gasen," *Anz. Akad. Wiss.*, (Vienna) 5 (1907): 72–73.

(e) "Über eine der Brownschen Molekularbewegung (etc.)," *Sitzungsber. Akad. Wiss., Math.-Naturw. Kl.* (Vienna) 116 (1907): 1139–49; dated July 11, 1907.

(f) "Über kolloidales Quecksilber," *Anz. Akad. Wiss.* (Vienna) 25 (1908): 513–14.

(g) "Eine Methode zur Messung der elektrischen Ladung kleiner Teilchen zur Bestimmung des elektrischen Elementarquantums," *Anz. Akad. Wiss.* (Vienna) 7 (1909): 72; dated March 4, 1909.

(h) "Eine Methode zur Bestimmung des elektrischen Elementarquantums, I," *Sitzungsber. Akad. Wiss., Math.-Naturw. Kl.* (Vienna), 118 (1909): 321–30; dated March 18, 1909, nominally appeared May 1, 1909.

(i) Same title, *Phys. Zs.* 10 (1909), 308–10; paper received April 10, 1909, nominally appeared May 1, 1909. (The next publication was a year later–see n. 108.)

It is unlikely that Millikan had easy access to these papers, partly because publication may have been delayed. Millikan hinted at this once, in passing (op. cit. [n. 68], p. 598): Referring to a paper in the *Sitzungsber. Akad. Wiss., Math.-Naturw. Kl.* (Vienna), dated May 12, 1910, he adds, "but this publication does not seem to have appeared till December 1910, at least it is not noted in '*Naturae Novitates*' before that date." Notices in the *Anzeiger* publication of the *Sitzungsber. Akad. Wiss., Math.-Naturw. Kl.* (*Vienna*) volumes bear out Millikan's observation concerning such delays.

77 Ibid., (g).

78 Ibid., (h).

79 Ibid., (g), (h), and (i).

80 Ibid., (h), p. 330; ibid., (i), p. 310.
The whole volume 118 of the *Sitzungsberichte* is a good indicator of the type and intensity of research in Austria circa 1909 and of the intellectual relations among individual researchers involved in this case. P. Frank writes an excellent pedagogic exposition of relativity, based on Einstein and Minkowski. A. Lampa writes on the effect of colloidal gold suspensions on light, drawing heavily on Hasenöhrl (1902) and Ehrenhaft (1903). Przibram, in a paper immediately following Ehrenhaft's, studies the mobility of ions of a great number of vapors, and parenthetically adopts the value $e = 4.65 \times 10^{-10}$ esu. Overlapping with this work, the longstanding interest of the Austrian physicists concerning "atmospheric electricity" yields five more lengthy *Beiträge* on the subject, for example, no. 30, by K. W. F. Kohlrausch, and no. 31, by

E. von Schweidler. Many others at that time were involved in the commission-sponsored, communal work – including Victor Hess. I intend to demonstrate, in a later contribution, the role that this rather pedestrian-appearing enterprise had in the unexpected discovery of cosmic rays within the following two years.

81 Or even into the report of the meeting, published by B.A.A.S. in 1910.

82 A letter from R. A. Millikan to his wife Greta conveys something of the meeting, and of the man (Millikan Archives, California Institute of Technology):

Monday, August 30, 1909

Dearest Greta

It is now 5.30 PM and I have been attending meetings all day. Had a fine letter waiting for me here when I came from the meeting hall to the official post office of the association. Crew had asked me to take dinner with him tonight at six o'clock at the Hotel Royal Alexandra so I shall have but a minute to tell you how much I love you, for the hotel is at least 15 minutes from here by the St. car. If we get through with our sessions I shall leave here tomorrow at 5.30 and be home Wednesday at 10.30 PM. It is barely possible however that I shall not be able to get away before the next night. This is staying away longer than I thought to do isn't it honey, but I am finding this meeting quite profitable. I haven't presented my results as yet and don't know that I shall do so, but if they want them tomorrow I am ready to give them. I made some lantern slides this AM so that they can be presented in short time. Goodnight honey I'll be back soon. Kiss the kiddies, and take a dozen of them for yourself.

Your own
Robert

83 *Report of the Seventy-ninth Meeting of the British Association for the Advancement of Science, Winnipeg, 1909* (London: John Murray, 1910).

84 Ibid., pp. 374–85. Rutherford's article was quickly republished in the *Phys. Zs.* 10 (1909): 762–71, under the title "Die neuesten Fortschritte der Atomistik." Millikan was probably in the audience to hear the paper, and he refers to it in his descriptions of the meeting, in *The Electron*, p. 75.

85 Rutherford, op. cit., n. 55.

86 Rutherford, op. cit. (n. 84), pp. 375, 381, and 385 (italics supplied).

87 Ibid., p. 376.

88 Ibid., pp. 379, 380.

89 Ibid., p. 381.

90 Ibid.

91 *Autobiography*, p. 75.

92 One of Larmor's students soon published a result. See E. Cunningham, *Proceedings of the Royal Society* (London) 83 (1910): 135. Millikan's new results are referred to twice.

93 Millikan reports this in the *Autobiography*, p. 75. A rather different account of the origin of the idea of using oil or mercury instead of water and alcohol was later given by Millikan's student Harvey Fletcher. See "Harvey Fletcher, Autobiographical Notes," correspondence with Fletcher, and a transcript (sixty-nine pages) of an interview conducted by Vern Knudsen, May 15, 1964, in the Center for History of Physics, American Institute of Physics, New York.

94 Millikan, "A New Modification of the Cloud Method of Measuring the Elementary Electrical Charge, and the Most Probable Value of That Charge," *Phys. Rev.* 29 (1909): 560–61; abstract of a paper given October 23, 1909, at the Princeton meeting of the American Physical Society.

95 Millikan, op. cit., n. 6.

96 Ibid., p. 227.

97 Ibid.

98 Ehrenhaft, op. cit., n. 76 (i); also, Millikan, op. cit. (n. 6), p. 226. Millikan's objections were: (a) In Ehrenhaft's method, Stokes's law is applied without modification to very small particles of doubtful sphericity; (b) the important velocity measurements are not made on one and the same particle but are mean values of observations on particles having speeds that can differ widely; (c) the radii are determined in a dubious way; (d) no provision was made for the possibility that multiple charges may be carried by some of the particles.

As the controversy heated up, additional and even more serious reservations came into view, such as the difficulty of knowing the density of the metal particles, and the role of Brownian movement in making measurements difficult. See H. Fletcher, "A Verification of the Theory of Brownian Movements and a Direct Determination of the Value of *Ne* for Gaseous Ionization," *Phys. Rev.* 33 (1911): 107–10; Millikan and Fletcher, "Ursachen der scheinbaren Unstimmigkeiten zwischen neuren Arbeiten über *e*," *Phys. Zs.* 12 (1911): 161–63; Millikan, "The Isolation of an Ion, a Precision Measurement of Its Charge, and the Correction of Stokes's Law," *Phys. Rev.* 32 (1911): 392–96; and Millikan, op. cit. (n. 68), pp. 595–625, largely repeated as ch. 8 of *The Electron*.

Ehrenhaft vigorously rebutted these and all other critics in numerous and extensive publications, for example, "Über die Quanten der Elektrizität," *Sitzungsber. Akad. Wiss., Math.-Naturw. Kl.* (*Vienna*) 123 (1914): 53–132. There were usually new results to report, even in his last lengthy paper on the subject, "The Microcoulomb Experiment," *Phil. Sci.* 8 (1941): 403–57.

99 Millikan, op. cit. (n. 6), p. 220.

100 Ibid.
101 Ibid.
102 Ibid., p. 219.
103 Ibid., p. 224. Millikan did continue to be concerned about this point, and he improved the Stokes's law calculations over the next years – for example, in his next paper (op. cit., n. 38), and in his 1912 report at the B.A.A.S meeting in Dundee.

Here is evidently a point where one would like to have access to Millikan's laboratory notebooks of 1909 with the data of his work of that period, for they might help us to see how his belief structure concerning the nature of electric charge aided him in deciding which observations were grounded in the nature of the phenomenon and which were not.

104 Ibid.
105 Heilbron and Kuhn, op. cit. (n. 66), p. 266. In Bohr's "On the Constitution of Atoms and Molecules," pt. 2, *Phil. Mag.* 26 (1913): 5, Bohr abandoned Rutherford's and Geiger's (1908) value of $e = 4.65 \times 10^{-10}$ esu and "adopts Millikan's value for *e*. . . ." It was a factor in improving Bohr's calculation of Rydberg's constant, bringing it from within 7 percent of the spectroscopic value, as in his previous work, to within 1 percent. Bohr actually writes $e = 4.7 \times 10^{-10}$ esu, without source; but it is equivalent to Millikan's value of 4.69×10^{-10} esu, given in his *Phys. Rev.* (1909) and *Phil. Mag.* (1910) papers (op. cit., n. 5 and n. 6, respectively).

Millikan's value of *e* changed as he continued his work. In *Science*, 1910, it became 4.9016×10^{-10}, "to less than ½%." By 1911, it was 4.891×10^{-10} esu, and in the grand *Phys. Rev.* paper of 1913, it is 4.774×10^{-10} esu – a value on which Millikan can finally rest, even to the 1924 reedition of his book, *The Electron*, p. 120. In 1912–13 Bohr evidently knew which value to keep his eye on.

106 Cf. H. Fletcher, "Einige Beiträge zur Theorie der Brownschen Bewegung mit experimentellen Anwendungen," *Phys. Zs.* 12 (1911): 202–8; Fletcher, op. cit. (n. 98), pp. 81–110. Space limitations forbid going into this field here, a necessity made more palatable by the fact that Millikan seems not to have been deeply involved in Fletcher's research.
107 A one-page abstract of Millikan's lecture of April 23, 1910 was published in July, "The Isolation of an Ion and a Precision Measurement of Its Charge," *Phys. Rev.* 31 (1910): p. 92. A lengthy "abridgment" of the paper was published on September 30, 1910 (op. cit., n. 38), pp. 436–48; it was republished in December in German, "Das Isolieren eines Ions, eine genaue Messung der daran gebundenen Elektrizitätsmenge und die Korrektion des Stokesschen Gesetzes," *Phys. Zs.* 11 (1910): 1097–09, and in French, "Obtention d'un ion isolé, mesure précise de sa charge; correction à la loi

de Stokes," *Le Radium* 7 (1910): 345–50 (abridged). There, Millikan writes that Fletcher and he "studied in this way between December [1909] and May [1910] from one to two hundred drops which had initial charges varying between the limits 1 to 150." Falling and rising oil drops are used; the values of e are now computed for each drop and for each run separately; the maximum electric field strength is now twice as large. But only one of the eleven drops for which actual data are given shows a unitary charge.

108 Ehrenhaft, "Über die kleinsten messbaren Elektrizitätsmengen. Zweite vorläufige Mitteilung der Methode zur Bestimmung des elektrischen Elementarquantums," *Anz. Akad. Wiss. (Vienna)*, no. 10 (April 21, 1910): 118–19.

A week later, in the *Anzeiger*, no. 11 (April 28, 1910): 175–76, Ehrenhaft's younger colleague K. Przibram reports briefly and mostly qualitatively about having repeated Millikan's experiment with water droplets. He acknowledges obtaining much the same results; but he adds a final, tortured sentence: "The few smaller values observed so far, i.e., smaller than e [which he takes to be about 3.5×10^{-10} esu], appear, in view of the certainty of measurement, to support the conclusion of Ehrenhaft that the deviation from the mean values, namely from 3×10^{-10} downwards, is considerably larger than the experimental error" (p. 176).

109 Ibid.

110 Ehrenhaft, "Über die Messung von Elektrizitätsmengen, die die Ladung des einwertigen Wasserstoffions oder Elektrons zu unterschreiten scheinen. Zweite vorläufige Mitteilung seiner Methode zur Bestimmung des elektrischen Elementarquantums," *Anz. Akad. Wiss. (Vienna)*, no. 13 (May 12, 1910): 215. A full paper appeared in the May 12, 1910, issue of *Sitzungsber. Akad. Wiss., Math.-Naturw. Kl. (Vienna)* 119 (1910): 815–66, under the title "Über die Messung von Elektrizitätsmengen, die kleiner zu sein scheinen als die Ladung des einwertigen Wasserstoffions oder Elektrons und von dessen Vielfachen abweichen." Nearly the same material appeared in "Über eine neue Methode zur Messung von Elektrizitätsmengen an Einzelteilchen, deren Ladungen die Ladung des Elektrons erheblich unterschreiten und auch von dessen Vielfachen abzuweichen scheinen," *Phys. Zs.* 11 (1910): 619–30 (received May 23, 1910). See *supra*, note 76, for Millikan's opinion on the date of actual publication.

111 This assumption was probably off by an order of magnitude: see H. Fletcher, op. cit. (n. 98), p. 108, and other arguments in the article.

112 See Ehrenhaft, *Sitzungsberichte*, op. cit. (n. 110), p. 866, and *Phys. Zs.*, op. cit. (n. 110), p. 630. See also Ehrenhaft, "Über eine neue Methode zur Messung von Elektrizitätsmengen, die kleiner zu

sein scheinen als die Ladung des einwertigen Wasserstoffions oder Elektrons und von dessen Vielfachen abweichen," *Phys. Zs.* 11 (1910): 940–52 (where Ehrenhaft announces he will replace Stokes's law "entirely by empirical formulas"), and articles of K. Przibram beginning in 1910, in the *Anzeiger, Sitzungsberichte,* and *Phys. Zs.*

113 *Sitzungsberichte,* op. cit., n. 110. Ehrenhaft showed that Millikan's method of treating data led to paradoxical situations. A drop with $q = 15.59 \times 10^{-10}$ esu had been placed among those assumed to be carrying three electrons, while another with charge $q = 15.33 \times 10^{-10}$ esu was among the drops assumed to be carrying four electrons.

114 Figure from Ehrenhaft, op. cit. (n. 110), Table I.

115 When such testimonies began to come in shortly afterward, they tended to favor Millikan's view. E. Regener, "Über Ladungsbestimmungen an Nebelteilchen. (Zur Frage nach der Grösse des elektrischen Elementarquantums)," *Phys. Zs.* 12 (1911): 135–41; R. Pohl, "Bericht über die Methoden zur Bestimmung des elektrischen Elementarquantums," *Jahrbuch der Radioaktivität und Electronik* 8 (1911): 406; A. Joffé, "Zu den Abhandlungen von F. Ehrenhaft: 'Über die Frage nach der atomistischen Konstitution der Elektrizität,'" *Phys. Zs.* 12 (1911): 268; L. M. McKeehan, "Die Endegeschwindigkeit des Falles kleiner Kugeln in Luft bei verminderten Druck," *Phys. Zs.* 12 (1911): 707–21.

Major support for Millikan's view concerning the quantum of charge came at the Solvay Congress of 1911. (P. Langevin and M. de Broglie, eds., *La Théorie du Rayonnement et les Quanta* [Paris: Gauthier-Villars, 1912], especially pp. 149, 150, 233–37, 251, 252.) Warburg, Rubens, Wien, Einstein, and especially Perrin spoke favorably of Millikan's results. Perrin took these views to a large public in his book *Les Atomes* (Paris, 1913, and later editions).

On the other hand, Ehrenhaft's supporters came chiefly from the circle of his students and collaborators in Vienna, for example, D. Konstantinowsky, F. Zerner, G. Laski, I. Parankiewicz. K. Przibram (cf. n. 108) was caught in a difficult position. A good researcher, he found his data used sometimes by one side, sometimes by the other. At the Solvay Congress in 1911 (Langevin and de Broglie, op. cit., p. 252), Hasenöhrl reported that Przibram was not joining in Ehrenhaft's conclusion regarding *e*; similarly, letters in the Caltech Archives from Przibram to Millikan (November 5, 1912, December 10, 1912) show he was then prepared to accept Millikan's results and method, and ended with "It is very gratifying to me that my researches on the subject, commenced with such a grievous mistake, are at last in fair agreement with your standard work."

116 Prior to publication, Millikan discussed his work on May 24, 1910,

at a Sigma Xi meeting in Chicago. An article "Substance of Address," which reads like a transcript of Millikan's remarks, was published by *The Daily Maroon* of the University of Chicago, May 25, 1910, under the heading, "Millikan Makes Great Scientific Discovery. Associate Professor in Physics Department Succeeds in Isolating Individual Ion. Holds It Under Observation. Proves Truth of Kinetic Theory of Matter – Result of Four Years of Research." Other newspapers also carried condensations of the talk: "Electric Secrets Located at Last. Prof. Millikan of Chicago University Proves Old Theories Long Advanced Will Be of Much Worth . . ." (*Chicago Daily Tribune*, 25 May 1910); "Electricity Is Defined. Problem of Centuries Solved . . ." (*Chicago Record-Herald*, 25 May 1910); "Proves the Theory of Electric Ions . . . Kinetic Theory Is Upheld . . ." (*The New York Times*, 25 May 1910); "Electrical Men Say They Give Credit to Milliken [*sic*]. Think That the Study of Single Ion Will Do Much for the Advance of Science" (*Chicago Inter Ocean*, 26 May 1910). The last of these quoted "electrical experts" as being unable to foresee practical applications: ". . . an expert on electricity said: 'I can't see that the discovery of the ion will have any effect on the use of the electricity . . .' L. T. Robinson of Schenectady, head of one of the largest electrical concerns in the world, says he is acquainted with Professor Milliken, and that if Milliken says he has discovered the ion he believes him. He thinks, however, that the discovery is of benefit only to science."

Millikan discussed his developing ideas on the oil drop method in "The Unit Charge in Gaseous Ionization," in *Transactions of the American Electrochemical Society* 18 (1910): 283–88. See also Millikan's unpublished lecture notes for summer 1910, in Folder 1.15, Millikan Archives, California Institute of Technology.

117 Millikan, op. cit., n. 107.

118 This and the next quotations are from Millikan, op. cit. (n. 6), p. 436. Italics supplied.

119 Ibid., p. 440; italics in original.

120 In Millikan's more extensive report of the same work (op. cit., n. 98), he adds: "Before we eliminated dust [in the viewing chamber] we found many drops showing these lower values of e_1 . . ." (p. 376).

121 In the 1911 version (op. cit., n. 98), Millikan omits only eight of the additional drops rather than ten, but with the same explanation (p. 382). See also Millikan and Fletcher, op. cit. (n. 98), pp. 161–63.

122 Millikan, "On the Elementary Electrical Charge and the Avogadro Constant," *Phys. Rev.* 2 (1913): 109–43, completed June 2, 1913. In *The Electron*, p. 106, he refers to the same results as having been "first presented before the Deutsche physikalische Gesellschaft in June 1912, and again before the British Association at Dundee, in September, 1912."

123 Ibid. (*Phys.* Rev. 2 [1913]), p. 139. The uncertainty in e claimed

at the end of the paper is "2 parts in 1000," with $e = 4.774 \pm 0.009 \times 10^{-10}$ esu.

A summary of the calculational path to e, using terms Millikan also employed, runs as follows:
v_1 is the constant speed of descent of the drop under the force of gravity (mg) equilibrated by the viscous force given by the (unmodified) Stokes's law expression ($6\pi\mu a v_1$).
v_2 is the constant speed of ascent as the drop, with frictional charge (q), rises in the electric field E.
Therefore,

$$\frac{v_1}{v_2} = \frac{mg}{Eq - mg}\text{, and } q = \frac{mg}{E}(v_1 + v_2)v_1^{-1} = \frac{\left(\frac{4}{3}\right)\pi a^3 \delta\, g}{E}(v_1 + v_2)v_1^{-1}$$

Replacing a^3 from Stokes's law gives

$$q = \frac{4}{3}\pi\left(\frac{9\mu}{2g}\right)^{3/2}\delta^{-1/2} \cdot \frac{g}{E}(v_1 + v_2)v_1^{1/2}$$

Thus if q is quantized, $e_{\text{frict}} = q/n$, or

$$e_{\text{frict}} \propto \left(\frac{v_1 + v_2}{n}\right)$$

Similarly, compute the charge picked from ions between successive ascents:

$$e_{\text{ionic}} = \frac{q_{n'} - q_n}{n'} = \frac{mg}{E}\left(\frac{v_2' - v_2}{n'}\right) \cdot v_1^{-1}$$

or

$$e_{\text{ionic}} \propto \left(\frac{v_2' - v_2}{n'}\right)$$

To take into account the breakdown of Stokes's law for small observed speeds v_1, assume

$$\frac{v_1}{(1 + Al/a)} = \frac{2}{9}\frac{g\, a^2 \delta}{\mu}$$

But since $e \propto v_1^{3/2}$, the "true" value of e is

$$e = \frac{e_1 \text{ (as "observed")}}{(1 + Al/a)^{3/2}}$$

So obtain e by plotting $e_1^{2/3}$ versus l/a for many runs, and read off the intercept, that is, $e_1^{2/3}$ when $l/a = 0$.

124 Or more precisely, deviates from the straight line on the graph of

$e_1^{2/3}$ against l/a or $1/pa$, where l = mean free path, a = radius of droplet, p = pressure in chamber.

125 *The Electron*, p. 106.

126 These protocols are in the Millikan Archives (Folders 3.3, 3.4 with "Oil Drop Experiments, R. A. Millikan," written on the covers).

127 This is followed by an entry made later, a second determination "by L. J. Lassalle, 10/31/11." Millikan frequently credited his students with participating in experiments, and in the laboratory notebooks not all the notations are in Millikan's handwriting.

128 Another example of this sort, based on the analysis by R. Bär, *Naturwissenschaften* 10 (1922): 344–45, shows how crucial it was to discover when the measurement of potential differences was being vitiated by changing voltages in the battery, changing calibration, and so on. Thus the ratio of successive charges on a droplet, n_1/n_2, will be given by the inverse ratio of the corresponding potential differences needed to suspend it against the pull of gravity, $(1/V_1 : 1/V_2)$. Thus $n_1/n_2 = V_2/V_1$. For example, if by experiment $V_1 = 47.5$ volts, and $V_2 = 71.1$ volts, then $n_1{:}n_2{:}{:}71.1{:}47.5$, or (to about two parts in one thousand) 3:2. Such a result would strongly support the hypothesis that droplets are charged in whole multitudes of one basic charge. But if errors produce a difference in the measurement of the relative value of V_1 or V_2 of only 1 percent, the case looks very different. Thus if V_1 were thought to be 47.0 volts, the ratio of n_1 and n_2 would be 71.1:47.0, or, to the nearest integers, 711:470 – by no means a convincing proof of the quantization of charge, and conversely evidence for unit charges much smaller than the charge of electron.

129 Despite the evident differences, I am using this phrase somewhat in the sense in which Coleridge, in *Biographia Literaria* (1817), applied it to the operation of the literary imagination. See also the letters of John Keats, December 28, 1817, and March 19, 1819.

For an example of the literature of the place of "belief" in scientific work, see Max Born, *Natural Philosophy of Cause and Chance* (Oxford University Press, 1951), pp. 123, 290.

130 K. Popper, "Autobiography," in P. A. Schilpp, ed., *The Philosophy of Karl Popper* (LaSalle, Ill.: Open Court Publ. Co., 1974), p. 29. Among challenges to the falsification theory, see, for example, E. Nagel, "What Is True and False in Science?" *Encounter* 29 (September 1967): 68–70.

131 In the presentation of the Nobel Prize for physics for 1923, Millikan's measurement of elementary charge was mentioned in considerable detail, whereas the investigation of the photoelectric effect was relegated to the last paragraph. However, as the chairman of the Nobel Prize Committee for Physics revealed, that work had helped to make not one but three decisions: "Without going into details I will only state that, if these researches of Millikan had

given a different result, the law of Einstein would have been without value, and the theory of Bohr without support. After Millikan's results, both were awarded the Nobel Prize for physics last year" (E. Gullstrand, *Nobel Lectures: Physics 1922–1941* [Amsterdam: Elsevier Publishing Co., 1965], p. 53).

132 R. A. Millikan, "A Direct Photoelectric Determination of Planck's 'h,'" *Phys. Rev.* 7 (1916): 384. A good, brief discussion of this episode is given in Roger H. Stuewer, *The Compton Effect: Turning Points in Physics* (New York: Science History Publications, 1975), pp. 72–77. For interesting material on the attitude of Millikan and others in the United States toward the quantum theory, see K. R. Sopka, "Quantum Physics in America, 1920–1935" (Ph.D. dissertation, Harvard University, 1976).

133 Millikan, op. cit. (n. 34), pp. 61–62. As late as 1920, Millikan was still not convinced: "The emission of electromagnetic radiation may or may not take place quantum-wise" (*Science* 51 [1920]: 505). In an address in December 1912, in which he declared that "the atomistic conception of matter has silenced the last of its enemies," he was struggling for some compromise that would avoid the photon. "That we shall ever return to a corpuscular theory of radiation I hold to be quite unthinkable." Similarly with the ether: "To deny the existence of this vehicle . . . is a bit of sophistry . . ." (Millikan, "Atomic Theories of Radiation," *Science* 37 [1913]: 133). Millikan was evidently able to adopt antithetical themata in different parts of his research, and he could overcome a deeply held thematic hypothesis when the experimental material would not coordinate with it.

134 F. Ehrenhaft, "Über eine neue Methode zur Messung von Elektrizitätsmengen, die kleiner zu sein scheinen als die Ladung des einwertigen Wasserstoffions oder Elektrons und von dessen Vielfachen abweichen," *Phys. Zs.* 11 (1910): 940–52. The paper contains a transcript of a discussion on the paper, with questions raised by various physicists.

From April 1910 to March 1911, Ehrenhaft and Przibram turned out the following publications, albeit there was considerable overlap in each case.

F. Ehrenhaft: *Anz. Akad. Wiss.* (*Vienna*), no. 10 (April 21, 1910), op. cit., n. 108; *Anz. Akad. Wiss.* (*Vienna*), no. 13 (May 12, 1910), op. cit., n. 110; *Sitzungsber. Akad. Wiss.* (*Vienna*) 119 (1910), op. cit., n. 110; *Phys. Zs.* 11 (1910), op. cit., n. 110; *Phys. Zs.* 11 (1910): 940–952; *Phys. Zs.* 12 (1911): 94–104; *Phys. Zs.* 12 (1911): 261–68.

K. Przibram: *Anz. Akad. Wiss.* (*Vienna*), no. 11 (April 28, 1910), op. cit., n. 108; *Anz. Akad. Wiss.* (*Vienna*), no. 17 (June 30, 1910): 262; *Sitzungsber. Akad. Wiss.* (*Vienna*) 119 (1910): 869–935; *Phys.*

Zs. 11 (1910): 630–32; *Sitzungsber. Akad. Wiss.* (*Vienna*) 119 (1910): 1719–53; *Phys. Zs.* 12 (1911): 260–61.

135 Ibid., first reference.

136 This is not to imply by any means that Millikan was sloppy or careless. On the contrary, he insisted on precision, where it counted. For example, he knew that it was crucial that accurate potential differences could be measured. See R. Bär, op. cit. (n. 128), pp. 344–45.

137 Millikan, op. cit. (n. 34), pp. 57–58.

138 Ehrenhaft, op. cit., n. 76(i).

139 Ehrenhaft, op. cit., n. 108.

140 Ehrenhaft, op. cit., n. 110.

141 Ibid. (*Sitzungsberichte* paper).

142 Ibid. (*Phys. Zs.* paper), p. 619.

143 Ehrenhaft, op. cit. (n. 134), p. 946.

144 Ehrenhaft, op. cit., n. 98 (*Sitzungsberichte* paper).

145 Ibid., p. 55.

146 Bär, op. cit., n. 128.

147 Schlick, op. cit., n. 23.

148 F. Ehrenhaft, "Ernst Mach's Stellung im wissenschaftlichen Leben," *Neue Freie Presse* (Vienna), Supplement, 12 June 1926, p. 12.

149 Ehrenhaft and his remaining followers occasionally revived the discussion, and then put the matter in a similarly ominous way. As late as 1934, Alfred Stein, *dozent* at the University in Vienna, reviewed the case, together with yet another set of Ehrenhaft's experiments, under the heading "Das Ende des Atomismus?" and the subheading "Ehrenhaft erschüttert den Aufbau der Welt." He concluded: "At any rate, the fight over the charge of the electron is still not decided – that war with heavy consequence, on whose outcome depends the existence of today's physics . . ." *Wiener Zeitung*, Beilage, 19 August 1934, p. 3.

150 Ehrenhaft, op. cit. (n. 108 and n. 110).

151 Letter of Lampa to Mach, in the Ernst Mach Archives, Freiburg.

152 The reference is to M. Planck, *Acht Vorlesungen über theoretische Physik* (Leipzig, 1910).

153 The appointment process continued to drag on. In *Phys. Zs.* 11 (June 15, 1910): 552, there appeared a note that Einstein had been proposed to fill the vacancy at Prague.

154 The "Monists," in particular, were jubilant about Perrin's work; thus Jacques Loeb linked his famous essay, "The Mechanistic Conception of Life" (1911), explicitly with Perrin's proof of the existence of molecules as the final vindication of the mechanistic philosophy.

155 Lampa also undertook to describe Ehrenhaft's work in the popular semimonthly, *Das Wissen für Alle* 11 (1911): 45–47.

Chapter 3. Dionysians, Apollonians, and the scientific imagination

1 Some psychodynamic reasons why scientists would tend to avoid such debates in any case are discussed in Chapter 7 of this book.

2 Don K. Price, "Purists and Politicians," *Science* 163, no. 3862, (1969): 25–31; the address was delivered on December 28, 1968. It is surely significant that Price was speaking as the first social scientist in many decades to be AAAS president.

3 One aspect of the cosmopolitan nature of the undercurrent of reaction against science, at least in the developed countries, was symbolized for me by a "Questionnaire on Science and Society," distributed by the editorial board of the Soviet journal *Literaturnya Gazetta* to Soviet as well as foreign scholars at an international congress in Moscow in the summer of 1971. Among the questions, in Russian and English, were these:

(a) Could, in your opinion, rapid development of science lead to some undesirable consequences?

(b) Do you think that some fields of scientific research may be "taboo" from the moral point of view? If so, which? Why?

(c) May there be any reasons for stopping a successful line of research? If so, what are they?

(d) Does scientific research by itself foster high moral qualities in men?

(e) Does it not seem to you that after a period of extreme popularity of the exact sciences among youths, a cooling off may set in?

4 Price, op. cit. (n. 1), p. 25.

5 Ibid., p. 31.

6 On aspects of Lewis Mumford's position, see Chapter 8 of this book.

7 The terms "new Dionysians" and "new Apollonians" are based not directly on the characteristic identifications with Greek mythological figures, but on the related usage given in Nietzsche's *Birth of Tragedy*. A. Szent-Gyorgyi, in *Science* 176 (1973): 966, has recently proposed the revival of the Nietzschean usage of Dionysian and Apollonian characteristics within science; he identifies the Dionysian elements with intuition, and the Apollonian element with the logical-rational. (Another useful discussion is in Y. Elkana, "The Problem of Knowledge in Historical Perspective," *Ellenike Anthropistike Etaireia* 24 [Athens, 1973]: 200–201.) Both these characteristic opposites have generally been credited with doing useful work within science; however, this does not apply to the *new* Dionysians and the *new* Apollonians as here defined, both of which are active outside science.

I should stress that I do not include among the new Dionysians the vastly larger and less sophisticated "public," which indeed provides the biggest popular following for and readership of the new Dionysian authors. That group may contain the largest share of the 32 million adults in the United States – more than one out of five – who said yes to the Gallup poll question "Do you believe in astrology?" (Gallup poll, released October 18, 1975); but it is also characterized by a set of favorable and sometimes even very positive beliefs about sciences and scientists that interlace with folklore, superstition, and so on, in a complex way. See Amitai Etzioni and Clyde Nunn, "The Public Appreciation of Science in Contemporary America," in Gerald Holton and William A. Blanpied, eds., *Science and Its Public: The Changing Relationship* (Boston: D. Reidel Publishing Co., 1976).

8 See for example the essays by George Basalla, Etzioni and Nunn, John A. Moore, Dorothy Nelkin, and Theodore Roszak, in Holton and Blanpied, op. cit., n. 7.

9 T. Roszak, *Where the Wasteland Ends* (New York: Doubleday, 1972), p. xxx.

10 T. Roszak, "Some Thoughts on the Other Side of This Life," *The New York Times*, 12 April 1973, p. 45.

11 Charles Reich, *The Greening of America* (New York: Bantam Books, 1970), pp. 241–42 (hereafter cited as *Greening*).

12 Philipp Frank, *Einstein, His Life and Times*, trans. George Rose, ed. and rev. Suichi Kusuka (New York: Alfred A. Knopf, 1947), p. 215. Frank was using these words to report on a characterization of Einstein's views in 1929.

13 A. Einstein, in an unpublished fragment apparently intended as an additional critical reply to one of the essays in P. A. Schilpp, ed., *Albert Einstein: Philosopher-Scientist* (Evanston, Ill.: Library of Living Philosophers, 1949).

14 A. Einstein, "Autobiographical Notes," in Schilpp, op. cit., n. 13. See also his essay "Motiv des Forschens" (1918), reprinted in an English translation as "Principles of Research," in Sonja Bargmann, trans. and rev., *Ideas and Opinions by Albert Einstein* (New York: Crown Publishers, Inc., 1954), pp. 224–27 (hereafter Bargmann's translation is cited as *Ideas and Opinions*). (I have used a revised translation of key passages here, however.) The differences between Reich's rather comfortable self to which he seems to repair gladly and Einstein's more pessimistic perception of the self are striking.

15 *Ideas and Opinions*, pp. 38–39.

16 *Greening*, p. 251.

17 Ibid., p. 266.

18 Ibid., pp. 279–80.

19 Ibid., p. 284.

20 Ibid., p. 285.

21 Ibid., p. 426.

22 See n. 14.

23 *Ideas and Opinions*, p. 226.

24 At least in its maturer form, readied from about the time of early general relativity theory. In Essay 9 ("Mach, Einstein, and the Search for Reality") of Holton, *Thematic Origins of Scientific Thought* (Cambridge, Mass.: Harvard University Press, 1973), I have treated the epistemological development of Einstein, from sensationism and empiricism to rational realism.

25 Letter from Einstein to Maurice Solovine, May 7, 1952, in A. Einstein, *Lettres à Maurice Solovine* (Paris: Gauthier-Villard, 1956), p. 120.

26 A. Einstein, *On the Method of Theoretical Physics* (Oxford, 1933) (The Herbert Spencer Lecture delivered at Oxford, June 10, 1933). Reprinted in *Ideas and Opinions*, pp. 270–76.

27 In Schilpp, op. cit. (n. 13), p. 673.

28 Ibid., pp. 21–25. All quotations in the next paragraphs are from that source unless otherwise indicated.

29 Einstein, Letter to M. Solovine, op. cit., n. 25.

30 This case is of considerable importance in throwing light on the actual working epistemology of Einstein, even at that early stage. It has been described in more detail in Holton, op. cit. (n. 24), pp. 189–190, 234–235.

It is interesting to note that in Karl Popper's "Autobiography" (in P. A. Schilpp, ed., *The Philosophy of Karl Popper*, The Library of Living Philosophers, [La Salle, Ill.: Open Court Publishing Co., 1974], vol. 14 bk. 1, pp. 28–29), Popper indicates that at its origin, his own falsifiability criterion owed much to what he perceived to be Einstein's example. He writes that, in 1919 (when Popper was seventeen years old and at about the time of the widely noted, successful test of the general relativity theory predictions by the eclipse expeditions), "I learned about Einstein; and this became a dominant influence on my thinking – in the long run perhaps the most important of all – Einstein gave a lecture in Vienna, to which I went; but I remember only that I was dazed." Popper explains that Mach's *Science of Mechanics* had contained an indication that a non-Newtonian theory was possible. But Mach "thought that before we could start on it we would have to await new experiences, which might come, perhaps, from new physical or astronomical knowledge about regions of space containing faster and more complex movements than could be found in our own solar system . . . Yet in spite of all this, Einstein had managed to produce a real alternative and, it appeared, a better theory, without waiting for new experiences . . .

"But what impressed me most was Einstein's own clear statement that he would regard his theory as untenable if it should fail in certain tests. Thus he wrote, for example: 'If the redshift of spectral lines due to the gravitational potential should not exist, then the general theory of relativity will be untenable' [A. Einstein, *Relativity: The Special and the General Theory, A Popular Exposition* (London: Methuen and Co., 1920), p. 132. Popper explains he has slightly improved upon the translation. In Austria, young Popper would have used the original edition of 1917.]

"Here was an attitude utterly different from the dogmatic attitude of Marx, Freud, Adler, and even more so that of their followers. Einstein was looking for crucial experiments whose agreement with his predictions would by no means establish his theory; while a disagreement, as he was the first to stress, would show his theory to be untenable.

"This, I felt, was the true scientific attitude. It was utterly different from the dogmatic attitude which constantly claimed to find 'verifications' for its favourite theories.

"Thus I arrived, by the end of 1919, at the conclusion that the scientific attitude was the critical attitude, which did not look for verifications but for crucial tests; tests which could *refute* the theory tested, though they could never establish it."

31 Albert Einstein, "Induktion und Deduktion in der Physik," *Berliner Tageblatt*, 25 December 1919.

32 *Greening*, p. 95.

33 As Reich puts it, "Consciousness III is deeply suspicious of logic, rationality, analysis, and of principle. Nothing is so outrageous to the Consciousness II intellectual as the seeming rejection of reason itself. But Consciousness III has been exposed to some rather bad examples of reason, including the intellectual justification of the Cold War and the Vietnam war. At any rate, Consciousness III believes it is essential to get free of what is now accepted as rational thought. It believes that 'reason' tends to leave out too many factors and values" (Ibid., p. 278).

34 Ibid., p. 414.

35 Ibid., p. 394.

36 C. Frankel, "The Nature and Sources of Irrationalism," *Science* 180 (1973): 930.

37 One's mind travels back to Johannes Kepler. Writing from Linz in October 1626 to Guldin, he tells what life was like when the peasants laid siege to the town. As luck would have it, Kepler (who was the local official mathematician) had been given an apartment in the County Hall, located in the city wall, with a fine view of the outskirts. Consequently, when the war reached the town "a whole company [of soldiers] descended on our house. Constantly the ear was attacked by gunfire, the nose by foul smoke, the eye by the flash of firing. All doors had to be kept open for the soldiers, who through their going back and forth ruined the sleep by night and my studies by day." Precisely in these terrible circumstances, Kepler concludes, he raised his eyes above the battle to the astro-

nomical heavens and settled down to work. (See also Chapter 7 of this book for psychological studies of motivations for scientific work.)

38 Bertrand Russell, "A Free Man's Worship" (1903), *Mysticism and Logic* (London: Longmans, Green, 1919), pp. 47–48.

39 The basic difference has been put well in an article, entitled "On Reading Einstein," by Charles Mauron, which T. S. Eliot himself took the trouble to translate in 1930 for publication in his journal, *Criterion*. Mauron compares the two basically antithetical epistemological approaches.

"The first holds that any profound knowledge of any reality implies an intimate fusion of the mind with that reality: We only understand a thing in becoming it, in living it. In this way Saint Theresa believed that she knew God; in this same way the Bergsonian believes that he knows at the same time his self and his world. The second type of opinion, on the contrary, holds that this mystical knowledge is meaningless, that to try to reach a reality in itself is vain, inasmuch as our mind can conceive clearly nothing but relations and systems of relations." (*The Criterion* 10 [1930]: p. 24).

40 K. R. Popper, "Normal Science and Its Dangers," in I. Lakatos and A. Musgrave, eds., *Criticism and the Growth of Knowledge* (Cambridge: Cambridge University Press, 1970), p. 56.

41 Ibid., p. 57.

42 Ibid., pp. 57–58. For another discussion of what he dismissed as the "subjectivist" approach, see K. R. Popper, *Objective Knowledge* (Oxford: Clarendon Press, 1974), p. 114 (hereafter cited as *Objective Knowledge*).

43 K. R. Popper, *The Logic of Scientific Discovery* (1934) (New York: Harper, 1965), p. 31.

44 See n. 30. When Popper and philosophers of that school do allow themselves remarks about the way the scientific imagination functions, one finds categorical but unsupported statements such as this one, on "the one really important difference between the method of Einstein and that of the amoeba": Einstein's "consciously critical attitude toward his own ideas" made it "possible for Einstein to reject, *quickly, hundreds of hypotheses* as inadequate before examining one or another hypothesis more carefully, if it appeared to be able to stand up to more serious criticism" (*Objective Knowledge*, n. 42, p. 247). Emphasis supplied.

45 *Objective Knowledge*, p. 179. However, in the last sentence of the same chapter a certain ambiguity emerges as to how seriously Popper took his proposal; he writes that his aim was to delineate "a theory of understanding which aims at combining an intuitive understanding of reality with the objectivity of rational criticism" (p. 190).

46 I. Lakatos, "Falsification and the Methodology of Scientific Research Programmes," in Lakatos and Musgrave, op. cit. (n. 40), p. 138 (hereafter cited as "Methodology"). Italics in original. See also I. Lakatos, "History of Science and Its Rational Reconstruction," in R. C. Buck and R. S. Cohen, eds., *Boston Studies in the Philosophy of Science* (Boston: D. Reidel, 1971), 8: 91–146, 174–82.

47 Lakatos, in Lakatos and Musgrave, op cit., p. 146.

48 A footnote at this point laconically adds: "This is a rational reconstruction. As a matter of fact, Bohr accepted this idea only in his [paper of] 1926."

49 The example has been analyzed by T. S. Kuhn, ibid., pp. 256–259, and by Y. Elkana, in "Boltzmann's Scientific Research Programme and Its Alternatives," in Y. Elkana, ed., *Some Aspects of the Interaction between Science and Philosophy* (New York: Humanities Press, 1975). See also Brian Easlea, *Liberation and the Aims of Science* (London: Chatto & Windus, 1973).

There is no space here to go further into the distorting effects of the rational reconstructionist's view of the progress of science. One or two short examples must suffice. Lakatos assures us that a theory must undergo "progressive problem shifts" to remain scientific. If the advance is made with the use of ad hoc proposals, the "progressiveness" is spoiled and such programs become "degenerating," so that one must "reject" them as "pseudoscientific." Since, however, ad hoc proposals frequently do figure in actual cases of what is widely acknowledged to be successful scientific work, he and his followers are forced into making the attempt to rescue Lorentz's ad-hoc-prone work from any possible charge that it might not be a single theory constantly undergoing "progressive problem shifts." To do this requires, however, new definitions of "ad hoc," "novel fact," and so on, which are patently ad hoc themselves; moreover, they, in turn, entail a number of distortions of well-known historical facts. For a definitive analysis of these failures in the work of Lakatos and followers such as E. Zahar, see A. I. Miller, "On Lorentz's Methodology," *British Journal of the Philosophy of Science* 25 (March 1974): 29–45; also E. E. Harris, "Epicyclic Popperism," *Brit. J. Phil. Sci.* 23 (1972): 55–67; R. V. Jones, "Methodologies and Myths," *Nature* 264 (1976): 121–123; and recent articles by S. Toulmin.

50 For example, see I. B. Cohen, "History and the Philosophy of Science," in F. Suppe, ed., *The Structure of Scientific Theory* (Urbana: University of Illinois Press, 1974), pp. 308–49, and his comments, pp. 351 ff.

51 "Methodology," p. 114.

52 Ibid., p. 116: emphases in original.

53 *Objective Knowledge*, p. 90. A strenuous attempt to "rescue"

Einstein from his own supposed erroneous ways has prompted Gary Gutting (in *Philosophy of Science* 39 [1972]: 51–68) to an attempt at an analysis which solemnly concludes that "any intuitions Einstein had . . . took their place in a logically coherent argument." The evidence from Einstein's own testimony that there were occasionally elements that did in fact not yield to conventional, logical analysis is dismissed by overamplification: To allow that, he holds, would amount to making a case that the discovery of relativity theory as a whole "was derived essentially from a private intuition."

54 "Methodology," p. 93.

55 Lakatos, "History of Science and Its Rational Reconstructions," op. cit. (n. 46), p. 133.

56 K. R. Popper, *Open Society and Its Enemies* (Princeton University Press, 1950), p. 410; 1966 ed., p. 224 of vol. 2. Also see the entire Chapter 24, "Oracular Philosophy and the Revolt Against Reason."

57 K. R. Popper, *Conjectures and Refutations* (New York: Basic Books 1963, 1965) p. 375.

58 "Methodology," p. 178. See Imre Lakatos, "Review of S. Toulmin's *Human Understanding*," *Minerva* 14 (Spring 1976): 128–29, for Lakatos's frank announcement that one ambition of his school is to "lay down *statute law* of rational appraisal," thereby capturing the direction of judgments concerning the evaluation and support of science.

59 In truth, those scientists who write textbooks all too often make matters worse by providing distortions of their own, namely by presenting severely rationalized versions of the scientific process – just as there implicitly hangs over the desks of all scientists, while they write up their research results for publication, the admonition attributed to Louis Pasteur: "Make it seem inevitable."

There are reasons why that is the case. But, in this respect, the hammer of the new Apollonians finds its most malleable materials in those who prepare didactic presentations. Usually, the latter provide indeed only "candles of information"; at worst, their products appear to have the character of unchallengeable, closed (and therefore essentially totalitarian) tracts. It may well be that the anarchic reconstructions of science on the part of the new Dionysians are largely reactions to the view of science that emanates from the rationalistic reconstructions of the new Apollonians *and* of science text writers – for those texts are perhaps all they (or most students) ever get to see of "science."

I have expanded some of these ideas and their consequences for the educational curriculum in science in Chapter 11 of this book, and in the essay "Science, Science Teaching, and Rationality," in S. Hook, P. Kurtz, and M. Todorovich, *The Philosophy of the Curriculum: The Need for General Education* (Buffalo, N.Y.: Prometheus Books, 1975), pp. 101–18.

60 P. B. Medawar, *Induction and Intuition* (Philadelphia: American Philosophical Society, 1969), p. 46.
61 Ibid., p. 57.

Chapter 4. Analysis and Synthesis as methodological themata

1 I shall presuppose the explanation of the role of themata in scientific work as given in G. Holton, *Thematic Origins of Scientific Thought: Kepler to Einstein* (Cambridge, Mass.: Harvard University Press, 1973) (hereafter cited as *Thematic Origins*), particularly Chapter 1 through 4; and in the first chapter of the present book.

I wish to acknowledge useful conversations on aspects of Chapter 4 with colleagues while a fellow at the Center for Advanced Study in the Behavioral Sciences at Stanford, California. An earlier draft was also circulated for discussion by the planners of a thematic encyclopedia for *Editore Einaudi* (Turin).

2 At least within limits; for, in synthesizing a peptide, for example, success with respect to proof of purity, amino-acid analysis, chemical and mass spectrometric evidence, and comparison of natural with synthetic material is always open to question and debate.

3 More generally, the word "analysis" in mathematics refers to the third and largest part of mathematics, that bordering on algebra and arithmetic on the one hand, and on geometry and topology on the other, although the boundaries are not sharp. It includes differential and integral calculus, ordinary and partial differential equations, and calculus of variations, Fourier series, complex analysis, and probability.

A discussion of analysis in modern philosophy can of course be found in almost any one of the standard texts. For a short but useful bibliography see also J. Passmore, *A Hundred Years of Philosophy*, 2nd rev. ed. (New York: Basic Books, Inc., 1966).

4 The pretensions of "holistic philosophies" were among the favorite targets of positivistic philosophers. In a chapter entitled "A Priori and the Whole," Richard von Mises wrote: "There is no lack of psychologic motives and expedient reasons for the use of the concept of "the whole" in various spheres of knowledge. But we see in it a model example of metaphysical postulates: an auxiliary concept, useful in many places, has been elevated to the rank of an 'absolutely valid idea,' which then is sold as a source of fundamental knowledge" (*Positivism* [Cambridge, Mass.: Harvard University Press, 1951], p. 286).

5 The fact that these are two members of a linked couple is tellingly indicated by the confusion ensuing upon the introduction of Kant's terminology "analytic" and "synthetic," sometimes in the way according with the Greek meaning and sometimes also in the

opposite manner. As a result, "analytic" and "synthetic" now are not infrequently found associated with deduction and induction respectively; but the opposite usage, more true to the original, will obtain in what follows, as it does in much of the literature. Cf. Colin Murray Turbayne, *The Myth of Metaphor*, rev. ed. (Columbia: University of South Carolina Press, 1970), ch. 2.

6 Cf. his letter to Maurice Solovine (May 7, 1952), discussed in Chapter 3.

7 The history of the general concept "element," from Aristotle to early modern chemistry, is given in Marie Boas Hall, "The History of the Concept of Element," in D. S. L. Cardwell, ed., *John Dalton and the Progress of Science* (New York: Barnes and Noble, Inc., 1968).

8 E. A. Burtt, *The Metaphysical Foundations of Modern Physical Science* (New York: Harcourt, Brace & Co., 1927), pp. 236–37.

9 One of the more readable exposés is in K. R. Popper's *Conjectures and Refutations* (New York: Basic Books, 1963, 1965), ch. 15.

10 E. O. Wilson, *Sociobiology: The New Synthesis* (Cambridge, Mass.: Harvard University Press, 1975). The discussion to page 134 is a condensation of my essay in M. Gregory and A. Silvers, eds., *Sociobiology and Human Nature* (San Francisco: Jossey-Bass, expected 1978).

11 Of the many accounts, I refer to the excellent brief one by Donald Fleming, in his introduction to the Harvard University Press reissue (1964) of Jacques Loeb's book *The Mechanistic Conception of Life*. See also Everett Mendelsohn, "Revolution and Reduction," in Y. Elkana (ed.), *The Interaction Between Science and Philosophy* (New York: Humanities Press, 1974), pp. 407–26.

12 Lucretius, *On the Nature of Things, Book I* (published by Henry Regnery Company for the Great Books Foundation, Chicago, Ill., 1969), pp. 2–5.

13 "Of Atoms, Mountains, and Stars: A Study in Qualitative Physics," *Science* 187 (1975): 605–12.

14 In *The Zinacantecos of Mexico: A Modern Maya Way of Life* (New York: Holt, Rinehart & Winston, 1970).

15 Ira R. Buchler, "Review of Hildred Geertz and Clifford Geertz, *Kinship in Bali*," *Science* 191 (January 9, 1976): 74–75.

16 Wheeler continued: "Matter, charge, electromagnetism, and other fields are only manifestations of the bending of space. Physics is geometry" (J. A. Wheeler, *Geometrodynamics* [New York: Academic Press, 1962], p. 225). In 1972 Wheeler revised the priority of the structure of space-time and announced that it is not basic to an understanding of the structure of elementary particles, but on the contrary can only be derived from the latter. This reversal, however, leaves untouched Wheeler's commitment to the method of monistic postulation. For references and discussion, see Adolf

Grünbaum, "The Ontology of the Curvature of Empty Space in the Geometrodynamics of Clifford and Wheeler," in P. Suppes, ed., *Space, Time and Geometry* (Dordrecht and Boston: D. Reidel Publishing Co., 1973), pp. 268–95.

17 "Analysis" and "synthesis" are two of the "abilities and skills" presented in D. R. Krathwohl, B. S. Bloom, and B. B. Masia, *Taxonomy of Educational Objectives: Classification of Educational Goals, Handbook II* (New York: David McKay Co., Inc., 1956), pp. 191–92.

18 *International Encyclopedia of Unified Science,* vol. 1, no. 1 (University of Chicago Press, 1938), p. 28.

19 William James characterized the monists as "optimists" and considered them "tender-minded," whereas he thought the pluralists to be pessimistic and "tough-minded." Ernest Gellner (*Legitimation of Belief* [Cambridge University Press, 1972], p. 2) suggests that James evidently supposed the tender-minded to be optimistic because they are monists and hence can rely "on the great all-embracing One to be ever there, in the background, guaranteeing a happy end."

20 For example, see Arthur Pap, *Elements of Analytical Philosophy* (New York: Hafner Publishing Co., 1949, 1972), ch. 16.

21 A. Einstein, "Notes on the Origin of the General Theory of Relativity," in Sonja Bargmann, trans. and rev., *Ideas and Opinions of Albert Einstein* (New York: Crown Publishers, Inc., 1954), pp. 288–89 (hereafter Bargmann's translation is cited as *Ideas and Opinions*). I have discussed the change in Einstein's epistemological credo in Chapter 8 of *Thematic Origins.*

22 "On the Method of Theoretical Physics," in *Ideas and Opinions,* p. 270.

23 In *Ideas and Opinions,* pp. 290–323.

24 In P. A. Schilpp, ed., *Albert Einstein: Philosopher-Scientist* (Evanston, Ill.: Library of Living Philosophers, 1949).

25 Ibid., p. 11.

26 Ibid., p. 674.

27 Ernst Mach, *Analysis of Sensations* (Chicago: The Open Court Publishing Co., 1910), p. 31.

28 It is in this sense that Bohr – whose favorite saying was the line from Schiller, "*Nur die Fülle führt zur Klarheit*" ("Only completeness [fullness, breadth] leads to clarity") – conceived the strength of the complementarity point of view. For a discussion of this point, see Chapter 4 in *Thematic Origins.*

29 Ernst Gombrich, *Art and Illusion,* 2nd ed. (Princeton, N.J.: Princeton University Press, 1969), p. 101. Going beyond the ontogeny of ideas in children, one is tempted to say that the child itself results from the repeated dichotomous differentiation of a single original cell.

Additional sources

Mortimer J. Adler, *The Great Ideas: A Syntopicon of Great Books of the Western World* (Chicago, Ill.: Encyclopaedia Britannica, Inc., 1952).

David Bohm, *Fragmentation and Wholeness* (New York: Humanities Press, 1976).

Ernest Gellner, *Legitimation of Belief* (London: Cambridge University Press, 1974).

Leroy E. Loemker, *Struggle for Synthesis: The Seventeenth Century Background of Leibniz's Synthesis of Order and Freedom* (Cambridge, Mass.: Harvard University Press, 1972).

Peter B. Medawar, *Induction and Intuition* (Philadelphia: American Philosophical Society, 1969).

Arthur Pap, *Elements of Analytic Philosophy* (New York: Hafner Publishing Co., Inc., 1949, 1972), particularly chs. 15, 16, 17.

D. C. Phillips, *Holistic Thought in Social Science* (Stanford, Calif.: Stanford University Press, 1976).

Philip Teitelbaum, "Levels of Integration of the Operant," in W. K. Honig and J. E. R. Staddon, eds., *The Handbook of Operant Behavior* (Englewood Cliffs, N.J.: Prentice-Hall, 1976).

Colin Murray Turbayne, *The Myth of Metaphor*, rev. ed. (Columbia: University of South Carolina Press, 1970).

UNESCO, *Science and Synthesis* (New York: Springer Verlag, 1971).

Chapter 5. Fermi's group and the recapture of Italy's place in physics

1 Quoted by S. C. Chandrasekhar, in *Collected Papers of Enrico Fermi* (Chicago: University of Chicago Press, 1965), 2: 927, and by Emilio Segrè in *Enrico Fermi, Physicist* (Chicago: University of Chicago Press, 1970), p. 80. The former will be abbreviated as *Collected Papers;* the latter, as *Enrico Fermi.* For other sources used throughout, see n. 143.

2 *Enrico Fermi*, p. 80.

3 Otto Hahn and Fritz Strassmann, "Nachweis der Entstehung aktiver Bariumisotope aus Uran . . . ," *Die Naturwissenschaften* 27, no. 6 (1939).

4 For example, the distinguished physicists in Florence included Gilberto Bernardini, Giuseppe Occhialini, Guilio Racah, and Bruno Rossi. The organization and contribution of this group merit a study of its own. There are analogies with the Fermi case – for example, Antonio Garbasso in Florence (physicist, director of the Physical Laboratory, senator, and mayor of Florence) was in some respects functionally similar to O. M. Corbino. Although the

groups in Rome and Florence grew up independently, they were linked to each other, primarily through contacts with B. Rossi. See also *Enrico Fermi,* passim, and G. Occhialini, American Institute of Physics (AIP) interview, passim (hereafter cited as AIP interview).

5 AIP interview, p. 158.

6 Emilio Segrè, biographical introduction to *Collected Papers*, p. 70.

7 Segrè, AIP interview, p. 51.

8 Sources for the History of Quantum Physics (SHQP) interview, p. 8 (hereafter cited as SHQP interview).

9 In his interview at the AIP, pp. 12–14, and in SHQP interview, pp. 20–21.

10 E. Amaldi, AIP interview, pp. 11–12.

11 "The Discovery of Fission," *Physics Today* 20, no. 11 (November 1967): 43.

12 *Enrico Fermi,* p. 71; Rasetti, F., in *Collected Papers,* p. 539.

13 See, for example, the testimony by B. Pontecorvo in "Review of the *Collected Papers of E. Fermi,*" *Nedelya* 25 (1969): 11.

14 Among Fermi's students, a remarkably large number reached distinction in their fields. For example, the list of those who passed through Fermi's classes in the United States includes H. Anderson, O. Chamberlain, M. Gell-Mann, M. Goldberger, D. Lazarus, T. D. Lee, J. Orear, M. N. Rosenbluth, A. H. Rosenfeld, W. Selove, J. Steinberger, A. Wattenberg, G. Weil, L. Wolfenstein, C. N. Yang, and others. See *Enrico Fermi,* p. 167 ff.

However, it is also important to see some limits within the obvious success of Fermi as physicist, teacher, and leader of groups of scientists. In retrospect it was evident that there were opportunities missed, for example, the emission of fission products from uranium atoms which had been bombarded by neutrons. There were also some simple mistakes made by the group, although the conservatism of Fermi's group, perhaps responsible for some of the missed opportunities, kept these to a very small number. Segrè points out: "Because Fermi loathed being in error, and error is occasionally unavoidable, he wanted to be in error only for having claimed too little" (ibid., p. 103).

15 Project Physics Course film transcripts, p. 11, of *The World of Enrico Fermi,* 16 mm black-and-white film, 47 min., distributed under number 075455-0 by Holt, Rinehart and Winston, New York (hereafter cited as Project Physics Fermi film transcripts).

16 See, for example, Laura Fermi, *Atoms in the Family* (Chicago: University of Chicago Press, 1954), p. 217.

17 Pontecorvo recounts that Rasetti once criticized an experimental apparatus, exclaiming to Fermi, "You do unpardonable things in the laboratory. Look at that electrometer. You would smear it with 'chicken's blood' [slang for a reddish, disgusting-looking paste used in the laboratory] if you thought it would help your results. Admit

it!" Fermi replied, calmly, "Of course, I would dunk all the electrometers in the laboratory in chicken blood if I thought we'd learn something essential" (op. cit. [n. 13]). There is evidence that Rasetti was considered a very elegant experimentalist, and that Fermi tended to exaggerate a little his own stylistic difference.

18 J. Chadwick in *Project Physics Course Reader, The Nucleus,* Unit 6 (New York: Holt, Rinehart and Winston, 1970), p. 28; reprinted from *Proceedings* of Tenth International Congress of the History of Science (Paris: Hermann et al., 1964).

19 C. Weiner, and Elspeth Hart, eds., *Exploring the History of Nuclear Physics,* American Institute of Physics Conference Proceedings, no. 7 (New York: American Institute of Physics, 1972), p. 185.

20 See my "The Roots of Complementarity," in G. Holton, *Thematic Origins of Scientific Thought: Kepler to Einstein* (Cambridge, Mass.: Harvard University Press, 1973), p. 115 ff.

21 Segrè, AIP Interview, p. 28. Amaldi similarly reported that Fermi delayed consciously introducing hypotheses which were not strictly necessary as "harmful if introduced into our mental picture. He insisted that one must proceed by reasoning with the observed experimental facts. The correct interpretation of the nature of the neutron groups would emerge as a necessary consequence of the data. He was afraid that a preconceived interpretation, however plausible, would sidetrack us from an objective appraisal of the phenomena that confronted us." E. Amaldi, in *Collected Papers* I: 808.

22 E. Persico, review of "Enrico Fermi: Collected Papers," in *Scientific American* 270, no. 5 (November 1962): 183.

23 "In experimental work, Fermi had a very personal style. His experimental ability did not consist in being able to build complicated devices or to perform measurements of high precision. It lay rather in the capacity to find out at the right moment which was the most important experiment to do, to project it in the simplest and most efficient way, and to carry it out with energy and patience without wasting time and labor on nonessentials. His experimental work was always intimately connected with his theoretical work, and he carried both of them through methodically and calmly, with great perseverence and an exceptional resistance to mental and physical fatigue." Persico, op. cit. (n. 22).

24 E. Segrè, "Fermi and Neutron Physics," *Review of Modern Physics* 27 (July 1955): 262 (hereafter cited as "Fermi and Neutron Physics").

25 One by-product was his textbook on atomic physics, *Introduzione alla Fisica Atomica* (Bologna: Zanichelli, 1928).

26 *Enrico Fermi,* p. 65.

27 AIP interview, p. 5.

28 "The work of Rutherford and his school was rather alien to us. Thus, the change to nuclear subjects cost us considerable effort. It was not a whim or a desire to follow a fashion, but the results of a deliberate plan that Fermi and his friends debated vigorously, even heatedly" (ibid.).

29 Rasetti did not want to return to Italy after his short stay at Columbia University, in view of the political developments of the time, namely, the Ethiopian war and the deterioration of the European political situation. Segrè was also in the United States that autumn and had begun to think of emigrating from Italy; when he did return, it was not to Rome but to Palermo, where he stayed until his immigration to the United States in 1938. Pontecorvo had left Italy for France. Only Amaldi and Fermi were then left in Rome ("Fermi and Neutron Physics," p. 261). Rasetti did return in 1936, but had to leave Italy again three years later.

30 See *Enrico Fermi*, pp. 65–68, and his introductory note to the translation by Fausta Segrè of Corbino's speech, "The New Goals of Experimental Physics," in Reports and Documents, *Minerva* 9, no. 4 (October 1971): 528–38.

31 Segrè, SHQP interview, p. 15.

32 Corbino, op. cit. (n. 30), p. 532.

33 Ibid., p. 535.

34 Ibid., p. 534.

35 Ibid., p. 534.

36 Ibid., p. 534.

37 Ibid., p. 535.

38 Ibid., p. 535.

39 Ibid., p. 536.

40 Ibid., p. 536.

41 Ernest Rutherford, James Chadwick, and C. D. Ellis, *Radiations from Radioactive Substances* (Cambridge: Cambridge University Press, 1930).

42 See the description of this seminar by E. Amaldi, *La Vita e l'Opera di Ettore Majorana, 1906–1938* (Rome: Accademia dei Lincei, 1966).

43 See E. Amaldi, lectures entitled "Recollection of Research," delivered at the International School of Physics, Varenna, 57th course, Summer 1972, published under the editorship of Charles Weiner, *History of Twentieth Century Physics* (New York: Academic Press, 1977), hereafter cited as *Varenna;* and Amaldi, AIP interview, p. 9. The transition was more marked in the case of Segrè, who was working on hyperfine structure. See Fermi and Segrè, in *Collected Papers*, items 75a and 75b.

44 ". . . what he [Fermi] tried to tackle at first – it's quite clear – he tried to go to [the] nucleus. The first thing he did was to try to go by hyperfine structure, which was a link to something he knew.

Then he tried to make gamma-ray spectra, gamma-ray spectroscopy, find nuclear levels . . ." Segrè, SHQP interview, p. 18.

45 *Collected Papers,* p. 548, comment by F. Rasetti.

46 Segrè, SHQP interview, p. 18; ibid., pp. 21–22. Amaldi, in his interview in the SHQP, stresses another aspect: "I went to Leipzig and I don't remember why. It was an extremely good place. Probably, also Segrè went to Stern because it was an extremely good place. But these trips – while I wouldn't say they were mistakes – were not planned. Fermi was encouraging us to go to other places. The group in Rome was so small that we needed to go to other places and talk with other people."

47 SHQP interview, p. 22.

48 *Collected Papers,* 1: 538. The conference also dealt to some extent with the other major field of current interest in Italy, cosmic rays; hence the list included Compton, Millikan, and Rossi.

A description of this conference and others that followed soon may be found in Charles Weiner, "Institutional Settings for Scientific Change: Episodes from the History of Nuclear Physics," published in A. Thackray and E. Mendelsohn, eds., *Science and Human Values* (New York: Humanities Press, 1976).

49 *Collected Papers,* 1: 548, comment by F. Rasetti.

50 Ibid.

51 Amaldi, First Lecture, *Varenna,* p. 297.

52 SHQP interview, p. 28. See also *Collected Papers,* item 78.

53 O. R. Frisch, "The Discovery of Fission," *Physics Today* 20 (November 1967): 44. Rome, however, was in a better position than most places with respect to sources.

54 Ibid., p. 45. See the description of a similar state of affairs in 1933 at the laboratory of Blackett, in the interview of Frisch at the AIP, p. 19.

55 SHQP interview, p. 18.

56 A valuable discussion of the state of the theory is to be found in Weiner and Hart, op. cit. (n. 19), pp. 121–93.

57 Ibid., p. 17. Rasetti adds: ". . . it must be emphasized that the idea of the neutrino had remained up to that time a rather vague hypothesis, while the construction of a formal theory had never been attempted . . . Apparently [Fermi] had some difficulty with the Dirac-Jordan-Klein method of the second quantization of fields, but eventually also mastered that technique and considered a beta-decay theory as a good exercise on the use of creation and destruction operators. He also made use of the isotopic spin formalism, recently invented by Heisenberg and later to prove of great usefulness in view of the charge independence of the strong interaction. The theory that he built on these foundations is remarkable for its ability to withstand almost unchanged two and a half decades of revolutionary advances in nuclear physics. One might say that

seldom was physical theory born in such definitive form" (*Collected Papers*, p. 539).

58 *Nuovo Cimento*, Supplement to vol. 2, series 10, no. 2 (1955), p. 482.

59 "They had known for many months before that aluminum bombarded with alpha particles emits positrons, but it had never occurred to them that this might be a delayed process. They had only observed the positrons during bombardment. Lawrence and his cyclotron people in California had made the same mistake" (Frisch, op. cit. [n. 53], p. 45).

60 Project Physics Fermi film transcripts, no. 11, p. 6.

61 "But when Joliot found the radioactivity by alpha particles, we were ready, in a certain sense, to jump on nuclear physics because we had started two years before to play with cloud chambers, to make counters, to make everything, so we had really the techniques. We had not yet found any problem to work on . . . Then came out the paper of Joliot, and Fermi immediately started to look for the radioactivity. He found it, and then everything started" (Amaldi, SHQP interview, p. 28).

62 Amaldi, SHQP interview, p. 26.

63 Radon sources had earlier been provided for Fermi's group by Trabacchi for use with the gamma-ray spectrometer. Hence they were familiar with the technique of radon preparation. On the importance of having good sources, see Occhialini, AIP interview, pp. 4–5.

64 Frisch, op. cit. (n. 53), p. 46.

65 Amaldi, SHQP interview, p. 26.

66 Amaldi notes that the leading position was lost partly as result of the deterioration of the political situation in Europe in general, and in Italy in particular, and partly because the means made available to them did not keep pace with their need, so that the – relatively small – Rome group could not continue to compete with other laboratories which had access to accelerators of various types, and hence to much better particle sources (*Varenna*, p. 324).

67 For the division of labor see Amaldi, *Varenna*, p. 301; Amaldi, AIP interview, p. 13; *Enrico Fermi*, pp. 78–80; "Fermi and Neutron Physics," pp. 258–59: "We organized our activities in this way: Fermi [in the spring of 1934] would do a good part of the experiments and calculations. Amaldi would take care of what we would now call the electronics, and I would secure the substances to be irradiated, the sources, etc. Now, of course, this division of labor was by no means rigid, and we all participated in all phases of the work, but we had a certain division of responsibility along these lines, and we proceeded at great speed. We needed all the help we could get, and we even enlisted the help of a younger brother of one of our students (probably 12 years old), persuading him that

it was most interesting and important that he should prepare some neat paper cylinders in which we could irradiate our stuff . . ."

68 Barbara Buck, in a survey of the Project Physics Fermi film transcripts.

69 Amaldi, AIP interview, pp. 38–40.

70 For insight into the stages in the rise of cosmic-ray and high-energy physics research after the war, see Amaldi, *Varenna*, 3, and his article "Italy," *Physics Today* 1, no. 1 (1948): 35–36.

71 *Ricerca Scientifica* 5, no. 1 (March 25, 1934): 283; *Collected Papers*, 1:639–46.

72 Amaldi, *Varenna*, p. 304.

73 Ibid., p. 13.

74 Thus, on April 23, 1934, Rutherford wrote Fermi: "Dear Fermi, I have to thank you for your kindness in sending me an account of your recent experiments in causing temporary radioactivity in a number of elements by means of neutrons. Your results are of great interest, and no doubt later we shall be able to obtain more information as to the actual mechanism of such transformations. It is by no means clear that in all cases the process is as simple as appears to be the case in the observations of the Joliots.

"I congratulate you on your successful escape from the sphere of theoretical physics! You seem to have struck a good line to start with . . . Congratulations and best wishes. Yours sincerely, Rutherford" (from copy of original, Cavendish Laboratory).

75 This phrase is used by Segrè, in "Fermi and Neutron Physics," p. 258.

76 E. Amaldi, "Production and Slowing Down of Neutrons," in S. Flügge, ed., *Handbuch der Physik* (Berlin: Springer Verlag, 1959), p. 64: "But in all these cases, increase of cross section with increase of velocity is mainly due to penetration of the electrostatic potential barrier. This is precisely what is absent for neutron bombardment" – as soon became clear. See also J. A. Wheeler, "Mechanism of Fission," *Physics Today* 20, no. 11 (November 1967): 49: "Every estimate ever made before [Fermi's slow-neutron experiments and the astonishing resonances that he had discovered. G. H.] indicated that a particle passing through a nucleus would have an extremely small probability of losing its energy by radiation and undergoing capture, if the current nuclear model was credible. Yet, directly in opposition to the predictions of this model, Fermi's [October 1934. G. H.] experiments displayed huge cross sections and resonances that were quite beyond explanation."

77 *Enrico Fermi*, p. 26.

78 Ibid., pp. 36–37.

79 Ibid., p. 10.

80 E. Segrè, a talk before the American Physical Society, Washington, D.C., April 25, 1972, transcript, p. 2.

81 There is evidence in the Fermi film interviews – with Castelnuovo's son and with Wick – that it was the great mathematician Guido

Castelnuovo, who having first met Fermi on a social occasion at his home and being impressed with Fermi on that occasion, drew Corbino's attention to the young man. Subsequently, Fermi met Castelnuovo's colleagues and fellow mathematicians of Rome in Castelnuovo's house; most of them had an interest in theoretical physics and could recognize Fermi's excellence. Wick believes that the idea of bringing Fermi to the University of Rome may have originated with the mathematicians. See Project Physics Fermi film transcripts, interview with Castelnuovo, no. 31, p. 4; with Wick, no. 11, p. 4.

82 *Collected Papers,* 1:120, translation quoted in *Enrico Fermi,* p. 26.

83 Ibid., p. 26.

84 Corbino certainly merits a full-scale biography, as Segrè properly insisted at the end of his interview, SHQP, pp. 60–61. He has been well described by Segrè in *Enrico Fermi* and by Laura Fermi in *Atoms in the Family* (op. cit., n. 16), as well as in many of the interviews, for example, Amaldi, SHQP. An important source is the autobiography of O. M. Corbino's brother, Epicarmo Corbino, *Racconto di una Vita* (Naples: Edizioni Scientifiche Italiane, 1972).

85 *Enrico Fermi,* p. 30. Several Italian physicists remarked on what they called the humane, "Sicilian" component of his personality. Amaldi adds that almost all of the younger generation even thought Corbino was sometimes "too much a Sicilian," for example, in the sense that in order to obtain a chair for the best possible man, "Corbino would apply practically the same method that one would apply to get the chair for a bad candidate" (SHQP interview, p. 11).

86 *Enrico Fermi,* p. 30.

87 See Laura Fermi, op. cit. (n. 16), pp. 71–73.

88 There is evidence that Corbino managed to divert the chair to physics from the general pool for all the sciences, and at the expense of biology. It has been remarked that things might have turned out differently if biology in Italy had had its Corbino.

89 Amaldi, SHQP interview, p. 13; Laura Fermi, who was also in the same class that day, recalls that Corbino referred to two new men – Fermi and Rasetti. See also, Laura Fermi, op. cit. (n. 16), p. 40.

90 E. Segrè, introductory note to F. Segrè, op. cit. (n. 30), p. 529.

91 *Enrico Fermi,* p. 31: Mrs. Fermi confirms this.

92 Corbino, quote from speech, in F. Segrè, op. cit. (n. 30), p. 532.

93 Quoted in *Enrico Fermi,* p. 45.

94 Samuel Goudsmit, Project Physics Fermi film transcripts, no. 10, p. 5. Others, however, point out that Fermi's capacity was so widely recognized that he was bound in any case to have reached a leading position in Italian cultural life.

95 Angelo Brofferio, *Storio del Piemonte dal 1814 ai nostri giorni* (Turin: Fontana, 1849–52), pt. III, p. 89; quoted in Barbara Buck, dissertation draft, p. 12.

96 Thus it would be natural for Senator Corbino to "take care that the

results [of Fermi's group in 1934 on the production of artificial radioactivity by neutron bombardment. G.H.] be properly advertised at the Italian Accademia dei Lincei in the solemn royal session" ("Fermi and Neutron Physics," p. 259).

Many prominent scientists and scholars were made senators, which was considered both an academic and a political honor. The list includes Blaserna, Cannizzaro, Cremona, Croce, Garbasso, Mossotti, Righi, and Volterra.

97 Corbino, quote from speech, in F. Segrè, op. cit. (n. 30), p. 536.

98 It is significant that Fermi did not hesitate to make that type of appeal himself when he visited the United States. The press release prepared by the Department of Public Information of Columbia University on August 3, 1936, quoted Fermi as follows: "The most obvious application of artificial radioactivity which can be foreseen is in the medicinal field. Radium, naturally radioactive, is used for the treatment of cancer. The completely new radioactive substances created in the laboratory should give medical men new tools, some of which may prove more efficient than radium." After the war, Fermi stressed the usefulness of nuclear reactors for the preparation of radioisotopes needed for medical research and therapy. One should add that before the discovery of fission, the medical uses of radioactivity were generally considered its most important application. Fermi and his collaborators often underlined these uses, and the patent taken out in October 1934 refers mainly to the use of slow neutrons for producing radioactive isotopes for practical applications.

99 Laura Fermi recalls that during their wedding trip in 1928, the possibility of being eventually considered for the prize was discussed by Enrico Fermi; however, he mentioned a list of physicists who would have to be given the prize first.

During the period 1901–52, Italy was credited with the following Nobel Prizes: a half-share of the prize in physics to Marconi (1909), together with F. Braun, for their services to the development of wireless telegraphy; the award of 1938 to Fermi, who was on his way to the United States as an emigrant; one-half of the prize for physiology and medicine for 1906 to Camillo Golgi, the other half going to Ramon y Cajal of Madrid; three prizes for literature: to Giosue Carducci of Bologna (1906), to Grazia Deledda (1926), and to Luigi Pirandello (1934). Also, in 1907, Ernesto Teodoro Moneta received the Nobel Prize for peace, with Louis Renault of Paris.

This excursion provides an opportune point to mention that the phrase "second-rank country, scientifically," used previously and often heard, needs of course some further specification. This has been attempted, for examples, by J. Ben-David. Insofar as Nobel Prizes are an indicator of a nation's productive effort – and they certainly are not a very good indicator – it may be useful to examine the figures and the relative standing of Italy to obtain a semi-quantitative estimate. Using – essentially for reasons of convenience – the year 1952 as the limit, one finds that Italy shares sixth place (with Sweden) in the number of individuals granted any part of a Nobel Prize in physics between 1901 and 1952 – assigned by country of nationality of the prize-winner. Considering all three categories of science prizes – physics, chemistry, and physiology and medicine – this standing becomes somewhat lower: Italy shares tenth place (with Canada). Making the same kind of calculation, but calculating the ranking by Nobel Prize winners per million of inhabitants, one finds that Italy shares sixteenth place with Argentina. Figures based on Ernst Meier, *Alfred Nobel, Nobel Stiftung, Nobelpreise* (Berlin: Duncker und Humblot, 1954).

100 Charles Weiner, "A New Site for the Seminar: The Refugees and American Physics in the '30s," in Donald Fleming and Bernard Bailyn, eds., *The Intellectual Migration: Europe and America, 1930–1960* (Cambridge, Mass.: Harvard University Press, 1969), pp. 190–234.

101 Ibid., p. 193.

102 Ibid., p. 195. See also Charles Weiner, "Institutional Settings for Scientific Change: Episodes from the History of Nuclear Physics," in Thackray and Mendelsohn, op. cit., n. 48.

103 AIP interview, p. 20.

104 *Enrico Fermi*, p. 10.

105 AIP interview, pp. 13–14.

106 Weiner, op. cit. (n. 100), passim.

107 The lack of flexibility in the Italian university system – indeed, for most appointments Italian citizenship was required – and the early forms of anti-Semitism on the part of Mussolini and others in policy-making positions may help to explain part of this phenomenon, but not all of it.

108 This is not, however, an attempt to "explain" the success of Fermi's group. Not everything is "explicable," least of all a phenomenon such as Fermi himself.

109 AIP interview, p. 1.

110 Amaldi agrees that Corbino "protected all of us from the criticisms of the traditional university environment" (*Varenna*, p. 317).

111 SHQP interview, pp. 23–24.

112 Ibid., p. 24.

113 See Joseph Ben-David, *The Scientist's Role in Society* (Englewood Cliffs, N.J.: Prentice-Hall, 1971), p. 140; also Daniel Bell, *The Coming of the Post-Industrial Society* (New York: Basic Books, 1973), pp. 350–51.

114 Ben-David, op. cit. (n. 113), p. 156.

115 Frisch, op. cit. (n. 53), p. 43. However, examples can be found –

the laboratories of Kamerlingh Onnes in Holland and R. A. Millikan at the California Institute of Technology – where analogies with aspects of Fermi's group may be discerned.

116 Rasetti, SHQP interview, p. 11.

117 Ibid., p. 13.

118 Segrè, SHQP interview, p. 7.

119 Segrè, AIP interview, p. 2.

120 Ibid.

121 See *Enrico Fermi,* pp. 7–8, and the appendix containing Fermi's early letters to Persico. Persico recalled: "I was supposed to become an engineer, and it was probably conversations with Fermi, who presented to me the possibility of becoming a professor of physics, that changed my direction" (Persico, SHQP interview, p. 10).

122 *Enrico Fermi,* p. 40.

123 Laura Fermi, op. cit. (n. 16), p. 42.

124 The word, though used rather casually, appears also in Laura Fermi, op. cit. (n. 16), p. 41. Referring to Amaldi, she writes: "He was soon accepted into the expanding family of physicists and physicists-to-be." Dr. J. Goodstein informs me that a similar, family-like relationship existed among the Italian mathematicians of that period.

One must of course guard against using the word "family" in more superficial contexts, for example, in describing the relationships in institutions, ranging from other laboratories to business enterprises, which are characterized merely by close personal ties and zeal.

125 *Enrico Fermi,* p. 8.

126 The "recruitment" of scientists by members of their immediate or wider family would make an interesting case study. Examples are Einstein's introduction to mathematics and science through his uncle; Frisch's account of the influence his aunt, Lise Meitner, had on his choice of career; and of course the recruitment of the scientists' children such as Curie, Bohr and Thomson into the same or similar fields of study. Many in Fermi's group were influenced early by relatives who were scientists. When he was ten years old, Segrè received a copy of Ganot's *Physics* from his uncle, a distinguished geologist; Amaldi is a mathematician's son, Rasetti the nephew of a well-known physiologist; Majorana was the nephew of a physicist.

There are negative aspects associated with this mechanism. One is discrimination against members of another class, either intellectually or indeed simply by class discrimination, which, in the long run, may also handicap the growth of science in a country. See, for example, "A Policy for Scientific Research: Proceedings of a Study Group Held in Rome, 2 and 3 December, 1961, Under the Auspices of the Central Office of Cultural Activities of the Christian

Democratic Party," reprinted in Reports and Documents, *Minerva* 2, no. 2 (Winter 1963): 210 ff. There it is pointed out that "only 6 per cent of the students [at the university level] come from the artisan class, 9 per cent from the working class, and a further 9 per cent from the peasantry. The offspring of entrepreneurs and persons in the liberal professions, who together constitute the 8.6 per cent of the male population, are disproportionately highly represented at the universities; they constitute 60 per cent of all students registered. On the other hand, offspring of white-collar workers are underrepresented" (p. 213).

Other negative aspects are indicated in that document, including the fragmentation and dispersal of the scientific effort into relatively many quite small enterprises throughout Italy; the apparent discrimination of research groups in favor of northern Italy and against the south, is indicated by Alessandro Alberigi Quaranta, in "Expenditure, Organization, and Policy-Making in Scientific Research," *Minerva* 2, no. 2 (Winter 1963): 217.

127 See Bruno Pontecorvo, *Fermi e la Fisica Moderna* (Roma: Editori Riunti, 1972), p. 30.

128 Rather, there was in the laboratory an informal, lighthearted atmosphere, "a certain playfulness, a naïve love of jokes and silly acting that they brought into their serious work" (Laura Fermi, op. cit. [n. 16], p. 45). See also Amaldi, SHQP interview, p. 27: "This was the great difference between the character of Fermi and probably of the Germans . . . We were so joking all the time and never taking seriously what we were doing . . . People [in Leipzig, which Amaldi had visited. G.H.] were extremely serious and had a certain style. They were conscious that they were scientists in an old tradition . . ."

In the group working at Florence under Garbasso, and possibly at other centers, the relationships in the groups were not greatly different. Occhialini has stressed the feudal and patriarchal model of organization in his AIP interview, adding, "in Rome they have a father image of Corbino, and in Florence they have a father image of Garbasso" (p. 18).

129 It may not be without significance that Corbino himself came from a large, closely knit family, one hierarchically organized in the old style, replete with periodic gatherings of the whole clan at their original home in Sicily for consultation. See Epicarmo Corbino, op. cit. (n. 84), *passim*.

130 Some years later, in 1937, Fermi, Amaldi, and Rasetti began work on a Cockcroft-Walton type of accelerator, to get increased neutron intensity from the $d+d$ reaction. Shortly before Fermi left Italy, work began on constructing a 1.1 Mev proton accelerator at the Istituto Superiore di Sanità, but it was far from operational in December 1938.

131 Segrè, AIP interview, p. 8.

132 Ibid., p. 10.

133 Similarly, one hears regrets about Fermi's distance and isolation with respect to cooperation with industrial enterprises and applied research to the possible detriment of Italian industry. On this point it would seem that Fermi's involvement with industry – for example, in connection with the construction of the 1.1 Mev accelerator – was on the rise during the period immediately before his virtually enforced departure, and it could well have become stronger if history had not prevented it.

134 Angelli Foundation Symposium, Torino, 1973, published in F. Cavazza and S. R. Graubard, eds., *Il Caso Italiano* (Milan: Garzanti, 1974).

135 OECD, *Report No. 25.651* (Paris: OECD, 1969).

136 Ibid., p. 20. Ben-David has confirmed this finding by observing that expenditure on research and development in Italy was 0.6 percent of the gross national product – a half or less of most others, with only Austria standing at the lower place than Italy (Ben-David, op. cit. (n. 113), p. 164.

137 OECD, op. cit., p. 21.

138 Ibid., p. 46.

139 Ibid., p. 58.

140 Ibid., p. 99.

141 Ibid., p. 106. However, some qualified observers in Italy have thought that this statement, and indeed the whole tenor of the OECD report, was somewhat more negative than a lengthier study would have warranted.

142 Ibid., p. 140.

143 I wish to acknowledge the generous help and support extended to me by Dr. Fabio Cavazza of the Agnelli Foundation and his colleagues, in arranging archival searches and obtaining documents. Much important material was found with their help, and will continue to be used in work on the same topic, by both myself and my students.

A large number of unpublished interviews with scientists and others close to Fermi and his colleagues were conducted under my general direction by a team headed by John Kemeny and Donald Brittain, in connection with making a documentary film on the life and work of Enrico Fermi, as part of the Project Physics Course development (n. 15). That part of the work was supported by funds from the Ford Foundation, and the analysis of those interviews was supported by a grant from the Section for the History and Philosophy of Science of the National Science Foundation. The transcripts of these interviews – including interviews with E. Amaldi, U. Fano, L. Fermi, E. Fubini, F. Rasetti, Bruno Rossi, E. Segrè, as well as most non-Italian chief collaborators and students of Fermi – are in my custody at the Jefferson Physical Laboratory, Harvard University, and copies have been deposited at the Center for History of Physics at the American Institute of Physics (AIP) in New York.

Another series of more extensive interviews, conducted chiefly by Charles Weiner with Amaldi, Beck, Cockcroft, Frisch, Goldhaber, Occhialini, Segrè, Wick, and others, is also on deposit at the Center for History of Physics at the AIP. They are referred to as AIP interviews. A third series of detailed interviews with Amaldi, Persico, Rasetti, and Segrè, among others, conducted chiefly by T. S. Kuhn, are in the archives of Sources for the History of Quantum Physics (referred to as SHQP), at the American Philosophical Society in Philadelphia, with copies at the other deposits of the SHQP project. The help extended to me by the directors and librarians of these archives is gratefully acknowledged – as is that of many of the persons who knew and worked in Fermi's circle.

In this paper I have stressed the information in these largely unpublished but accessible interviews. Where possible I have cited them rather than the more widely available books. But the latter have of course not been neglected. Foremost among them are, naturally, the *Collected Papers of Enrico Fermi* (n. 1), and the two biographies of Fermi, by Emilio Segrè (n. 1) and Laura Fermi (n. 16).

I profited from the three lectures by Edoardo Amaldi on "Recollection of Research," given at Varenna (n. 43). The basic physics of the work discussed in this article is authoritatively discussed in two publications: One is Franco Rasetti, *Elements of Nuclear Physics* (New York: Prentice-Hall 1936), English edition of F. Rasetti's *Il Nucleo Atomico* (Bologna: Zanichelli 1936), which gives an interesting contemporaneous view. The other, more extensive and written from a more recent vantage point, is the contribution by E. Amaldi, "Production and Slowing Down of Neutrons," in vol. 38, pt. II of the *Handbuch der Physik* (n. 76).

Another basic resource is the book *Exploring the History of Nuclear Physics* (n. 56). This is the record of two conferences of major participants in nuclear physics. Appendix 2 and Appendix 3 are useful surveys of data, original source materials, and other resources for doing research on the history of nuclear physics. A set of essays on the discovery of the neutron and its consequences, which I arranged for presentation at the Tenth International Congress of the History of Science at Ithaca, 1962, included the contributions of Emilio Segrè, E. M. Purcell, and S. Goudsmit (n. 18).

Use has been made of the archival collection of Enrico Fermi's own papers at the University of Chicago; of the archives and library of the Institute of Physics at the University of Rome; and of the archives of the Accademia dei Lincei.

In addition to my own visits to archives, I was also enormously helped by Professor Giorgio Tabarroni of Modena. Through his intercession, we obtained valuable documentary material from the archives of the Academy of Italy, the Archivio Centrale dello Stato, Rome, and other resources; these illuminated not only the work of the Fermi group, but also the early history and role of the Società Italiana di Fisica and the Società Italiana per il Progresso delle Scienze in the rise of physics in Italy. This material will continue to be of importance in our ongoing work on the history of the physical sciences in Italy. My thanks are also due to Dr. Tullio Derenzini of the Domus Galilaeana in Pisa, where laboratory notebooks and other documents of the Fermi group are kept and were made accessible to me.

Edoardo Amaldi, Laura Fermi, Emilio Segrè, Bruno Rossi, and Charles Weiner read a draft of the paper and advised me on it. I was also helped by discussions with Barbara Buck, David J. Taylor, and Kenneth Swartz.

Chapter 6. Can science be measured?

1 *Science Indicators 1972: Report of the National Science Board 1973* (Washington, D.C.: National Science Board [NSB], National Science Foundation [NSF], 1973) (hereafter cited as *S.I. 72*). Obtainable from the Superintendent of Documents, U.S. Government Printing Office, Washington, D.C., 20402; Stock Number 3800-00146. The report was the fifth in a series of annual reports to the President and to Congress, as mandated by Congress in its 1968 amendment of the National Science Foundation Statute (Public Law 90-407).

The congressional charge to the board was that the report focus on "the status and health of science and its various disciplines [including] an assessment of such matters as national scientific resources . . . progress in selected areas of basic scientific research, and an indication of those aspects of such progress which might be applied to the needs of American society." As quoted by Norman Hackerman, Chairman of the NSB, in U.S. Congress, *Measuring and Evaluating the Results of Federally Supported Research and Development,* Science Output Indicators – Part I, Hearings Before the Subcommittee on Domestic and International Scientific Planning and Analysis, Committee on Science and Technology, 94th Cong., 2nd sess.; House, 19 and 26 May 1976 (Washington, D.C.: Government Printing Office, 1976), p. 7.

The intention to issue such science indicator reports every second year may now be modified by the plan to absorb such information in the projected annual science and technology reports of the new Office of Science and Technology Policy (OSTP).

2 *S.I. 72*, p. iii.

3 *Science Indicators 1974: Report of the National Science Board 1975* (Washington, D.C.: National Science Foundation, 1975) (hereafter cited as *S.I. 74*). For sale by the Superintendent of Documents, U.S. Government Printing Office, Washington, D.C., 20402; Stock Number 038-000-00253-8.

4 Release by the Office of the White House Press Secretary, February 23, 1976.

5 *S.I. 72*, p. vii.

6 *The New York Times*, 14 March, 1976, front page.

7 Hackerman, op. cit. (n. 1), p. 8.

8 Ibid., p. 19. See also *Science* 191 (March 12, 1973): 1033.

9 The NSB has in fact invited such efforts and has, through NSF grants, already assisted several studies (e.g., in "limitations to scientific literature citations, investment in innovation, technology transfer, economics of publishing scientific literature, social indicator utilization by executives, international scientific activities, and many others") (Roger W. Heyns in the hearing cited in n. 1).

It is also significant that *S.I. 74*, p. viii, expresses appreciation for the convening of an invitational seminar, cosponsored by the NSF, at the Center for Advanced Studies in the Behavioral Sciences at Stanford, California, centering on the *S.I. 72* volume. The papers given there are being published in a book entitled *Toward a Metric of Science*, edited by Y. Elkana, J. Lederberg, R. K. Merton, A. Thackray, and H. Zuckerman (New York: Wiley-Interscience, scheduled for 1977). The present paper is based on an earlier, briefer version given at that meeting; I wish to acknowledge gratefully helpful comments by the editors of that volume, by S. D. Ellis, C. S. Gillmor, N. J. Smelser, and B. A. Stein, as well as partial support for research by a grant from the National Science Foundation.

Among other efforts devoted to the same end is the establishment by the Social Science Research Council of a Subcommittee on Science Indicators under the Advisory Committee on Social Indicators. Its long-range objectives were announced to be "to advance the quantitative description and analysis of science as a social phenomenon and to improve the empirical base for science policy." See *Social Science Research Council Items* 30, no. 1 (March 1976): 9.

It is also relevant that even in the original planning in the NSB which led to the first *Science Indicators* volume, one of the major purposes was "to stimulate social scientists' interest" in this area (Roger W. Heyns, p. 10 in the hearing cited in n. 1).

10 Albert Einstein, "Ernst Mach," *Physikalische Zeitschrift*, 17 (1916): 2.

11 A recent example is the discussion in population biology concerning the relative merits of special "tactical" models with scaled graphs, and abstract "strategic" models with scaleless diagrams. Cf. R. M.

May, "Review of 'The Mathematical Theory of the Dynamics of Biological Populations,'" *Science* 183 (March 22, 1974): 1188–89; and R. Mitchell and R. M. May, Letters to the Editor, *Science* 184 (June 14, 1974): 1131.

12 Lynn White, "Technology Assessment from the Standpoint of a Medieval Historian," *American Historical Review* 79 (1974): 1–13.

13 Harvey Brooks, "The Physical Sciences: Bellwether of Science Policy," in J. A. Shannon, ed., *Science and the Evolution of Public Policy* (New York: Rockefeller University Press, 1973), p. 125.

14 Based on an idea suggested in private conversation by E. M. Purcell.

15 See, for example, H. Inhaber and K. Przednowek, "Quality of Research and the Nobel Prizes," *Social Studies of Science* 6 (1976): 33–50; and Harriet Zuckerman, *Scientific Elite* (New York: The Free Press, 1977).

16 Hackerman, op. cit. (n. 1), p. 15. See also *S.I. 72*, p. 7, which reports that "an effort was made to estimate the relative 'quality' or 'significance' of the literature" by calculating the average number of citations per published article in selected fields for different countries. The result was the "upbeat" conclusion that the United States ranked first by a large margin in physics and geophysics, molecular biology, systematic biology, psychology, engineering, and economics; higher than all others except the USSR, by a relatively narrow margin, in mathematics; and second only to the USSR, by about the same narrow margin, in chemistry and metallurgy. In *S.I. 74* the results are not very different, except that economics was dropped, and the USSR now was first in both mathematics and chemistry.

17 On the last page of *S.I. 74*, one member of the NSB, in "Supplementary Comments," spelled out his dissatisfaction on these points. He noted that scientists in the USSR have in most fields by far the highest self-citation index; that a small-scale, high-quality effort such as that in certain fields in France would be swamped in a citation index; and that in some countries the editorial and space limitations on publications allow little room for citations.

18 *S.I. 74*, p. 106. The innovations themselves "were selected by an international panel of experts who rated the innovations based on their technological, economic, and social importance" (Hackerman, op. cit. [n. 1], p. 16).

19 L. J. Anthony et al., *Reports on Progress in Physics* 32 (1969), pt. II, pp. 764–65.

20 Cf. Statement of Robert Park at the May 26, 1976, hearing before the Subcommittee on Domestic and International Scientific Planning and Analysis, Committee on Science and Technology, U.S. House of Representatives (cf. n. 1).

21 In fact, the baseline for a great deal of "spectroscopy" (data on publication, funds, manpower, etc.) for the various subsections of fields was available in publications such as *Physics in Perspective*

(Washington, D.C.: National Academy of Science, 1972) and *Physics Manpower 1973* (New York: American Institute of Physics, 1973), but they were neither used nor even referred to bibliographically in *S.I. 72*, and little in *S.I. 74*.

22 Y. Elkana, *The Discovery of the Conservation of Energy* (London: Hutchinson, 1974).

23 *S.I. 72*, p. iii; emphasis supplied.

24 *S.I. 72*, p. vii.

25 *S.I. 72*, p. 59.

26 A. C. Nixon, Letter to the Editor, *Science* 184 (June 7, 1974): pp. 1028–30.

27 *Work in America*, Report of Special Task Force to the Secretary [Elliot Richardson], Department of Health, Education, and Welfare (Cambridge, Mass.: MIT Press, 1973), pp. 155–56.

28 Many other evidences point to a need for more detail in order to arrive at an accurate picture of the employment situation for scientists, especially in the 1970s. Unfortunately, the support for an annual science manpower survey, long provided by NSF, was discontinued some years ago just when the job crunch became notable, and professional societies had to assume the support to continue the enterprise just when many of them were least in a position to do so. Thus an American Institute of Physics (AIP) survey of new physicists in the summer of 1972 (American Institute of Physics Publication No. R-207.5 [New York: American Institute of Physics, 1972]) showed that of the Ph.D.'s who had graduated the year before, 7 to 8 percent were still seeking jobs, and an additional small percentage was changing professions. Among more recent graduates, 16 percent of the Ph.D.'s and 19 percent of the terminal M.A.'s were still seeking employment. Of the 1970–71 B.A. recipients in physics surveyed by AIP in December 1971 who did not go on to graduate work or into military service, "only 14 percent found jobs that make extensive use of the physics training received in college" (AIP Publication No. R-211.3 [New York: American Institute of Physics, 1972]).

Similarly, a survey of the American Astronomical Society (Editorial, "Current Trends in Ph.D. Production and Employment Among Astronomers," *Bulletin of the American Astronomical Society* 6, no. 2 [1974]: 233–34) made in 1973–74 showed unemployment was both serious and still rising. In the 1974–75 academic year, about 200 newly trained astronomers were vying for about 20 faculty openings.

29 *S.I. 74* had more details on enrollments, although not on employment.

30 AIP, *Physics Manpower 1973* (New York: American Institute of Physics, 1973), p. 47. See also *Newsletter of the Forum on Physics and Society* 4, no. 1 (February 1975), published by AIP.

31 *S.I. 72*, p. vii.

32 *Work in America*, op. cit. (n. 27), emphasis supplied. The "full employment" movement is evidently becoming stronger in the United States. See, for example, E. Ginzburg, ed., *Manpower Goals for the American Democracy, Report of the American Assembly* (Englewood Cliffs, N.J.: Prentice-Hall, 1976).

33 The "quality of life" of scientists and engineers, including that of the younger ones who did find new jobs or held on to makeshift arrangements in a distressed employment market, would also have been worth some attention. Fine-structure indicators can well be developed; for example, the American Institute of Physics Placement Register of jobseeking physicists at the (Spring) American Physical Society meetings shows that the ratio of employees to registrants present had dropped to 0.07 for 1972, roughly one-seventh of what it had been six or seven years earlier.

It is significant that Chapter 6 of the National Academy of Sciences (NAS) report, *Physics in Perspective*, vol. 1 (Washington, D.C. NAS-NRC, 1973), is entitled "The Consequences of Deteriorating Support" – with a discussion of the decline of federal support for research in Chapters 4, 5, and 10; the effects "on the livelihood of highly trained people" is the subject of detailed data in chapter 12 (which is devoted to manpower).

I have used many examples from physics, although I am aware that this field was in some ways the one hardest hit by the deterioration of science support from the late 1960s on. My choice is necessitated chiefly by the fact that I am far more familiar with this field; but I agree with the remark on this point in the NAS report: "To a considerable degree, of course, these effects [of deteriorating support] apply to any field of scientific research and are not unique to physics. Although our discussion is rooted in physics, the problems we address have much broader relevance" (*Physics in Perspective*, ibid., p. 452).

34 NSF, "Undergraduate Enrollments in Science and Engineering" (Washington, D.C.: U.S. Government Printing Office, January 5, 1972), NSF 71-42. More recently, the NSF has published the report, "Women and Minorities in Science and Engineering" (Washington, D.C.: U.S. Government Printing Office, 1977), NSF 77-304. This report relies largely on data supplied by the NSF's Manpower Characteristics System, which was instituted in July 1972 and became fully operable in 1974.

35 B. Henderson, "Unemployment Among Scientists," *Geotimes* 16 (September 1971): 16.

36 For the text of Malek's memoranda, see "New Federalism," Record of Hearings of Committee on Government Operations, House of Representatives, January 29, 1974. Also see *The New York Times*, 13 August 1974.

37 A recent report by the National Board on Graduate Education pointed out that while the proportion of new Ph.D.'s having no

specific job prospects rose from 4.5 percent in 1968 to 17.2 percent in 1973, the figures for new Ph.D.'s in English who had no job increased from 3.9 percent to 21.5 percent during that period (*The New York Times*, 21 January 1976).

38 W. Heisenberg, *Physics and Beyond* (New York: Harper & Row, 1971), pp. 62–63. In a letter to this author, elaborating further on this account, Heisenberg referred to his Principle of Indeterminacy, formulated the year after this exchange with Einstein. Heisenberg noted wryly that the principle added a refinement to Einstein's dictum, for it tells what one *cannot* observe.

39 Harvey Brooks, "Models for Science Planning," *Public Administration Review* 31 (May/June 1971): 367–68. According to R. Heyn's testimony (see the congressional hearing cited in n. 1), the NSB had established an internal Committee on Science Indicators by 1971.

40 Brooks, op. cit., n. 13.

41 *S.I.* 74, p. 71. The level may be even lower if one uses the correct deflator for basic research, one that properly includes the increasing complexity of research tools and management. The NSF publication "A Price Index for Deflation of Academic R&D Expenditures" (NSF 72-310) shows that during the decade of 1960–70, the increases in prices of academic research and development were considerably higher than price increases in the economy at large. They were as much as twice the rate given by the GNP deflator, even if little allowance is made for increases in R&D costs owing to increases in complexity and sophistication of basic equipment.

42 Hackerman, op. cit. (note 1), p. 13.

43 Imre Lakatos, "Review of S. Toulmin's *Human Understanding*," *Minerva* 14 (Spring 1976): 128–29.

44 Brooks, op. cit. (n. 13), p. 127.

45 This section was dropped in *S.I.* 74, although some of the old subsections are included in the new section "Industrial R&D and Innovation."

46 Here, too, fine structure is of interest. The American Physical Society (APS) and the American Association of Physics Teachers are both essential societies with somewhat overlapping memberships and mandates; but the latter has for some years been on the verge of financial disaster, while the former has enjoyed a handsome income (total net income for 1972 was $400,000; for 1973, $244,000; the accumulated Income and Reserve Fund of the APS stood at $2.38 million at the end of 1973). See "Summary of American Physical Society Income and Expense," *Bulletin of the American Physical Society* 19, ser. 2 (November 1974): 1069–71.

47 The distinction is suggested by Arnold Thackray, "Measurements in the Historiography of Science," paper presented at the annual meeting of the AAAS in Boston, February 1976.

48 *S.I.* 72, p. 73.

49 *S.I.* 72, p. 84.
50 *S.I.* 72, p. 17.
51 The issue of *Physical Review Abstracts* being analyzed here is vol. 5, no. 15, August 1, 1974. As an aside to citation-index scholars, I note that under the masthead is the injunction: "This journal is not to be cited."
52 *S.I.* 74, p. 8.
53 *S.I.* 72, pp. 25 and 111.
54 *S.I.* 72, p. 26. The field of educational research would also benefit from the development of *qualitative* indicators of the research being performed.
55 *S.I.* 72, p. 85.
56 *S.I.* 72, p. 88; see also the list of fifteen suggested "currently or potentially important graduate educational changes," on p. 93.
57 *S.I.* 72, pp. 97, 98. In *S.I.* 74 the results are slightly more positive toward science. On the subject of simultaneously held but contrary public opinions about science, see the essays by A. Etzioni and Clyde Nunn and by Dorothy Nelkin in G. Holton and W. Blanpied, eds., *Science and Its Public: The Changing Relationship* (Dordrecht and Boston: D. Reidel, 1976).
58 *S.I.* 72, pp. 88, 90 and 91.
59 *S.I.* 72, p. 92.
60 For example, *National Assessment of Educational Progress, Science Results* (Education Commission of the States, Denver, Colorado, 1970 and after).
61 Even if all *applied* research support by NSF is included, the 5 percent figure rises only to 8½ percent. This and similarly disturbing data are in the report *Social and Behavioral Science Programs in the National Science Foundation* (Washington, D.C.: NAS-NRC, 1976). For example, the report points out that nearly 24 percent of the nation's scientists are engaged in basic research in these fields, but that they receive only about "one-third as much support as those in other sciences" (p. 14).
62 For example: Are not the existing divisions of science and of current candidates for financing taken too much for granted, so that research in agriculture and nutrition, fundamental learning theory, or certain aspects of energy research continue to receive less attention than they need? Should one not try to find indicators of how large a fraction of scientific research is being done on small, "safe," and relatively unadventurous programs? If this is too large a fraction, as I think would be found, may it not be argued that new institutions of funding are needed to encourage the kind of research more risky to the researcher's career, more long term and less likely to guarantee early results – and yet possibly far more important in terms of attending to areas of scientific ignorance perceived to be at the base of social problems? At this time, such research would have to be called "altruistic."

A listing of additional concerns might include the following: Were the expectations for more social science research support actually fulfilled on an adequate scale? How adequate do scientists think the current process is by which intended publications and research-support applications are reviewed? Should science indicators be extended to cover also the fields which the NSF does not traditionally support? Would it be useful to include such indications of the general benefits of scientific research as the rapidly decreasing U.S. death rates for specific diseases, or the rise in the yields of specific crops? Cf. Nestor E. Terleckyj, *State of Science and Research: Some New Indicators: A Chart Book Survey* (Washington, D.C.: National Planning Association, 1976); and N. E. Terleckyj, *The State of Science and Research: Some New Indicators* (Boulder, Colo.: Westview Press, in press).

63 An example of the dubious use of support data is the finding in Figure 4-14 (*S.I. 74*, p. 101) that the share of the "major innovations" in the United States has been by far the lowest, and constantly dropping since 1953, for the group of those companies that have 5,000 to 9,999 employees (now down to a mere 3 percent of the major innovations), whereas it has been by far the largest, and constantly rising, for the group of companies having 10,000 employees or more (now about 45 percent). However, if one calculates the ratio of Innovations per Employee (*I/E*), the picture looks very different, as indicated by these figures (kindly made available to me by Barry Stein):

Employment size of firm	*I/E* (all industry)	*I/E* (manufacturing industry only)
1–99	4.5 ($\times 10^{-6}$)	21.4 ($\times 10^{-6}$)
100–999	10.4 ($\times 10^{-6}$)	18.5 ($\times 10^{-6}$)
1,000–9,999	9.2 ($\times 10^{-6}$)	12.9 ($\times 10^{-6}$)
$>$ 10,000	9.5 ($\times 10^{-6}$)	11.8 ($\times 10^{-6}$)

(Data for Employees (*E*) from *General Report on Industrial Organizations, 1967, Part 1*, Census Bureau; omits CAOs and support [nonproductive activities]. Data for Innovations (*I*) from final report by Gellman Research Associates, Inc., *Indicators of International Trends in Technological Innovation* [Jenkintown, Pa.: Gellman Research Assoc., April 1976], cited as the source of the data on Innovations in *S.I. 74* itself.) Thus, far from supporting the simple but erroneous conclusion one might draw from *S.I. 74* that major innovations are correlated with large company size, the data when properly presented point to a very different conclusion – as has been previously shown by the work of J. Jewkes, J. Smookler, and others. See Barry A. Stein, *Size, Efficiency, and Community Enterprise* (Cambridge, Mass.: Center of Community Economic Development, 1974), ch. 2, particularly pp. 31–37, and the bibliography given by E. Layton, "Conditions of Technological Development," in I. Spiegel-Rösing and D. deS. Price, eds., *Science, Technology and Society* (London and Beverly Hills, CA: Sage Publications, 1977), esp. pp. 214–215. Layton also gives a concise account (pp. 206–209) of the ideological and professional shortcomings that doomed Project Hindsight.

Chapter 7. On the psychology of scientists, and their social concerns

1 This chapter is condensed from an essay entitled "Scientific Optimism and Societal Concerns: A Note on the Psychology of Scientists," which appeared in Marc Lappé and Robert Morison, eds., *Ethical and Scientific Issues Posed by Human Uses of Molecular Genetics*, vol. 265 of the *Annals* of the New York Academy of Sciences (December 1975). The essay was based on a paper delivered at the Conference on Ethical and Scientific Issues Posed by Human Uses of Molecular Genetics, organized by the New York Academy of Sciences and the Institute of Society, Ethics, and the Life Sciences; the latter organization published a version of the paper in *The Hastings Center Report* (December 1975), 5: 39–47.

I am happy to acknowledge partial support of a National Science Foundation grant for research on which this chapter is based, as well as helpful comments on an earlier draft by Jonathan Cole, David C. McClelland, Anne Roe, and Vivien Shelanski.

2 A. Einstein, "Motiv des Forschens," republished in English translation in Sonja Bargmann, trans. and rev., *Ideas and Opinions of Albert Einstein* (New York: Crown Publishers, Inc., 1954), pp. 224–27. The somewhat revised translation used here is from G. Holton, *Thematic Origins of Scientific Thought: Kepler to Einstein* (Cambridge, Mass.: Harvard University Press, 1973), pp. 376–78 (hereafter cited as *Thematic Origins*).

3 An implied claim of moral superiority of the scientist engaged in pure research without direct commercial applications can be discerned in other confessions of this sort. Thus, writing about the university physicist, a scientist in the 1920s said, "Financially his lot is not a wealthy one, nor socially is he high . . . [but close] to nature and by it closer and closer every day to the Almighty God who made him, what matters it to the physicist if the days be dull or neighboring man uncouth? His soul is more or less aloof from mortal strife" (R. Hamer, *Science* 61 [1925]: 109–10; quoted by S. Weart, *Nature* 262 [July 1, 1976]: 15).

4 B. Kutznetsov, "Nonclassical Science and the Philosophy of Optimism," in R. McCormmach, ed., *Historical Studies in the Physical Sciences,* (Princeton, N.J.: Princeton University Press, 1975), 4: 192–231.

5 For example, G. Holton and W. A. Blanpied, eds., *Science and Its Public: The Changing Relationship* (Boston: D. Reidel Publishing Co., 1976); also the *Newsletters* of the Harvard University Program on Public Conceptions of Science (now the *Newsletter on Science, Technology and Human Values*). Such ambivalence is characteristic not only of the public. In its Bicentennial issue of April 1975, *Fortune* magazine ran excerpts of an interview with the President of the United States in which he was quoted as saying (p. 80): "The System [of our government] seemed to be flexible enough and viable enough to meet whatever the problems were then, and I believe that the System is sufficiently adaptable today to meet the problems that we are facing, either domestically or internationally. The one thing that makes us a little less confident is the tremendous impact of science and technology."

6 A. Szent-Györgyi, *American Journal of Physics* 43, no. 5 (1975): 427. The distinguished editor of the journal *Science,* Philip H. Abelson, has drawn attention to an ironic consequence that causes interference even with the attempt of scientists to communicate among *themselves:* "Feeling their status threatened, and being dependent on the support of others, scientists have increasingly talked of the need to increase the public understanding of science. The majority cannot even effectively convey scientific information to each other. This is true of verbal presentations, in which decade after decade most scientists use slides that cannot be read beyond the front row of the audience. This illustrates not only technological incompetence but, more seriously, a blind spot – a failure to take into account the other person's requirements. In general, people who cannot or do not customarily analyze and respond to the needs of others cannot communicate" ("Communicating with the Publics," editorial, *Science* 194, no. 4265 [1976]: 565).

7 Marshall Bush, in an interesting essay, "Psychoanalysis and Scientific Creativity," abridged and reprinted in Bernice T. Eiduson and Linda Beckman, eds., *Science as a Career Choice* (New York: Russell Sage Foundation, 1973), pp. 243–57 – the whole book (hereafter cited as *Science as a Career Choice*) is an excellent source to which we shall often refer – properly reminds us that the notion of a single entity called "the creative scientist" and of a "scientific life style" is not only suspicious from commonsense arguments but has been seriously challenged by research results. Still, he does find useful generalizations to hold. See also F. Barron, *Creative Person and Creative Process* (New York: Holt, 1969), ch. 9.

8 B. T. Eiduson, "Psychological Aspects of Career Choice and

Development in the Research Scientist," in *Science as a Career Choice*, p. 15; italics supplied.

9 A. Einstein, "Autobiographical Notes," in P. A. Schilpp, ed., *Albert Einstein: Philosopher-Scientist* (New York: Harper and Brothers, 1959), p. 81.

10 D. C. McClelland, "On the Psychodynamics of Creative Physical Scientists" (1962), reproduced in abridged form in *Science as a Career Choice*, pp. 187–95. The quotations are from pp. 188 and 189, respectively.

11 Eiduson, op. cit. (n. 8), p. 15. In the same volume, Robert K. Merton's essay "Behavioral Patterns of Scientists" (1969) (abridged and reprinted on pp. 601–11), makes a point that may well parallel this change toward less interpersonal conflict: "Among the multitude of multiple discoveries in the history of science, Elinor Barber and I have examined a sample of two hundred and sixty-four in detail and have found, among other things, that there is a secular decline in the frequency with which multiples are an occasion for intense priority-conflicts. Of the thirty-six multiples before 1700 that we have examined, ninety-two percent were strenuously contested; the figure drops to seventy-two percent in the eighteenth century; remains at about the same level in the first half of the nineteenth century and declines to fifty-nine percent in the second half, reaching the lowest level of thirty-three percent in the first half of this century. Perhaps the culture of science today is not so pathogenic as it once was" (p. 606).

12 The only trait not fully matched in the list given earlier is Fermi's "familial" and close personal relations with those who worked with him (see Chapter 5). Even with respect to scientists who did not belong to this close group of co-workers, the applicability of David McClelland's description to this case can be documented.

13 L. S. Kubie, "Some Unresolved Problems of the Scientific Career," *American Scientist* 41 (1953): 596–613, and 42 (1954): 104–12. See also L. S. Kubie, "The Fostering of Creative Scientific Productivity," *Daedalus* 91 (1962): 304 ff.

14 For example, in David Shapiro, *Neurotic Styles* (New York: Basic Books, 1965), as cited in D. J. Taylor, "Enrico Fermi: the Psychology of Scientific Style" (unpublished doctoral thesis, Yale University School of Medicine, 1975). Taylor asks the interesting question why there have been so few psychoanalytic studies on scientists ever since Sigmund Freud's theories of neurosis.

He notes that Freud published studies of many creative persons – philosophers, artists, writers – but never of a scientist other than himself (though the work on Leonardo perhaps came close); moreover, the work of those who followed Freud has had largely the same bias. It may well be because science, of which Freud thought psychoanalysis to be a special example, was considered by Freud to be endowed with "impersonal" objectivity and therefore that there was a relative paucity of emotional goals and of fantasies in the work of scientists. Only in *other* fields would the subjectivity of the creators show itself sufficiently for them to be interesting subjects for scientific (i.e., psychoanalytic) studies.

Ironically, the claims of psychoanalysis to scientific status were long denied by other scientists. Einstein, for one, was quite skeptical; see his exchanges with Freud, for example, in O. Nathan and H. Norden, eds., *Einstein on Peace* (New York: Schocken Books, 1960), pp. 185–88. The reluctance of scientists to this day, by and large, to take interest in the psychological study of the scientific imagination itself is also relevant.

15 E. Durkheim, *The Rules of Sociological Method* (Glencoe, Ill.: The Free Press, 1950; originally 1895), p. 104.

16 Anne Roe, *The Making of a Scientist* (New York: Dodd, Mead & Co., 1952).

17 I have discussed the coherence of intellect and character in the case of highly achieving scientists in Chapter 10 of *Thematic Origins*. Many of the findings of Roe (n. 16) are of course parallel to others, made over the years. Compare, for example, Francis Galton's passage in *Hereditary Genius* (New York: Macmillan, 1869), p. 33: "By natural ability, I mean those qualities of intellect and disposition, which urge and qualify a man to perform acts that lead to reputation. I do not mean capacity without zeal, nor zeal without capacity, nor even a combination of both of them, without an adequate power of doing a great deal of very laborious work. But I mean a nature which, when left to itself, will, urged by an inherent stimulus, climb the path that leads to eminence and has strength to reach the summit – on which, if hindered or thwarted, will fret and strive until the hindrance is overcome, and it is again free to follow its labouring instinct."

18 Thus Roe (n. 16) notes that, whatever the reason, the people who "become physical and biological scientists early found special interest and special satisfactions away from personal relations." It is reminiscent of the famous passage in Einstein's *Autobiographical Notes*, concerning his own childhood and early adolescence: "It is quite clear to me that the religious paradise of youth, which was thus lost [at the age of twelve], was a first attempt to free myself from the claims of the 'merely-personal,' from an existence which is dominated by wishes, hopes and primitive feelings. Out yonder there was this huge world, which exists independently of us human beings and which stands before us like a great, eternal riddle, at least partially accessible to our inspection and thinking. The contemplation of this world beckoned like a liberation, and I soon noticed that many a man whom I had learned to esteem and to admire had found inner freedom and security in devoted occupation

with it. The mental grasp of this extra-personal world within the frame of the given possibilities swam as highest aim half consciously and half unconsciously before my mind's eye" (A. Einstein, in Schilpp, op. cit. [n. 9], p. 5).

I have discussed the methodological themes of projection (or externalization) and retrojection (or internalization) in the essay "Thematic and Stylistic Interdependence" (particularly pp. 101–12), in *Thematic Origins*.

19 Additional support was provided later from the Project TALENT data bank, computed in 1968. For example, those whose occupational goals (measured at age twenty-three) were in the sciences had shown, when the variable was measured at age eighteen, to test markedly low on the measure "desire to serve the public." The mean quartile rank (with all scores based on a 1 to 4 classification, where 1 corresponds to the lowest quartile, and 4 to the highest), were as follows for selected sciences: physics, 1.69; mathematics, 2.12; chemistry, 2.09; engineering, 2.17; computer science, 2.06; biological sciences, 2.02; medicine, 2.53. By comparison, some scores from outside the physical and biological sciences are as follows: business administration, 2.48; psychology, 2.76; social sciences, 2.22; law, 3.08; arts and humanities, 2.37 (from *Physics Manpower 1968* [New York: American Institute of Physics, 1969], p. 41).

See also M. Bush, in *Science as a Career Choice*, pp. 244–45; J. A. Chambers, in *Science as a Career Choice*, p. 350; and M. J. Moravcsik, "Motivation of Physicists," *Physics Today* 20 (October 1975): 9. The results of his survey, Moravcsik writes, "can be summarized by stating that the *internal* motivations (satisfaction from learning, release of curiosity, conversion of talent, and satisfaction from discovery) play a considerably more important role than the *external* motivations (priority, peer recognition, competition, prestige, influence, finances, or social service)."

20 Anne Roe, *The Psychology of Occupations* (New York: John Wiley & Sons, 1956), p. 319. See also the research results in Anne Roe and Marvin Siegelman, *The Origin of Interests*, American Personnel and Guidance Association Inquiry Study No. 1 (Washington, D.C.: APGA, 1964), particularly pp. 3–7, 45, 64–67.

21 W. W. Cooley and Paul R. Lohnes, *Predicting Development of Young Adults*, Project TALENT, 5-year Follow-up Studies, Interim Report 5 (Palo Alto, Calif.: Project TALENT, American Institutes for Research, and Pittsburgh: School of Education, University of Pittsburgh, 1968).

Though warning that sharp predictability is seldom obtained, that "all predictions of occupational choice are necessarily probabilistic," and that "personality is only one broad factor in the decisions made at any occupational choice point," Roe, in another study, found that we are dealing with significant and deep-going relations between personality traits and career choices: "The assumption that there are differences in various aspects of personality between those following different occupations has been upheld whenever it has been subjected to research." See Anne Roe, "Personality Structure and Occupational Behavior," in H. Borow, ed., *Man in a World at Work* (Boston, Mass.: Houghton Mifflin, 1964), p. 211.

22 Cooley and Lohnes, op. cit. (n. 21), p. 4.25.

23 Cooley and Lohnes, ibid., p. 4.44. The cited passage has further references to supporting work, that is, that of Roe and Siegelman, op. cit. (n. 20); Cooley, op. cit. (n. 27); and D. E. Super et al., *The Psychology of Careers* (New York: Harper & Row, 1957).

A schematic "career development tree" derived by Cooley and Lohnes from the Project TALENT data for boys is given on p. 4.57 in Cooley and Lohnes. It shows that 56 percent of the male children make the initial "thing" (or "science-technology") choice by early junior high school. After many other branchings (e.g., college versus noncollege) this has led, five years after high school, to the following choices: 1.0 percent – biomedical researchers, physicians; 2.3 percent – dentists, pharmacists, biologists; 0.9 percent – research scientists, mathematicians; 7.5 percent – applied scientists, engineers; 15.0 per cent – skilled and technical workers; and 9.2 percent – miscellaneous, and laborers.

Of the 44 percent who made the early "people" choice, the career choices five years after high school have yielded the following: 5.2 percent of the total original sample – lawyers, social scientists; 12.2 percent – counselors, teachers, welfare workers, artists, musicians; 1.4 percent – managers, executives; 10.1 percent – businessmen, auditors; 18.6 percent – accountants, office equipment operators, salesmen; 16.6 percent – clerks, office and service workers.

R. B. Cattell and J. E. Drevdahl, in "A Comparison of the Personality Profile (16P.F.) of Eminent Researchers with That of Eminent Teachers and Administrators, and of the General Population" (1955), abridged and reprinted in *Science as a Career Choice*, p. 178, broaden the dichotomy to "things *and* ideas" versus "people." They add, concerning their results of measurements of the psychological and social operations of personality factors on research scientists, "It is easy to see that the schizothymic preoccupation with things and ideas, rather than people; the self-sufficiency which favors creativity and independence of mind; the dominance which gives satisfaction in mastery of nature for its own sake; and the emotional instability which permits radical restructuring and creativity, would all be vital to the best kind of basic research performance – though perhaps unpleasant in an administrator and inapt in a businessman."

24 J. A. Davis, *Undergraduate Career Decisions* (Chicago: Aldine

Publishing Co., 1965), especially pp. 10–13, 50–63, 132–39, 152–65. See also, for example, Bernice T. Eiduson, *Scientists, Their Psychological World* (New York: Basic Books, 1962). Her story of forty research scientists reports their "chief interest in childhood" in descending order as follows: reading, 33; science, 23; sports, 22; artistic, 16; mechanical, 7; social, 4.

Among other publications, the following contain interesting work along similar lines: Robert S. Albert, "Toward a Behavioral Definition of Genius," *American Psychologist* 30 (1975): 140–51 (includes a good bibliography); Charles C. Gillispie, "Remarks on Social Selection as a Factor in the Progressiveness of Science," *American Scientist* 56 (1968): 439–50; Howard E. Gruber, Glenn Terrell, and Michael Wertheimer, eds., *Contemporary Approaches to Creative Thinking* (New York: Atherton Press, 1962), especially the essay by Allen Newell, J. C. Shaw, and Herbert A. Simon, "The Processes of Creative Thinking," pp. 63–119; Liam Hudson, *Contrary Imaginations* (London: Pelican, 1966); Ian I. Mitroff, *The Subjective Side of Science* (Amsterdam: Elsevier, 1974); Ian I. Mitroff and Ralph H. Kilman, "On Evaluating Scientific Research: The Contribution of the Psychology of Science," *Technological Forecasting and Social Change* 8 (1975): 163–74; Michael Mulkay, "Some Aspects of Cultural Growth in the Natural Sciences," *Social Research* 36 (1969): 22–52; and Barry F. Singer, "Toward a Psychology of Science," *American Psychologist* 26 (1971): 1010–15 (includes a good bibliography).

25 At least a brief remark is needed here on the deeper reasons for the existence of the "people/thing" split. It seems to me based in the Cartesian subject/object split, and may be among the earliest thematic antitheses with which the growing child must struggle. This is supported also in Freud's evolutionist view of recapitulatory development. Possibly influenced by the Comtean tradition, Freud saw the evolution of human views of the universe, historically and individually, as a three-stage sequence: e.g., in *Totem and Taboo* (1912): ". . . the *animistic* phase would correspond to narcissism both chronologically and in its content; the *religious* phase would correspond to the state of object-choice of which the characteristic is a child's attachment to his parents; while the *scientific* phase would have an exact counterpart in the stage at which an individual has reached maturity, has renounced the pleasure principle, adjusted himself to reality and turned to the external world for the object of his desires."

In the development of the future scientist there may thus occur a more pronounced externalization, and hence a more pronounced Cartesian split, than in the development of nonscientists. Cf. Philip Rieff, "Freud and the Authority of the Past," in R. J. Lifton and E. Olsen, eds., *Explorations in Psychohistory* (New York: Simon and Schuster, 1975), pp. 81–82.

26 Cf. W. O. Hagstrom, *The Scientific Community* (New York: Basic Books, 1965).

27 W. W. Cooley, *Career Development of Scientists: An Overlapping Longitudinal Study, Cooperative Research Project No. 436* (Cambridge: Graduate School of Education, Harvard University, 1963).

28 See, for example, the discussion in Bernice T. Eiduson, "Scientists as Advisors and Consultants in Washington," *Science and Public Affairs, Bulletin of the Atomic Scientists* 22 (1966): 26–31. She cites a typical reaction of a scientist to the "Washington Experience": "I was becoming a government committee figure. I was really big business. I found myself doing unpleasant things with people I despised and even before I knew it, I was finding myself trying to rival them. Then I realized I was really selling out my soul to them; and that if I were going to do this, I might as well go into the Hollywood business, or do something else that would be more interesting and profitable.

"This attitude came to me one day when, at a committee meeting . . . a man leaned over the table, and said to me, 'Is this what we went to college for?' I felt nauseous, and almost had a blackout, and decided to get out right then and there . . . It's an emotionally revolting and emotionally draining experience, and I've done my duty."

29 J. Primack and F. von Hippel, *Advice and Dissent* (New York: Basic Books, 1974).

30 Philip Boffey, *The Brain Bank of America* (New York: McGraw-Hill Book Company, 1975).

31 The term "introversion" may be confusing; for Freud, for example, the introvert is on the way to neurosis "if he does not find other outlets for his pent-up libido" (Freud, *General Introduction to Psychoanalysis*; 1920). As H. J. Eysenck uses the term, for example, in *The Structure of Human Personality* (New York: John Wiley & Sons, 1953), the conception of introversion–extraversion is a dimension considered to be orthogonal to one called neuroticism. Cooley finds the lower scores on the PSPs on "persuasive interests" significant of introversion; the low scores on "social interests" are not thought to be so significant by Cooley.

Cooley and Lohnes, op. cit. (n. 21), p. 2.7, note that "science-oriented scholasticism" (a function combining the motives of scholastic interest, science interest, mathematical ability, and verbal knowledge which they hold to be "the major explanatory and predictive personality function in our theory of the antecedents of adult development") is positively correlated with the motive-domain factor "introspection" (rather than introversion) – although more for boys than girls – and negatively with "sociability" – though more for girls than for boys. See also pp. 2.4 and 4.74.

32 Cooley, op. cit. (n. 27), p. 108.

33 Albert Einstein, in *Ideas and Opinions* (n. 2), p. 9.

34 Cooley, op. cit. (n. 27), pp. 113–14.

35 The only proper "test" will always be wise and compassionate actions rather than the score on some "ethics readiness" test. One cannot, for example, shake off the suspicion that the persons in Nixon's White House who brought this nation to the brink of disaster, with their public relations, law, and other such backgrounds, would have scored quite well on tests concerned with "sociability," "extraversion," and others on which scientists as a group do poorly. Conversely, in their early years, a Leo Szilard, a James Franck, or a Niels Bohr would quite possibly have produced a low score on such a test.

36 See the discussion of the program in Chapter 11. One of the studies showing that a conscious attempt to change erroneous, negative notions held about science by students can be built into a sound science program is in the discussion by A. Ahlgren and H. J. Walberg, "Changing Attitudes Toward Science Among Adolescents," *Nature* 245 (1973): 187–90. They conducted a survey of students in three different types of science courses that had been separately found to produce about the same final level of understanding of the conceptual scientific content. The courses differed markedly, however, in their attitudes to the epistemological, social, and other contexts of science in general and physics in particular. They found that the content and context of the science course "can have a marked effect on the student's view" of science, and that "on the basis of the test data on student attitudes" one can indeed hope to have "succeeded in restoring the social, humanitarian, and artistic aspects that had been lost" in other curricula (p. 189).

37 This is closely allied with Elting Morison's identification of a chief need for our educational efforts today, the "rational coordination of impulses and thought"; see E. Morison, *From Know-how to Nowhere: The Development of American Technology* (New York: Basic Books, 1974), particularly chs. 1 and 9.

38 I have intentionally not been dealing with "applied research" or "development," where existing scientific knowledge is put to use. The benefits one expects from these efforts is different from what one may expect from basic scientific research and researchers who are able to go beyond the barriers of current ignorance on fundamental matters.

39 See, for example, Dorothy S. Zinberg, "Education Through Science: The Early Stages of Career Development in Chemistry," *Social Studies of Science* 6 (May 1976): 215–46. However, of her sample of students in a second-tier university in England, less than 10 percent were headed for academic careers.

See also K. J. Gilmartin, D. H. McLaughlin, L. L. Wise, and R. J. Rossi, *Development of Scientific Careers: The High School Years* (Palo Alto, Calif.: American Institutes for Research, 1976,

Report AIR-48200). This is an extensive, TALENT-related study, with the intention to increase understanding of career guidance factors that affect the development of high school students' scientific potential. Special attention was paid to women and minorities. The findings include: (1) More than three times as many female high school students in the 1975 sample as in the 1960 TALENT sample were planning science careers. (2) Students of oriental background showed the largest proportion of students planning science careers. (3) Only 12 percent of males and 6 percent of females who had planned a science career in 1960 persisted in this plan over the subsequent twelve years. Over half of the women in science in 1972 – twice the proportion for men – had had nonscience career plans in high school in 1960.

40 See H. Zuckerman and J. R. Cole, "Women in American Science," *Minerva* 13 (1975): 82–102; and two useful earlier studies: Roe and Siegelman, op. cit., n. 20; and H. S. Astin, *The Woman Doctorate in America* (New York: Russell Sage Foundation, 1966), particularly pp. 26–27, giving data on rates of marriage and divorce quite analogous to those found by Roe for male scientists. It would also be instructive to compare data from the United States with that from societies in which women in science have had a rather different position, for example in France and in the USSR.

41 Cooley and Lohnes, op. cit. (n. 21), pp. 1.29, 4.42. See also p. 4.48. The bias among high school counselors to define science as a "man's job," and the consequent problems raised for female students, are discussed in a study presented in *FAS* [*Federation of American Scientists*] *Professional Bulletin,* no. 2 (October 1973).

42 *Factors Influencing the Science Career Plans of High School Students* (West Lafayette, Ind.: Purdue University, June 1975), p. 15.

43 Ibid., p. 19. Revenna Helson, in "Women Mathematicians and the Creative Personality" (1971), in *Science as a Career Choice*, pp. 563–74, shows that a group of women who matured before the current generation of students have evidently faced even more intense career-formation problems and psychological barriers than those that characterized the men of their days – as one would expect. "The traits most characteristic of the creative woman would seem to [include] these: rebellious independence, narcissism, introversion, and a rejection of outside influence . . . These traits have all been ascribed to the creative person, regardless of sex, but they appear more clearly in creative women mathematicians than they do in creative men mathematicians" (p. 570).

Similar results are given in Louise M. Bachtold and Emmy E. Werner, "Personality Profiles of Gifted Women: Psychologists" (1970), abridged and reprinted in *Science as a Career Choice*, pp. 551–63.

Chapter 8. Lewis Mumford on science, technology, and life

1 Mumford, *The Myth of the Machine II: The Pentagon of Power* (New York: Harcourt Brace Jovanovich, Inc., 1970; London: Secker and Warburg). This essay appeared as a book review in *Minerva* 9, October 1971: 568–75.

2 For another example of Mumford's *ad hominem* response to what he regarded as an adversary position, see his letter to the editor, *The New York Times Book Review,* 10 January 1971.

Chapter 9. Frank Manuel's Isaac Newton

1 This chapter appeared in *Methodology and Science* 8, no. 23 (July 1975): 68–74, largely as reprinted from *The New Republic* 172, no. 9 (1975): 26–28.

Chapter 10. Ronald Clark and Albert Einstein

1 Ronald W. Clark, *Einstein: The Life and Times* (New York and Cleveland: The World Publishing Company, 1971). This chapter is closely based on a review that appeared in *The New York Times Book Review,* vol. 76, September 5, 1971.

2 Martin J. Klein, *Paul Ehrenfest, Vol. 1, The Making of a Theoretical Physicist* (Amsterdam: North-Holland Publishing Company, 1970), 303–4.

Chapter 11. On the educational philosophy of the Project Physics Course

1 This chapter is an expansion of an invited talk at Congreso Nacional de Enseñanza de la Física, Sociedad Mexicana de Física, Mexico City, July 17, 1975. When published in its original form in the journal *Physics Education* 11, no. 5 (1976): 330–35, the address carried an introductory, explanatory note by one of the journal's editors, John Harris:

"Science teaching at secondary level in the United States is quite different from anything in secondary schools in Britain. The most common pattern is for biology, chemistry and physics to be offered in the last three years (10th, 11th, 12th grades of high school respectively). Physics is thus an optional subject for students (the word 'pupil' has only optical connotations in the U.S.) in their last year of secondary schooling. And most of these students will not have taken any physics before 12th grade.

Professor Holton describes the interests and abilities of the students that the Project Physics Course is intended for in this article. One must remember that there is less specialization in the American schools than in British schools and that the students staying at school until 12th grade are from a wider ability than 'traditional' British sixth formers.

The Project Physics Course materials include a wide range of material: laboratory equipment, 16mm film, 8mm film loops, projectable transparencies, etc., as well as printed material for students and teachers.

The main course is organized into six (basic) units: (1) Concepts of motion, (2) Motion in the heavens, (3) The triumph of mechanics, (4) Light and electromagnetism, (5) Models of the atom, (6) The nucleus."

2 F. G. Watson, "Why Do We Need More Physics Courses?" *The Physics Teacher* 5 (May 1967): 210–12; G. C. Bates, "More on the Problem of Physics Enrollments," *Science Teacher* 42 (October 1975): 29–30.

3 During the question period, a member of the audience asked to know the relation between the Project Physics Course and my earlier college texts (*Introduction to Concepts and Theories in Physical Science* [Reading, Mass.: Addison-Wesley Publishing Co., 1952, 1973], and the text *Foundations of Modern Physical Science*, with D. H. D. Roller [Reading, Mass.: Addison-Wesley Publishing Co., 1958]). The answer was that these texts are, in a sense, the grandfathers of the Project Physics Course. This happened in the following manner. One day in 1960 a high school teacher from California came into my office at Harvard. He said he had been using my college text, which was first published in 1952, as a text for the seniors in his high school class. It worked well except that the sentences were often too lengthy for his students. He suggested I rewrite the text for that audience, and I in turn urged him to write a suitable textbook himself, possibly with a grant from the National Science Foundation. His name was James Rutherford.

When the NSF failed to fund that idea, Rutherford, Fletcher Watson (supervisor of the Ph.D. thesis Rutherford was then completing at Harvard), and I decided to undertake the work with a modest starter grant from the Carnegie Corporation, beginning in 1962. All this was in the back of my mind at the October 1963 NSF conference, and the start we had already made gave us the courage to expand our plans considerably into the Project Physics Course for schools and colleges, from 1964 on. Support was provided from the U.S. Office of Education, NSF, Carnegie Corporation, Ford Foundation, Sloan Foundation, and the publisher (Holt, Rinehart and Winston, New York City).

The word Harvard was originally used in the title, since the project headquarters were there; but between the development of the prototype course and the issuing of the publisher's version for national use, the name Project Physics Course was adopted (and with it came, inevitably, the use of the abbreviation PPC).

4 The components in the U.S. version are not *all* required for each student taking a Project Physics Course but were designed to provide a rich variety of materials from which teachers or students could choose. For example, when studying vectors, most students find it very helpful to supplement the discussion printed in a textbook by other media that use other cognitive channels. Therefore, we made a group of film loops about motions and vectors in the real world (airplanes going in a wind, boats crossing a river), as well as a set of programmed instruction booklets and some laboratory activities.

Also, there is much material for students to select if their class is run so as to encourage them to read on their own and to do special projects. Many teachers treat the passages on history and philosophy of science as reading assignments. Others use the *Readers* as resources for technical, historical, and sociological discussions about science – not assigned as classroom work that must be done, but as material from which to choose a project. In many classes, students are invited to choose a special project every few months. It can be mathematical, a laboratory project, or a report on historical reading. The teacher does not have to become a polymath, but can leave the selection of materials open to the students' choice, in accord with their own interests. Even if by training and inclination the teacher is chiefly interested in the "purely scientific" parts only, as long as he or she shows respect for and does not downgrade the other cultural elements, the students will find their own way.

5 A book-length account of the total evaluation process and results has been prepared by F. G. Watson, W. Welch, and H. Walberg. In addition, some forty articles have been published in various professional journals by members of the evaluation group of the project, using the extensive test results. A list of those, and of the theses, is obtainable from F. G. Watson, Graduate School of Education, Longfellow Hall, Harvard University, Cambridge, Mass., 02138.

6 A. Ahlgren and H. J. Walberg, "Changing Attitudes Toward Science Among Adolescents," *Nature* 245 (1973): 187–90. For a brief summary, see n. 36 in the notes to Chapter 7 of this book.

7 This and two of the other teaching styles often used in Project Physics classes are the subject of three of the twenty-one teacher-briefing films obtainable from Holt, Rinehart and Winston, or rentable from Modern Talking Picture Service, Inc., in New York City. The film dealing with the method here described is called

Teaching Styles II (listed in the Holt catalog of Project Physics materials as No. 084020-1).

8 Earlier discussions include my essays "Project Physics: A Report on Its Aims and Current Status," *The Physics Teacher* 5 (May 1967): 198–211; "The Relevance of Physics," *Physics Today* 23, no. 11 (November 1970): 40–47; and Chapter 15 in G. Holton, *Thematic Origins of Scientific Thought: Kepler to Einstein* (Cambridge, Mass.: Harvard University Press, 1973).

9 Actual articles are contained in the seven *Readers*. Annotated bibliographies are given in several "Resource Letters" (contained in the students' *Handbook*) reprinted from *The American Journal of Physics*. Further bibliographies are in the instructors' *Resource Book*. An extensive bibliography of books and articles in physics, history of science, philosophy of science, and so on, directed to the same kind of students, is given on pp. 555–70 of G. Holton, *Introduction to Concepts and Theories in Physical Science*, 2nd ed., with S. Brush (Reading, Mass.: Addison-Wesley, 1973).

ACKNOWLEDGMENTS

Material for many of the chapters in this book is based on the following
Introduction: "How a Scientific Discovery Is Made: A Case History," *American Scientist* 84 (July–August 1996): 364–375.

Chapter 1: "On the Role of Themata in Scientific Thought," *Science* 188 (April 25, 1975): 328–34. Copyright 1975 by the American Association for the Advancement of Science.

Chapter 3: "On Being Caught Between Dionysians and Apollonians," *Daedalus* 103, no. 3 (Summer 1974): 65–81. "Mainsprings of Scientific Discovery," in Owen Gingerich, ed., *The Nature of Scientific Discovery* (Washington, D.C.: Smithsonian Institution Press, 1975), pp. 199–217.

Chapter 4: "Analysis and Synthesis as Methodological Themata," *Methodology and Science* 10, no. 1 (Spring 1977): 3–33.

Chapter 5: "Striking Gold in Science: Fermi's Group and the Recapture of Italy's Place in Physics," *Minerva* 12, no. 2 (April 1974): 158–98.

Chapter 7: "Scientific Optimism and Societal Concerns," *Annals of the New York Academy of Sciences* 265 (January 1976): 82–101. Reprinted by permission of the New York Academy of Sciences.

Chapter 8: "On Science, Technology and Life: A Review of Lewis Mumford's *The Myth of the Machine II*," *Minerva* 9, no. 4 (October 1971): 568–75.

Chapter 9: "Review of Frank E. Manuel's *The Religion of Isaac Newton*," *The New Republic* 172, no. 9 (March 1, 1975): 26–28.

Chapter 10: "Einstein, the Life and Times," *The New York Times*, 5 September 1971. © 1971 by The New York Times Company. Reprinted by permission.

Chapter 11: "The Project Physics Course—Notes on its Educational Philosophy," *Physics Education* 11 (July 1976): 330–35.

INDEX